METHODS IN MOLECULAR BIOLOGY™

John M. Walker, SERIES EDITOR

Protein and Peptide Analysis
by Mass Spectrometry

METHODS IN MOLECULAR BIOLOGY™

Protein and Peptide Analysis by Mass Spectrometry

Edited by

John R. Chapman

Manchester, UK

Humana Press ✳ Totowa, New Jersey

© 1996 Humana Press Inc.
999 Riverview Drive, Suite 208
Totowa, New Jersey 07512

All authored papers, comments, opinions, conclusions, or recommendations are those of the author(s), and do not necessarily reflect the views of the publisher.

This publication is printed on acid-free paper. ∞
ANSI Z39.48-1984 (American Standards Institute) Permanence of Paper for Printed Library Materials.

Cover illustration: Fig. 1 from Chapter 10, "Protein Secondary Structure Investigated by Electrospray Ionization," by Carol V. Robinson.

Photocopy Authorization Policy:
Authorization to photocopy items for internal or personal use, or the internal or personal use of specific clients, is granted by Humana Press Inc., provided that the base fee of US $5.00 per copy, plus US $00.25 per page, is paid directly to the Copyright Clearance Center at 222 Rosewood Drive, Danvers, MA 01923. For those organizations that have been granted a photocopy license from the CCC, a separate system of payment has been arranged and is acceptable to Humana Press Inc. The fee code for users of the Transactional Reporting Service is: [0-89603-345-7/96 $5.00 + $00.25].

Printed in the United States of America. 10 9 8 7 6 5 4 3 2

Library of Congress Cataloging in Publication Data

Main entry under title:

Methods in molecular biology™.

Protein and peptide analysis by mass spectrometry / edited by John R. Chapman.
     p. cm. — (Methods in molecular biology™)
     Includes index.
     ISBN 0-89603-345-7 (alk. paper)
     1. Proteins—Analysis. 2. Peptides—Analysis. 3. Mass spectrometry. I. Chapman, John R.
II. Series: Methods in molecular biology™ (Totowa, NJ).
QP551.P6956 1996
574.19'245—dc20   96-25738
                                                            CIP

Preface

The purpose of the preface is to explain the book's objectives and how to use it; give warnings, disclaimers, and the like.*

The main objective of *Protein and Peptide Analysis by Mass Spectrometry* is quite straightforward—to present authoritative, up-to-date, and practical accounts of the use of mass spectrometry in the analysis of peptides and proteins.

How to use it? Every reader will have their own particular interests and will surely be drawn toward the chapters that cover these interests. Within the remaining chapters, however, techniques are described with analytical possibilities that such a reader can then only guess at. So, read the book fully.

Again, as is customary in the *Methods in Molecular Biology* series, the chapter format (Introduction, Materials, Methods, and Notes) allows the authors to introduce the techniques, to explain their relevance and applicability, and, above all, to provide detail—detail that represents each author's accumulated experience and enables the reader to use and benefit from these methods. So, read the book fully, and read it diligently.

Warnings and disclaimers: Mass spectrometry today offers the protein chemist ready access to a wealth of information that is otherwise available only with great difficulty, or perhaps not at all. With this goal in sight, any warnings and disclaimers will almost surely be ignored. So, a warning anyway; the use of mass spectrometry might be habit forming.

Finally, the book has a second objective—to honor the memory of Mickey Barber, a good friend and an excellent scientist. In this respect, I can do no better than recommend Peter Roepstorff's chapter, which sets out his personal view of Mickey's innovative contributions to mass spectrometry.

John R. Chapman

*White, J. V. (1988) *Graphic Design for the Electronic Age.* Watson-Gupthill, New York.

Dedication

Professor Michael Barber, FRS
November 3, 1934–May 8, 1991

Contents

Contributors

PHILIP C. ANDREWS • *Department of Biological Chemistry, University of Michigan, Ann Arbor, MI*

BOBBY L. BARNETT • *Miami Valley Laboratories, The Procter and Gamble Company, Cincinnati, OH*

MARK D. BAUER • *Miami Valley Laboratories, The Procter and Gamble Company, Cincinnati, OH*

ALMA L. BURLINGAME • *Mass Spectrometry Facility, Department of Pharmaceutical Chemistry, University of California, San Francisco, CA*

MICHAEL P. CARRIER • *Academic Information Services, University of Salford, Salford, UK*

BRIAN T. CHAIT • *The Rockefeller University, New York, NY*

JOHN R. CHAPMAN • *Manchester, UK*

JOHN S. COTTRELL • *Finnigan MAT Ltd., Hemel Hempstead, Hertfordshire, UK*

THOMAS COVEY • *PE Sciex, Concord, Ontario, Canada*

NATALIE DALES • *Department of Chemistry, University of Wisconsin, Madison, WI*

DOMINIC M. DESIDERIO • *The Charles B. Stout Neuroscience Mass Spectrometry Laboratory, Departments of Neurology and Biochemistry, University of Tennessee, Memphis, TN*

CHRISTOPHER R. ERWIN • *The Children's Hospital Medical Research Foundation, Cincinnati, OH*

DAVID C. GALE • *Chemical Methods and Separations Group, Chemical Sciences Department, Pacific Northwest National Laboratory, Richland, WA*

BRIAN N. GREEN • *Micromass UK Ltd., Altrincham, Cheshire, UK*

DAVID J. HARVEY • *Oxford Glycobiology Institute, Department of Biochemistry, University of Oxford, UK*

THERESE HUTTON • *Micromass UK Ltd., Altrincham, Cheshire, UK*

THOMAS W. KEOUGH • *Miami Valley Laboratories, The Procter and Gamble Company, Cincinnati, OH*

MARTIN P. LACEY • *Miami Valley Laboratories, The Procter and Gamble Company, Cincinnati, OH*

RACHELE R. OGORZALEK LOO • *Department of Biological Chemistry, University of Michigan, Ann Arbor, MI*

RONALD W. A. OLIVER • *Biological Materials Analysis Research Unit, University of Salford, Salford, UK*

GEERT JAN RADEMAKER • *Department of Mass Spectrometry, University of Utrecht, The Netherlands*

CAROL V. ROBINSON • *Oxford Centre for Molecular Sciences, Oxford, UK*

PETER ROEPSTORFF • *Department of Molecular Biology, Odense University, Odense, Denmark*

KEVIN L. SCHEY • *Department of Cell and Molecular Pharmacology, Medical University of South Carolina, Charleston, SC*

BRENDA SCHWARTZ • *Environmental Molecular Sciences Laboratory, Pacific Northwest National Laboratory, Richland, WA*

CHRISTINE A. SETTINERI • *Mass Spectrometry Facility, Department of Pharmaceutical Chemistry, University of California, San Fransisco, CA*

MARSHALL M. SIEGEL • *Wyeth-Ayerst Research, Lederle Laboratories, Pearl River, NY*

RICHARD D. SMITH • *Chemical Methods and Separations Group, Chemical Sciences Department, Pacific Northwest National Laboratory, Richland, WA*

BERNARD SPENGLER • *Institute of Laser Medicine, University of Dusseldorf, Germany*

YIPING SUN • *Miami Valley Laboratories, The Procter and Gamble Company, Cincinnati, OH*

CHRIS W. SUTTON • *Finnigan MAT Ltd., Hemel Hempstead, Hertfordshire, UK*

JANE THOMAS-OATES • *Department of Mass Spectrometry, University of Utrecht, The Netherlands*

SERGE N. VINOGRADOV • *Department of Biochemistry, School of Medicine, Wayne State University, Detroit, MI*

YOSINAO WADA • *Department of Molecular Medicine, Osaka Medical Center and Research Institute for Child and Maternal Health, Osaka, Japan*

ELLEN S. WANG • *Miami Valley Laboratories, The Procter and Gamble Company, Cincinnati, OH*

RONG WANG • *The Rockefeller University, New York, NY*

JOSEPH ZAIA • *Department of Chemistry and Biochemistry, University of Maryland, Baltimore, MD*

1

Mass Spectrometry in the Analysis of Peptides and Proteins, Past and Present

Peter Roepstorff

When the editor, John Chapman, asked me to write the introductory chapter to this volume and told me that it would be dedicated to the late Michael "Mickey" Barber, I felt very honored and also humbled because I have always considered Mickey to be one of the most outstanding pioneers in the field of mass spectrometry (MS) of proteins. Most younger scientists associate Mickey's name with the invention of ionization of nonvolatile compounds by fast-atom bombardment (FAB) in 1981 *(1)*. It is also true that this invention had a major impact on the practical possibilities for mass spectrometric analysis of peptides and proteins. However, only a few of the present generation of scientists involved in MS of peptides and proteins know that MS of peptides was already an active field 30 years ago and also that Mickey's career in many ways reflects the development in the field through all these years.

In the 1960s, a number of groups had realized and actively investigated the potential of MS for peptide analysis. The only ionization method available was electron impact (EI), which required volatile derivatives. This necessitated extensive derivatization of the peptides prior to mass spectrometric analysis. The following groups were pioneers in investigating MS: the group headed by M. Shemyakin at the Institute for Chemistry of Natural Products of the USSR Academy of Sciences, which worked with acylated and esterified peptides *(2)*; K. Biemann's group at Massachusetts Institute of Technology, which used acylation followed by reduction of the peptides to amino alcohols followed by trimethylsilylation and gas chromatography (GC)/MS *(3)*; and E. Lederer's group at the Institute for Chemistry of Natural Substances in Gif sur Yvette, France, which studied natural peptidolipids among other compounds. Mickey Barber, who at that time worked at AEI in Manchester, got involved in the

From: *Methods in Molecular Biology, Vol. 61: Protein and Peptide Analysis by Mass Spectrometry*
Edited by: J. R. Chapman Humana Press Inc., Totowa, NJ

work of E. Lederer. The French group had isolated a peptidolipid called fortuitine, which was analyzed by Mickey on the MS9 mass spectrometer. Fortuitine appeared to be a very fortuitous compound. It was N-terminally blocked with a mixture of fatty acids, was naturally permethylated, and contained an esterified C-terminus. Mickey Barber obtained a perfect EI spectrum of this 1359-Da peptide and was able to interpret the spectrum *(4)*. I believe that this, at that time, was the largest natural compound ever analyzed by MS. The achievement contains many of the elements now considered to be only possible with the most contemporary MS techniques. Thus, heterogeneity both in the sequence and the secondary modifications was determined and the fragment ions, always present in EI spectra, allowed sequencing. The realization of the effect of *N*-methylation on the volatility of the peptidolipid resulted in development of the permethylation procedure for peptide analysis by MS *(5)*.

For me, personally, the fortuitine paper has also been very fortuitous. Shortly after its publication, I became involved in peptide synthesis by the Merrifield method, which, unfortunately, did not always yield the expected product. Modifications, which researchers had no practical method to analyze, were frequent. The fortuitine paper inspired me to investigate the possible use of MS for analysis of these modifications. Since then, MS has been my main tool in protein studies. A few years later, I first met Mickey during a visit to AEI in Manchester. I was very fascinated by his lively engagement in the subject, and also realized over a pint of beer in a nearby pub that MS was not his only scientific interest. He was deeply involved in surface science and had been also very active in the development of equipment for photo-electron spectroscopy, as well as in exploring the possibilities of electron spectroscopy for chemical analysis (ESCA). Shortly after we first met, he moved from AEI to a lectureship at the University of Manchester Institute of Science and Technology, where he became a full professor in 1985. In the same year, he was elected Fellow of the Royal Society.

In the 1970s, MS of peptides progressed slowly and, although its potential was demonstrated by a number of applications to structure elucidation of modified peptides, the field was stagnating by the end of the decade. Two new ionization methods, chemical ionization and field desorption, appeared during that period. They created new hopes for improvements in peptide analysis by MS, but unfortunately, they did not result in a real breakthrough.

In that period, I had no real contact with Mickey Barber. It is my impression that he, intuitively or consciously, was realizing that the opening of new possibilities for mass spectrometric analysis had to come from surface science. Anyway, among other subjects, he started to investigate surface analysis by secondary-ion mass spectrometry (SIMS). This led to his discovery of a new technique for desorption and ionization of involatile organic compounds. He termed the technique FAB because, instead of the primary ions used in SIMS, he used a beam of 3–10 keV argon atoms to effect desorption and ionization. The choice of atoms instead of ions was mainly determined by a desire to avoid surface charging phenomena, which could disturb ion focusing in the sector instrument used. It was later realized that primary ions work just as well as atoms, and the fast atom gun is now frequently replaced by a cesium gun creating 20–30 keV primary cesium ions.

The concept of SIMS of organic solids was not new. Benninghoven et al. *(6)*, at the University of Münster in Germany, had already, some years earlier, demonstrated mass spectra of organic solids, including amino acids, by SIMS. The spectra, however, were only transient because the surface was quickly destroyed by the high flux of primary ions. As a matter of fact, the real discovery by Mickey Barber was the use of a liquid matrix and the technique is now often termed liquid secondary-ion mass spectrometry (LSIMS) when a primary beam of cesium ions is used instead of fast atoms. The trick of using a matrix was that the matrix surface was continuously replenished with sample so that secondary ions could be produced continuously over a long period of time. This feature also made the technique directly compatible with scanning mass spectrometers. In fact, an important reason for the immediate success of FAB was that it could readily be installed on existing sector field and quadrupole mass spectrometers. In my own laboratory, for example, we installed FAB in 1982 simply by replacing the standard solids inlet probe with a simple rod and by placing an ion gun in the place of the GC inlet on a Varian MAT 311A double-focusing sector instrument. To my knowledge, Mickey was the first to introduce the use of a matrix in MS and, as is described later in this chapter, the use of a matrix is essential in all mass spectrometric techniques used for analysis of peptides and proteins.

Another technique that allowed the analysis of large, involatile organic molecules was plasma-desorption mass spectrometry (PDMS) developed as early as 1974 by Torgerson et al. *(7)*. The technique had been shown to be capable of the analysis of large underivatized peptides *(8)*, and, soon after, of proteins, by the demonstration of the first mass spectrum of insulin in 1982 *(9)* and of a number of snake toxins with molecular masses up to 13 kDa *(10)*. Instrumentation for PDMS became commercially available a few years later, and the real breakthrough for this technique came 2 years later with the simultaneous discoveries of the advantages of using nitrocellulose as support *(11)* or reduced glutathione as matrix *(12)*. Shortly after the publication of the PD-spectrum of insulin, FAB mass spectra of insulin were also published *(13,14)*, and in the following years, mass spectra of proteins as large as 25 kDa were published using both techniques, concomitantly with the gradual acceptance of the potential of mass spectrometric analysis by a number of protein chemists.

In that period, Mickey Barber visited my laboratory for a period during which he "played" with our plasma-desorption mass spectrometers to get a personal feeling of the potential of this technique compared to FABMS. Mickey's enthusiastic engagement made his stay a great inspiration for me as well as for my students, and we had many long discussions about the status and future perspectives of the field. It seemed to us that both techniques had fundamental limitations that would prevent their full acceptance among protein chemists. The major limitations were that it was difficult to extend the mass range beyond 25 kDa, and that this range could only be attained for a few ideally behaved proteins. The sensitivity, which was in the low- to mid-picomole range, was also not as good as desired, and finally, mixture analysis was subject to considerable selectivity owing to suppression effects.

In 1988, two new mass spectrometric techniques, which dramatically extended the potential of MS for protein analysis, were published, and it was soon appreciated that they were able to overcome most of the limitations mentioned. At the American Society for Mass Spectrometry (ASMS) conference in June in San Francisco, John Fenn from Yale University gave a lecture on the application of a new ionization technique, termed electrospray ionization (ESI), for protein analysis *(15)*. Those of us who attended the lecture walked away with the feeling that we had witnessed a real breakthrough for the mass spectrometric analysis of large biomolecules. A few months later at the International Mass Spectrometry Conference in Bordeaux, France, Franz Hillenkamp gave a lecture describing another new ionization technique, matrix-assisted laser desorption/ionization (MALDI) *(16)*. In this lecture, he showed molecular ions of proteins up to 117 kDa using time-of-flight analysis. MALDI seemed at least as promising for protein analysis as ESI.

A few years later, commercial instruments were available for both techniques, and the question was really which of the two techniques would be the future method of choice in the protein laboratory. In fact, at present, I consider the two techniques to be highly complementary. They have both dramatically improved the perspectives for the application of MS in protein chemistry to such an extent that a protein chemistry laboratory without access to these two techniques or at least one of them cannot be considered up to date. A common feature of both techniques is that, if the idea of a matrix is considered in its widest sense, both can be considered to be matrix-dependent just as FAB and PD. MALDI is, as indicated by its name, a matrix-dependent method. However, ESI, in spite of an entirely different ionization mechanism, can be considered to be matrix-dependent, since a prerequisite is that the analyte is dissolved in an appropriate solvent prior to ion formation in the electrospray process.

Mickey, unfortunately, died in May 1991, and therefore did not get the chance to see how his dream about the role of MS in protein studies and his concept of using a matrix have made their triumphal progress during the past 5 years. We in the scientific community have been deprived of the possibility to obtain his interpretation of which elements are common to the four techniques that have created the progress in MS applied to protein chemistry. In the absence of his interpretation, I will try to use the way of thinking and arguing I have experienced in my discussions with Mickey to outline what I consider to be the main function of the matrix. This is independent of whether it is the nitrocellulose support in PD, the liquid matrix in FAB, the solid matrix in MALDI, or the solvent in ESI.

A common feature is that the matrix/support is present in a large molar excess relative to the analyte. This indicates that a prime purpose of the matrix is to isolate the protein molecules and prevent aggregation. Next, the matrix must create a platform that can be removed, leaving the single analyte molecules free in the gas phase. This is effected by different means in the four techniques. In PDMS, the nitrocellulose support most likely decomposes on high-energy impact, so that the analyte molecules are left free and pushed off the surface by the resulting pressure wave. In FAB, the analyte molecules are sputtered from the liquid matrix surface, maybe still partly solvated in microdroplets of liquid matrix, followed by desolvation by multiple collisions just above the matrix surface. In MALDI, the solid matrix absorbs most of the laser energy, and decomposes or evaporates leaving the analyte molecules free in the expanding matrix plume. Finally, in ESI, the microdroplets created in the electrospray process by combined evaporation and coulombic explosions are subdivided until each droplet contains only one or a few analyte molecules, which, on final desolvation, leaves the analyte molecules free. The last step is that the analyte molecules must be ionized. Several different ionization mechanisms are without doubt operative in the different techniques and also within a single technique. In PD, FAB, and MALDI, chemical ionization is most likely to be the dominant ionization mechanism, although preformed ions, as well as other mechanisms, may also play a role. The ionization mechanism in ESI is still controversial, and it is outside the scope of this introductory chapter to enter this debate.

In summary, the matrix serves to isolate single analyte molecules, to create a removable platform from which the analyte molecules can be brought into the gas phase, and to create a medium that can ionize the analyte molecules. To be able to create ions of the proteins is, however, not sufficient to make the techniques usable in the protein laboratory. They must be compatible with the procedures generally used in protein chemistry in terms of sensitivity and acceptance of solvents, detergents, and buffers. They must be able to handle impure samples and complex mixtures. Last, but not least, the information gained must be of sufficient value to justify the effort and cost needed to obtain it. The numerous applications of MS to protein studies published during the last decade and the rapid acceptance of MS in the protein community clearly show that these conditions are now fulfilled. The following chapters in this volume describe these mass spectrometric techniques in detail and demonstrate a wide variety of applications: The use of MS for protein identification in combination with high-resolution separation techniques, such as 2D-PAGE, its compatibility with buffers and detergents, and its use in combination with HPLC are illustrated. Several methods for the sequencing of peptides, determination of disulfide bonds, and different methods for the localization and structure determination of secondary modifications, including glycosylation, are described. Even examples of tasks considered very difficult, such as the analysis of very hydrophobic proteins (e.g., membrane proteins), the highly specific quantitation of biologically active peptides, and studies of noncovalent interactions between proteins or between proteins and low-mol-wt ligands, are now within reach. In my mind, there is no doubt that MS will be an essential technique in all protein studies in the future.

References

1. Barber, M., Bordoli, R. S., Sedgwick, R. D., and Tyler, A. N. (1981) Fast atom bombardment of solids (FAB): a new ion source for mass spectrometry. *J. Chem. Soc. Chem. Commun.* **1981,** 325–327.
2. Shemyakin, M. M., Ovchinnikov, Yu. A., Kiryushkin, A. A., Vinogradova, E. I., Miroshnikov, A. I., Alakhov, Yu. B., et al. (1966) Mass spectrometric determination of the amino acid sequence of peptides. *Nature* **211,** 361–366.
3. Biemann, K. and Vetter, W. (1960) Separation of peptide derivatives by gas chromatography combined with mass spectrometric determination of the amino acid sequence. *Biochem. Biophys. Res. Commun.* **3,** 578–584.
4. Barber, M., Jolles, P., Vilkas, E., and Lederer, E. (1965) Determination of amino acid sequence in oligopeptides by mass spectrometry. I. The structure of fortuitine, an acyl-nonapeptide methyl ester. *Biochem. Biophys. Res Commun.* **18,** 469–473.
5. Das, B. C., Gero, S. D., and Lederer, E. (1967) *N*-methylation of *N*-acyl oligopeptides. *Biochem. Biophys. Res. Commun.* **29,** 211–215.
6. Benninghoven, A., Jaspers, D., and Sichtermann, W. (1976) Secondary-ion emission of amino acids. *Appl. Phys.* **11,** 35–39.
7. Torgerson, D. F., Skowronski, R. P., and Macfarlane, R. D. (1974) New approach to the mass spectrometry of non-volatile compounds. *Biochem. Biophys. Res. Commun.* **60,** 616–621.
8. Macfarlane, R. D. and Torgerson, D. F. (1976) Californium-252 plasma desorption mass spectrometry. *Science* **191,** 920–925.
9. Håkansson, P., Kamensky, I., Sundqvist, B., Fohlman, J., Peterson, P., McNeal, C. J., and Macfarlane, R. D. (1982) 127-I plasma desorption mass spectrometry of insulin. *J. Am. Chem. Soc.* **104,** 2948,2949.

10. Kamensky, I., Håkansson, P., Kjellberg, J., Sundqvist, B., Fohlman, J., and Petterson, P. (1983) The observation of quasi molecular ions from a tiger snake venom component (MW 13309) using 252-Cf plasma desorption mass spectrometry. *FEBS Lett.* **155,** 113–116.
11. Jonsson, G. P., Hedin, A. B., Håkansson, P. L., Sundqvist, B. U. R., Säve, G. S., Nielsen, P. F., et al. (1986) Plasma desorption mass spectrometry of peptides and proteins absorbed on nitrocellulose. *Anal. Chem.* **58,** 1084–1087.
12. Alai, M., Demirev, P., Fenselau, C., and Cotter, R. J. (1986) Glutathione as matrix for plasma desorption mass spectrometry of large peptides. *Anal. Chem.* **58,** 1303–1307.
13. Dell, A. and Morris H. R. (1982) Fast atom bombardment—high field magnetic mass spectrometry of 6000 Dalton polypeptides. *Biochem. Biophys. Res. Commun.* **106,** 1456–1461.
14. Barber, M., Bordoli, R. S., Elliot, G. J., Sedgwick, R. D., Tyler, A. N., and Green, B. N. (1982) Fast atom bombardment mass spectrometry of bovine insulin and other large peptides. *J. Chem. Soc. Chem. Commun.* **1982,** 936–938.
15. Meng, C. K., Mann, M., and Fenn, J. B. (1988) Electrospray Ionization of Some Polypeptides and Small Proteins. *Proceedings of the 36th ASMS Conference on Mass Spectrometry and Allied Topics,* San Francisco, CA, June 5–10, pp. 771,772.
16. Hillenkamp, F. (1989) Laser desorption mass spectrometry: mechanisms, techniques and applications, in *Advances in Mass Spectrometry,* vol. 11 (Longevialle, P., ed.), Heyden and Sons, London, pp. 354–362.

2

Mass Spectrometry

Ionization Methods and Instrumentation

John R. Chapman

"... There's jasmine! Alcohol there! Bergamot there! Storax there! Grenouille went on crowing, and at each name he pointed to a different spot in the room, although it was so dark that at best you could only surmise the shadows of the cupboards filled with bottles." (Patrick Süsskind, *Perfume*).

1. Introduction

Mass spectrometry (MS) *(1)* is one of the most important physical methods in analytical chemistry today. A particular advantage of MS, compared with other molecular spectroscopies, is its high sensitivity, so that it provides one of the few methods that is entirely suitable for the identification or quantitative measurement of trace amounts of chemicals. A mass spectrometer, in its simplest form, is designed to perform the following three basic functions:

1. Produce gas-phase ions from sample molecules. This is accomplished in the ion source and, at one time, would normally have required these neutral molecules to be already in the vapor state. New ionization techniques have, however, extended this process to neutral molecules which are essentially in a solid (condensed) state or in solution.
2. Separate gas-phase ions according to their mass-to-charge (m/z) ratio. This takes place in the analyzer.
3. Detect and record the separated ions.

Conventionally, the process of ion formation, just like ion analysis and detection, takes place in a vacuum. Some more recent methods, however, use instruments in which ions are produced in a source that operates at atmospheric pressure, although analysis and detection still require a vacuum environment.

From: *Methods in Molecular Biology, Vol. 61: Protein and Peptide Analysis by Mass Spectrometry*
Edited by: J. R. Chapman Humana Press Inc., Totowa, NJ

Table 1
Ionization Methods

Ionization method	Sample preparation for ionization	Thermal input associated with ionization	Method category
EI	As vapor	Relatively high	Thermal
CI	As vapor	Relatively high	Thermal
FAB	Dissolved in matrix such as glycerol	Virtually none	Energetic particle bombardment
MALDI	Mixed with matrix, such as sinapinic acid	Virtually none	Energetic particle bombardment
TS	Dissolved in solvent	Moderate	Field desorption (in some cases)
ES (or ion spray)	Dissolved in solvent	Virtually none	Field desorption
APCI	Dissolved in solvent	Moderate	Thermal

A large number of different instrumental configurations can be used to perform these three functions. For example, there are different sample inlet systems, different methods of ionization, and different mass analyzers. This chapter first looks at the various methods of ion formation that are available, in particular those that are applicable to macromolecules. The remainder of this chapter then deals with mass analyzers.

2. Ionization Methods

The ionization methods *(1)* that are generally available are summarized in Table 1.

2.1. Electron Ionization

Electron ionization or electron impact (EI) was the first ionization method to be used routinely and is still the most widely employed method in MS overall. Although of only marginal relevance in peptide and protein analysis, EI can conveniently be used to describe the main features of a mass spectrum. The EI source is a small enclosure traversed by an electron beam that originates from a heated filament and is then accelerated through a potential of about 70 V into the source. Gas-phase molecules entering the source interact with these electrons. As a result, some of the molecules lose an electron to form a posi-

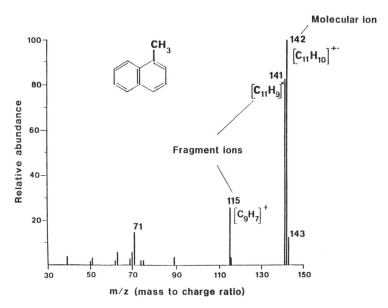

Fig. 1. EI spectrum of methylnaphthalene.

tively charged ion whose mass corresponds to that of the original neutral molecule: This is the molecular ion (Eq. [1]). Many molecular ions then have sufficient excess energy to decompose further to form fragment ions that are characteristic of the structure of the neutral molecule (Eq. [2]).

$$A - B \rightarrow [A - B]^{+\bullet} + e^- \tag{1}$$

$$[A - B]^{+\bullet} \rightarrow A^+ + B^\bullet \tag{2}$$

Thus, the molecular ion gives an immediate measurement of the molecular weight of the sample, whereas the mass and abundance values of the fragment ions may be used to elicit specific structural information. Taken together the molecular and fragment ions constitute the mass spectrum of the original compound (Fig. 1). Overall, an EI spectrum of an organic compound may be used as a fingerprint to be compared with existing collections of mass spectra. The principal collections of reference spectra, which are available in data system-compatible form, represent some 200,000 separate compounds. Computer-based searching of library data has other applications in MS, some of which (e.g., Chapter 6) are of more immediate relevance to peptide and protein analysis.

One other aspect of the spectrum in Fig. 1 that should be noted is the existence of isotope peaks, which correspond to the molecular ion and to each fragment ion. For example, although the molecular ion, at m/z 142, has the composition $C_{11}H_{10}$, its isotope ion, at m/z 143, has the composition $C_{10}{}^{13}CH_{10}$.

A more detailed consideration of the effects of isotope peaks is presented in the Appendix section of the book.

EI is suitable for the analysis of a large number of synthetic and naturally occurring compounds, but is limited by the need for sample vaporization prior to ionization. Thus, the conventional thermal vaporization routines that are used in conjunction with EI mean that this technique is quite unsuitable for the labile, involatile compounds that are encountered in biological work. On the other hand, the coupling of EIMS with capillary gas chromatography (GC/EIMS) is certainly the most widely used analytical technique in organic MS today.

2.2. Chemical Ionization

In chemical ionization (CI) MS *(2)*, ions that are characteristic of the analyte are produced by ion–molecule reactions rather than by EI. Again, just as with EI, the direct relevance of CI to peptide and protein analysis is small, but the basic CI process is very likely an integral part of more relevant ionization methods (*vide infra*).

CI requires a high pressure (approx 1 torr) of a so-called reagent gas held in an ion source that is basically a more gas-tight version of the EI source. EI of the reagent gas, which is present in at least 10,000-fold excess compared with the sample, eventually produces reagent ions (*see* Eqs. [3] and [4] for typical reactions of methane reagent gas), which are either nonreactive or react only very slightly with the reagent gas itself, but which react readily, by an ion–molecule reaction (Eq. [5]), to ionize the sample.

$$CH_4 \rightarrow CH_4^{+\bullet}, CH_3^+, CH_2^{+\bullet} \qquad (3)$$

$$CH_4^{+\bullet} + CH_4 \rightarrow CH_5^+ + CH_3^{\bullet} \qquad (4)$$

$$M + CH_5^+ \rightarrow MH^+ + CH_4 \qquad (5)$$

Thus, although many compounds fail to give a molecular ion in EI, the energetically mild ion–molecule reactions in CI afford an intense quasimolecular ion, indicative of molecular weight, with the same sample. For example, reagent ions from isobutane or methane ionize sample molecules by a proton transfer process (Eq. [5]), which leads to a positively charged quasimolecular ion at a mass that is 1 u higher than the true molecular weight. In addition, unlike EI, which produces only positive ions, CI can be used to produce useful ion currents of positive or negative ions representative of different samples.

Examination of the CI spectrum of histamine (Fig. 2) illustrates both the advantage (mol-wt information) and the possible disadvantage (little or no fragment ion information) of CI. Just as with EI, however, a major disadvantage of CI is the need for sample vaporization prior to ionization, which again rules

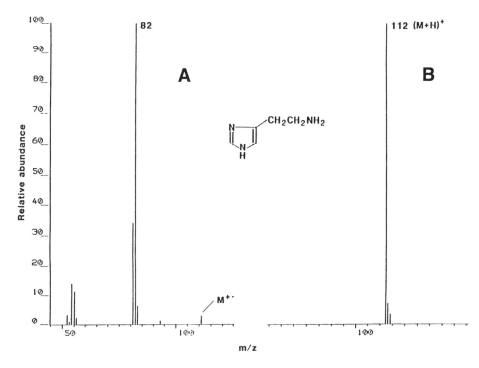

Fig. 2. (A) Electron impact and **(B)** CI spectra of histamine (M_r 111).

out any application to higher mol-wt, labile materials. On the other hand, CI processes, for example, the formation of $(M + H)^+$ ions, are implicated in a number of ionization techniques, such as fast-atom bombardment (FAB) (Section 2.3.) and matrix-assisted laser desorption/ionization (MALDI) (Section 2.4.), which are used for the analysis of macromolecules, as well as in techniques such as thermospray (TS) (Section 2.7.) and atmospheric pressure chemical ionization (APCI) (Section 2.6.), which can be used for the liquid chromatography (LC)/MS analysis of relatively labile molecules.

2.3. Ionization by Fast-Atom Bombardment (FAB)

The introduction of FAB *(3)* as an ionization method marked the first entry of an energetic-particle bombardment method (Table 1) into routine analysis, as well as the effective entry of MS into the field of biopolymer analysis. In such methods, the impact of an energetic particle initiates both the sample volatilization and ionization processes so that separate thermal volatilization is not required.

In the FAB source, a beam of fast moving neutral xenon atoms, (a) in Fig. 3, directed to strike the sample (b) which is deposited on a metal probe tip (c),

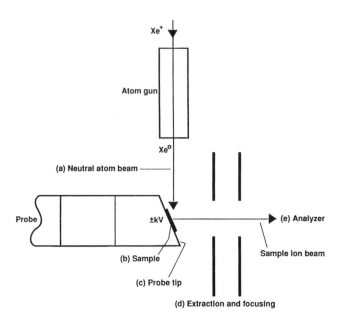

Fig. 3. FAB ion source.

produces an intense thermal spike whose energy is dissipated through the outer layers of the sample lattice. Molecules are detached from these surface layers to form a dense gas containing positive and negative ions, as well as neutrals, just above the sample surface. Neutrals may subsequently be ionized by ion–molecule reactions within this plasma. Depending on the voltages used, positive or negative ions may be extracted into the mass analyzer (e). Subsequently, the neutral primary beam was replaced by a beam of more energetic primary ions, such as Cs^+, and this technique was named liquid secondary-ion mass spectrometry (LSIMS) *(4)*.

With a dry-deposited sample, there is a rapid decay in the yield of sample ions owing to surface damage by the incident beam. In FAB, however, the sample is routinely dissolved in a relatively involatile liquid matrix, such as glycerol. The use of a liquid matrix, which is an absolutely crucial element in the success of this method, provides continuous surface renewal, so that sample ion beams with a useful intensity may be prolonged for periods of several minutes. In addition, the matrix behaves, in the vapor phase, in the same way as a reagent gas in CI, for example by protonating the analyte (to form an $[M + H]^+$ ion) in the positive-ion mode. Most other methods for the analysis of involatile and/or labile materials (e.g., MALDI [Section 2.4.] and electrospray (ES) ionization [Section 2.5.]) also use a matrix in some form.

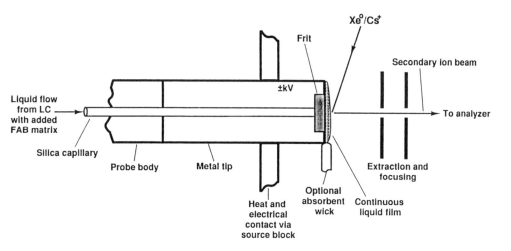

Fig. 4. Continuous-flow FAB ion source and liquid inlet.

The introduction of FAB saw the immediate extension of MS to the analysis of a wide range of thermolabile and ionic materials, as well as to biopolymers, such as peptides, oligosaccharides, and oligonucleotides. FAB is a relatively mild ionization process, so that fragment ions are generally of low abundance, or, particularly with analytes of higher molecular weight, absent altogether. Useful fragmentation can, however, be deliberately introduced by the use of MS/MS techniques (Section 3.1.).

FAB is also the basis of an effective coupling technique for LC/MS, viz. continuous-flow FAB (CF-FAB) *(4)*. In this technique (Fig. 4), a liquid flow, which is typically 5–10 µL/min split from the LC effluent, is directed toward a gently heated FAB source via a narrow fused-silica capillary. The flow, to which ~0.5% glycerol matrix has been added, enters the source through a small metallic frit interposed between the the capillary exit and the mass spectrometer vacuum. Not only is the metal surface of the frit easily wetted, but the thermal conductivity of the metal surface and the narrow orifices in the frit also encourage a stable liquid-evaporation process. As a result, the liquid flow forms a continuous film on the probe tip in which previously eluted sample is continually removed from the area where FAB takes place. An additional technique, sometimes used to promote a stable liquid film, is to place a layer of absorbent material against the edge of the probe to remove liquid from the tip.

2.4. Matrix-Assisted Laser Desorption/Ionization (MALDI)

Another obvious source of energetic particles for sample bombardment is the laser. Just as with FAB, the successful use of a laser was found to depend on the provision of a suitable matrix material with which the sample is admixed.

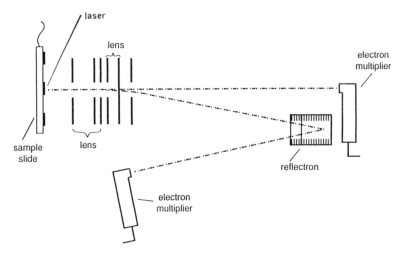

Fig. 5. MALDI source installed on a time-of-flight analyzer (Section 3.). Both linear and reflectron analysis modes are shown.

In this way, the technique of MALDI (Fig. 5) was conceived. In MALDI *(5)*, the matrix material (generally a solid and again present in large excess) absorbs at the laser wavelength and thereby transforms the laser energy into excitation energy for the solid system, a process that leads to the sputtering of surface molecular layers of matrix and analyte. A typical MALDI matrix compound (Appendix) displays a number of desirable properties:

1. An ability to absorb energy at the laser wavelength, whereas the analyte generally does not do so.
2. An ability to isolate analyte molecules within some form of solid solution.
3. Sufficient volatility to be rapidly vaporized by the laser in the form of a jet in which intact analyte molecules (and ions) can be entrained.
4. The appropriate chemistry so that matrix molecules excited by the laser can ionize analyte molecules, usually by proton transfer.

Using this technique, ionized proteins with molecular masses in excess of 200 kDa are readily observed—considerably greater than anything previously achieved. MALDI is, in fact, applicable to a wide range of biopolymer types, e.g., proteins, glycoproteins, oligonucleotides, and oligosaccharides. Again, with an appropriate choice of matrix, MALDI is more tolerant than other techniques toward the presence of inorganic or organic contaminants (Appendix VI). Thus, compared with FAB, MALDI has enormously extended the mol-wt and polarity range of samples for which an MS analysis is possible while providing an analytical technique that is easy to use and can be more tolerant of the difficulties encountered in purifying biochemical samples.

The principal ion seen in most MALDI spectra is an $(M + H)^+$ "molecular" ion. As the analyte molecular weight increases, however, doubly charged ions, $(M + 2H)^{2+}$, also will be detected, and triply charged analogs, and so forth, may be detected at still higher molecular weights. In addition, signals that correspond to molecular clusters, e.g., dimers $(2M + H)^+$ and trimers $(3M + H)^+$, may also be found in MALDI spectra. Negative-ion MALDI spectra are just as easily recorded and also show an equivalent range of molecular-ion types. Any of these "molecular" ions can provide a direct measurement of the molecular weight of the analyte, but useful fragmentation is usually absent. Recent developments (Chapter 4), however, have provided methods by which structurally meaningful fragment ions may be recorded in MALDI.

Unlike any of the ionization techniques mentioned so far, MALDI differs insofar as it is a discontinuous ionization technique. Analyte ions are only produced, for a very short period, each time the laser is fired. For this reason, conventional scanning analyzers, which operate on an inappropriate time scale, are replaced by a time-of-flight mass analyzer (Section 3.) for MALDI. Alternatively, the use of an integrating detector, with a magnetic sector analyzer for example, is also suitable for MALDI (Chapter 18).

2.5. Electrospray (ES) and Ion-Spray Ionization

In the ES ionization process *(6)*, a flow of sample solution is pumped through a narrow-bore metal capillary held at a potential of a few kilovolts relative to a counterelectrode ("filter" in Fig. 6B). Charging of the liquid occurs, and as a result, it sprays from the capillary orifice as a mist of very fine, charged droplets. This spraying process takes place in atmosphere and the whole ionization process is, in fact, a form of atmospheric pressure ionization (API).

The charged droplets, with a flow of a warm drying gas to assist solvent evaporation, decrease in size until they become unstable and explode (Coulomb explosion) to form a number of smaller droplets. Finally, at a still smaller size, the field due to the excess charge is large enough to cause the desorption of ionized sample molecules from the droplet. These ions, which are field desorbed (Table 1) from the droplets at atmospheric pressure, are then sampled through a system of small orifices with differential pumping, into the vacuum system for mass analysis.

ES ionization is a very mild process, with little thermal input overall, by which analyte ions may again be derived from molecules with molecular weights in excess of 100 kDa. Fragmentation is virtually absent, and only mol-wt information is available from the spectra. Useful fragmentation can, however, be deliberately induced when using ES, ion-spray (*vide infra*), or APCI (Section 2.6.), either by MS/MS techniques (Section 3.1.) or by an increase in the voltage between the cone plates in Fig. 6B. Thus, with a lower

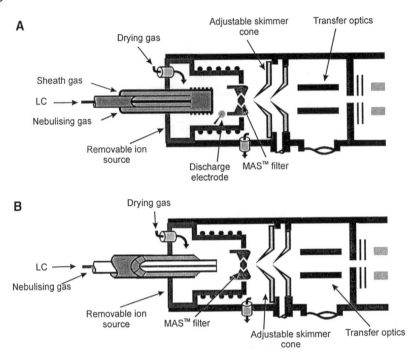

Fig. 6. (A) APCI source and **(B)** ES/ion-spray ion source. (Courtesy VG Instruments, Manchester, UK.)

voltage difference, ions will pass undisturbed through the intermediate pressure region between these plates. On the other hand, an increase in the voltage difference causes sample ions to undergo more energetic collisions with gas molecules in this region, and, as a consequence, the ions can dissociate into structurally significant fragment ions. Unlike the use of MS/MS techniques, however, this in-source collision-induced dissociation (CID) is not selective and has some effect, at a given voltage difference, on a significant proportion of all the different ion types passing through the region.

A particular feature of ES ionization spectra is that the molecular ions recorded are multiply charged, $(M + nH)^{n+}$, in the positive-ion mode, or $(M - nH)^{n-}$ in the negative-ion mode, and also cover a range of charge states. On average, one charge is added per 1000 Da in mass. Since mass spectrometers separate ions according to their mass-to-charge (m/z) ratio, rather than their mass, this means that, for example, an ion of mass 10,000 which carries 10 charges will actually be recorded at m/z 1000, thereby reducing the m/z range required from any analyzer. This is an especially convenient feature, since ES ionization is, by this means, able to generate, from relatively massive molecules, ions that can readily be analyzed using a simpler "low-mass" analyzer, such as a quadrupole mass filter (Section 3.).

In the ion-spray technique (Fig. 6B), a flow of nebulizing gas in an annular sheath, which surrounds the spraying needle, is used to input extra energy to the process of droplet formation. Using this technique, the practical upper limit for the liquid flow that can be sprayed to provide a stable ion current is increased from perhaps 10 μL/min in earlier ES sources to approx 1000 μL/min in newer ion-spray sources. Ion-spray also offers increased tolerance toward the presence of higher water levels in the solvent flow as well as being less affected by the presence of electrolytes and, as a result, provides a more robust system that is overall more suitable than ES for LC/MS.

With either ion-spray or ES, the coupling of low-flow chromatographic systems requires the use of a make-up flow, in the form of a liquid sheath around the spraying needle, to increase the liquid flow rate to a suitable value and/or to provide an overall solvent composition that is suitable for electrospraying. In general, LC/MS operation benefits from the very much simpler interfacing afforded by the use of an atmospheric pressure ion source. In particular, source access is excellent, and there are no vacuum effects that might affect the performance of lower flow columns.

ES and ion-spray ionization offer methods for the mol-wt determination of proteins and other biopolymers with very acceptable accuracy, and also offer convenient access, particularly by in-source dissociation, to some fragment-ion information. Again, as techniques in which ionization takes place in the liquid phase, they are highly compatible, especially in the ion-spray configuration, with LC/MS operation. On the other hand, ES is somewhat more susceptible to the presence of impurities and therefore perhaps less useful than MALDI as a "first-pass" analytical method.

2.6. Atmospheric Pressure Chemical Ionization (APCI)

In APCI *(7)*, the liquid flow, which carries the sample and which enters the atmospheric pressure source through a narrow tube, is nebulized by a coaxial stream of gas and by thermal energy from an adjacent heater (Fig. 6A). This process produces a vapor that contains both sample and solvent, and which is subsequently ionized by a corona discharge established, still in atmosphere, from an electrode located just after the nebulizing inlet.

Ionization of the sample takes place by means of a chemical ionization process, at atmospheric pressure, in which the solvent vapor is initially ionized by the discharge and then, very efficiently, ionizes sample molecules. This last step is an ion–molecule reaction with, for example, proton transfer from species such as $H(H_2O)^+$ in the positive-ion mode or reaction with solvated O^{2-} in the negative-ion mode. A stream of drying gas removes most clustering solvent molecules from the sample ions and prevents solvent neutrals from entering the mass analyzer. As with ES, the products of these atmospheric pressure processes are sampled, via a system of orifices, into the vacuum system of the mass spectrometer for mass analysis.

APCI provides a good basis for practical LC/MS interfacing. The APCI source can handle 1–2 mL/min of most solvents, and the atmospheric pressure configuration simplifies interfacing and provides a source that is very often more sensitive than alternative techniques, such as TS (Section 2.7.). The large overall size of the source, together with careful routing of gas flows, means that wall collisions are minimized during the ionization process and, as a result, relatively labile molecules may be analyzed routinely by LC/APCI-MS. Despite these advantages, however, the use of thermal volatilization (cf Table 1) prior to ionization means that APCI is not directly applicable to the analysis of proteins or peptides.

2.7. Thermospray (TS) Ionization

In the TS ionization process *(8)*, a volatile electrolyte, usually ammonium acetate, is added to an aqueous or partly aqueous solution of the sample, which flows through a heated capillary, introduced into the TS source. The capillary heating vaporizes most of the liquid flow so that the remainder, with the sample and ammonium acetate still in solution, is sprayed from the capillary exit, by the vapor, into the ion source. Although the solution is overall electrically neutral, statistical fluctuations ensure that each of these tiny droplets bears a slight excess positive or negative charge from the added ammonium acetate.

As in ES ionization (Section 2.5.), the charged droplets decrease in size owing to evaporation until they become unstable and explode to form much smaller droplets. Under appropriate conditions, the desorption of intact ionized sample molecules from the smallest highly charged droplets is possible. Alternatively, since the ion source, unlike ES, operates at an elevated temperature, more volatile neutral analyte molecules may be transferred directly to the gas phase during the droplet evaporation process and then ionized by an ion-molecule process. If the TS source is used with a mainly organic solvent, insufficient ionization occurs unless an auxiliary source of ions, such as a heated filament or discharge electrode, is used. TS ionization is a relatively mild ionization process, so that, in many cases, only ions indicative of the molecular weight are seen and structurally informative fragment ions are absent. Again, however, MS/MS may be used as an ancillary technique.

TS is also a practical LC/MS interfacing technique that, owing to the rotary pump attached to the ion source, will readily accept flow rates of 1–2 mL/min. This pump is able to remove most of the solvent vapor so that only a small fraction has to be pumped via the source housing. The same source, with the assistance of an additional discharge electrode or filament, will also accept a

Table 2
Upper *m/z* Value and Resolving Power (RP)
for Various Mass Analyzers

Mass analyzer	m/z^a	RP^a
Quadrupole	4000	Unit mass
Double-focusing sector	10,000	50,000
Time-of-flight	1,000,000	500
Ion trap	2500	Generally low
FTMS	High, but see RP	High, e.g., >>50,000, but depends on *m/z*

[a]These figures relate to routine use and do not represent an absolute limit to technical capabilities.

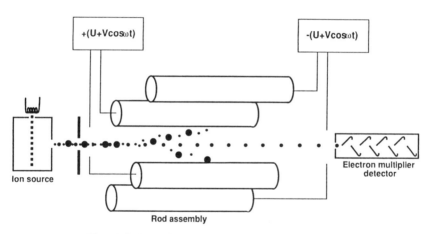

Fig. 7. Schematic of a quadrupole mass analyzer.

wide range of solvent polarities. TS has been used for the analysis of many different solute types, including small peptides, but shows poor sensitivity with higher mol-wt analytes, and has not been used in a practical manner for the analysis of peptides or proteins.

3. Mass Analyzers

The function of the mass analyzer (Table 2) is to separate ions according to their mass-to-charge (*m/z*) ratio. Figure 7 shows one of the most commonly used analyzers—the quadrupole mass filter *(9)*. In this device, a voltage made up of a DC component U and an RF component Vcos ωt is applied between adjacent rods of the quadrupole assembly, whereas opposite rods are connected electrically. With a correct choice of voltages, only ions of a given *m/z* value

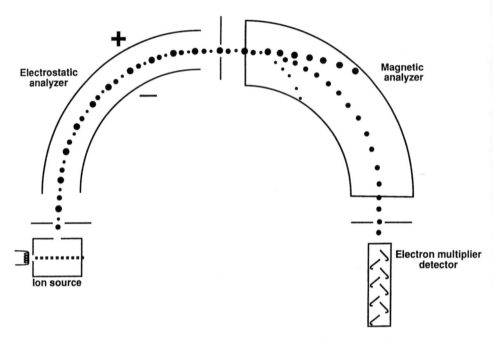

Electrostatic analyzer

Magnetic analyzer

Ion source

Electron multiplier detector

Fig. 8. Schematic of a double-focusing magnetic sector analyzer.

can traverse the analyzer to the detector, whereas ions having other m/z values collide with the rods and are lost. By scanning the DC and RF voltages, while keeping their ratio constant, ions with different m/z ratios will pass successively through the analyzer. In this way, the whole m/z range may be scanned and a complete mass spectrum recorded.

In a magnetic sector analyzer (Fig. 8), accelerated ions are constrained to follow circular paths by the magnetic field. For any one magnetic field strength, only ions with a given m/z ratio will follow a path of the correct radius to arrive at the detector. Other ions will be deflected either too much or too little. Thus, by scanning the magnetic field, a complete mass spectrum may be recorded, just as with a quadrupole analyzer. When the magnetic analyzer is operated in conjunction with an electrostatic analyzer (Fig. 8), the instrument then provides energy as well as direction focusing and is capable of attaining much higher mass resolution. This double-focusing magnetic sector instrument *(10)* is also much more suitable than the quadrupole analyzer for the analysis of ions of higher m/z ratio. Detection of ions is accomplished, in all the instruments discussed so far, by a device, such as an electron multiplier, placed at the end of the analyzer. The output from the electron multiplier is then directed toward some kind of recording facility, usually a data system.

For routine analyses, the foregoing analyzers can be operated in one of two modes. The first of these is the scanning mode where the mass analyzer is scanned over a complete mass range, perhaps from m/z 1000 to m/z 40 and usually repetitively, in order to record successive full spectra throughout an analysis. This mode of operation provides a survey analysis where the spectra provide information on every component that enters the ion source during the analysis. The other mode is called selected-ion monitoring (SIM). In this case, the instrument is set to successively monitor only specific m/z values, chosen to be representative of compounds sought. This type of analysis detects only targeted compounds, but does so with a much higher sensitivity because of the longer monitoring time devoted to the selected m/z values compared with the scanning mode.

Magnetic sector analyzers have been used in high-mol-wt analysis for some time because of their high specifications for m/z range and mass resolution (Table 2) and because of their versatility, e.g., as part of more complex MS/MS instruments (Section 3.1.) or used with an integrating detector (Chapter 18). Quadrupole analyzers now provide equally useful facilities in the high-mol-wt area through the analysis of multiply charged ions from ES ionization (Section 2.5.). In addition, although certainly of lower specification, quadrupoles demand a less sophisticated approach to operation and can be more tolerant of operation at high pressure. Again, the quadrupole analyzer is also an integral part of many MS/MS instruments (Section 3.1.).

A third analyzer system, which is increasingly commonly used, notably with MALDI, is the time-of-flight analyzer *(11)* (Fig. 5). In this simple device, ions are accelerated down a long, field-free tube to a detector. The m/z ratio of each ion is calculated from a measurement of the time from their start, e.g., the ion-formation laser pulse in MALDI, to the time at which they reach the detector. Unlike the quadrupole and magnetic sector analyzers, ions with an m/z ratio other than that which is being currently recorded are not rejected; all ions that leave the ion source can, in principle, reach the detector. With this type of instrument, therefore, there is no real distinction between scanning and SIM modes. A time-of-flight analyzer of improved mass resolution, the so-called reflectron instrument, uses an electrostatic mirror to compensate for energy differences among the ions.

As mentioned in Section 2.4., a time-of-flight analyzer is ideally suited to the analysis of ions that are created on a discontinuous basis, e.g., as the result of a laser pulse in MALDI. Time-of-flight analyzers have a very high sensitivity and a vitually unlimited m/z range (Table 2), but generally have not offered a particularly good mass resolution, although recent developments in this area *(12)* are very encouraging.

Another analyzer that does not distinguish between the scanning and SIM modes is the ion-trap analyzer *(13)* (Fig. 9). Ions are either made within the

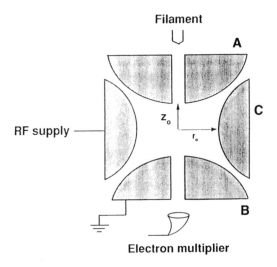

Fig. 9. Schematic of a quadrupole ion-trap analyzer.

trap, e.g., by EI, or injected into the trap from an external ion source, over a short period of time. These ions are then maintained in orbits within the box-like trap by means of electrostatic fields. After the ion formation period, ions within the trap are ejected, in order of m/z ratio, so that a conventional spectrum is recorded. Again, all ions that are formed within the ion source are eventually recorded. The ion trap is particularly of interest because of its suitability for a range of MS/MS experiments (Section 3.1.). The Fourier transform-ion cyclotron resonance (FTMS) instrument *(14)* (Fig. 10) is also based on a trapping analyzer, in this case, located within the solenoid of a superconducting magnet. A particular feature of FTMS instruments is their very high mass resolution (e.g., Chapter 9). A further advantage, as with the ion trap, is its suitability for a range of MS/MS experiments (Section 3.1.).

3.1. MS/MS Instruments

Tandem mass spectrometry or MS/MS (sometimes written MS^2) is an important technique that is proving to be increasingly useful in many areas of analysis *(15)*. Most MS/MS instruments consist of two mass analyzers arranged in tandem, but separated by a collision cell (Fig. 11). In an MS/MS instrument, sample ions of a specified m/z value can be selected by the first analyzer and then directed into the collision cell where they collide with neutral gas molecules. The use of a collision cell means that ion fragmentation is induced deliberately and in a specific region of the instrument. For example, in a triple quadrupole instrument (Fig. 11), the first analyzer is a conventional quadru-

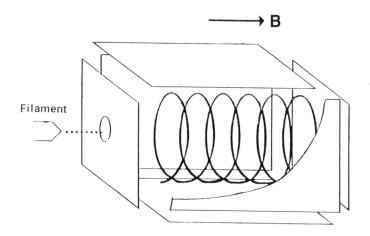

Fig. 10. Schematic diagram of FTMS instrument showing a circular ion path (B = magnetic field).

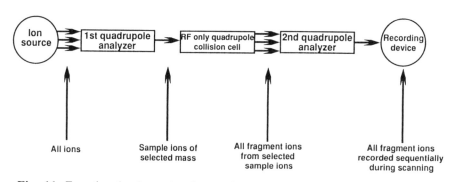

Fig. 11. Functional schematic of a tandem mass spectrometer based on a triple quadrupole instrument.

pole analyzer, set to transmit ions of the required m/z value, whereas the collision cell is another quadrupole analyzer that holds collision gas and to which only an RF voltage is applied to transmit fragmentation products of whatever m/z value. Scanning the second mass analyzer, in this case a third quadrupole, which follows the collision cell, will then record all those fragment ions that originate from fragmentation of the precursor ion selected by the first analyzer.

Other types of mass analyzers may be used in tandem to give alternative forms of MS/MS instrumentation. In particular, a collision cell may be interposed between a double focusing magnetic sector analyzer and a quadrupole analyzer to give what is known as a hybrid instrument. Alternatively, the collision cell may be interposed between two double-focusing magnetic sector analyzers to give a four-sector instrument. A four-sector instrument is particularly appropriate when CID experiments on ions with an *m/z* ratio much in excess of 1000 are to be carried out. As mentioned in Section 2.4., the specific use of time-of-flight instrumentation to study fragmentation processes in MALDI is dealt with in Chapter 4.

It should be emphasized that a number of the ions formed in the ion source will fragment in any case, without the intervention of a collision process. These less stable ions, which dissociate between the source and the detector, are known as metastable ions. CID is used, nevertheless, since it provides increased sensitivity, confines the observed fragmentation process to a specific part of the instrument, and can open up new, analytically useful, fragmentation routes.

Ion-trap and FTMS instruments are suitable for MS/MS experiments without being combined with other instrumentation and can provide facilities for MS/MS experiments that are based on the manipulation of an ion population in a single physical location. Thus, unlike the previous experiments, the steps of a trapping MS/MS experiment are separated in time, but not in space. Typically, all ions, except those of a selected *m/z* value, are ejected from the trap while the remaining, selected ions are energized and then made to collide with neutral gas molecules in the trap. A spectrum of the product ions from these collisions is then recorded in the usual manner. In particular, ion trapping lends itself very well to the implementation of sequential MS/MS steps or so-called (MS)^n experiments.

MS/MS experiments have found increasing use in conjunction with soft ionization techniques, such as FAB, ES or ion-spray, MALDI, and TS. For example, a conventional spectrum recorded by any of these techniques will quite often show mainly a quasimolecular ion, indicative of molecular weight, such as the ion at *m/z* 1233 in Fig. 12A. If, however, this same ion is selected for fragmentation by CID, the product ions may be detected using MS/MS techniques. The record of these fragment ions (Fig. 12B) is then a source of useful structural information (cf Chapter 3), which is unavailable in the original spectrum.

Another application of MS/MS techniques is to remove background ions, which may, for example, originate from the solvent in LC/MS or be ions resulting from unresolved components in the sample. Thus, if the first analyzer is set to transmit the quasimolecular ion of interest, then none of the background ions or ions resulting from unresolved components will reach the collision cell. Because of this, the collision-induced spectrum contains only information that relates to the component of interest, while other interfering ions have been eliminated. Different operating modes allow other valuable experiments to be carried out with MS/MS instrumentation. For example, a characteristic fragmentation may be monitored to detect a particular compound or members of a

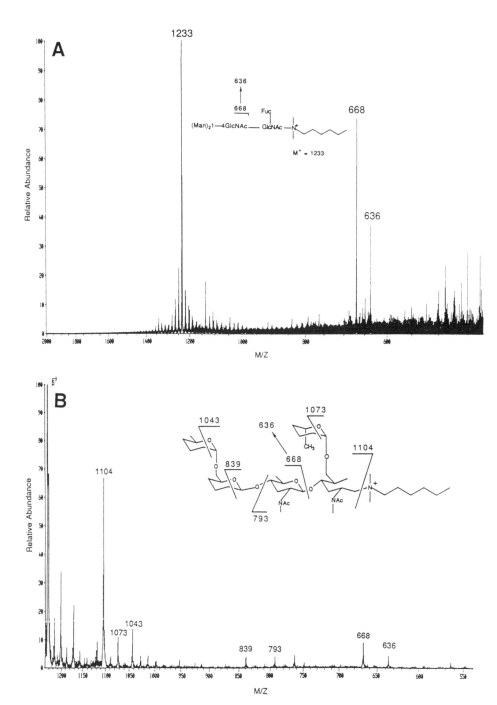

Fig. 12. (A) LSIMS spectrum of oligosaccharide after reductive amination and permethylation and **(B)** CID spectrum of the ion at *m/z* 1233 showing fragmentation (methyl esters have been omitted in the formula for clarity). Reproduced with permission from ref. *16*.

particular class of compounds. This technique, using the first analyzer to transmit a chosen precursor ion or ions and a second analyzer to transmit the chosen fragment ion or ions, is usually referred to as selected-reaction monitoring (cf SIM, Section 3.).

References

1. Chapman, J. R. (1993) *Practical Organic Mass Spectrometry,* 2nd ed. Wiley, Chichester, UK.
2. Harrison, A. G. (1983) *Chemical Ionization Mass Spectrometry.* CRC, Boca Raton, FL.
3. Barber, M., Bordoli, R. S., Elliott, G. J., Sedgwick, R. D., and Tyler A. N. (1982) Fast-atom bombardment mass spectrometry. *Anal. Chem.* **54,** 645A–657A.
4. Caprioli, R. M. (ed.) (1990) *Continuous-Flow Fast Atom Bombardment Mass Spectrometry.* Wiley, Chichester, UK.
5. Hillenkamp, F., Karas, M., Beavis, R. C., and Chait, B. T. (1991) Matrix-assisted laser desorption/ionization of biopolymers. *Anal. Chem.* **63,** 1193A–1203A.
6. Fenn, J. B., Mann, M., Meng, C. K., Wong, S. F., and Whitehouse, C. M. (1990) Electrospray ionization—principles and practice. *Mass Spectrom. Rev.* **9,** 37–70.
7. Bruins, A. P. (1991) Mass spectrometry with ion sources operating at atmospheric pressure. *Mass Spectrom. Rev.* **10,** 53–77.
8. Arpino, P. (1990) Combined liquid chromatography mass spectrometry. Part II. Techniques and mechanisms of thermospray. *Mass Spectrom. Rev.* **9,** 631–669.
9. Dawson, P. H. (1976) *Quadrupole Mass Spectrometry and Its Applications.* Elsevier, New York.
10. Duckworth, E., Barber, R. C., and Venkatasubramanian, V. S. (1988) *Mass Spectroscopy,* 2nd ed. Cambridge University Press, Cambridge, UK.
11. Cotter, R. J. (1992) Time-of-flight mass spectrometry for the structural analysis of biological molecules. *Anal. Chem.* **64,** 1027A–1039A.
12. Vestal, M. L., Juhasz, P., and Martin, S. A. (1995) Delayed extraction matrix-assisted time-of-flight mass spectrometry. *Rapid Commun. Mass Spectrom.* **9,** 1044–1050.
13. March, R. E. and Hughes, R. J. (1989) *Quadrupole Storage Mass Spectrometry.* Wiley, New York.
14. Asamoto, B. (ed.) (1991) *FT-ICR MS: Analytical Applications of Fourier Transform Ion Cyclotron Resonance Mass Spectrometry.* VCH, Weinheim, Germany.
15. Busch, K. L., Glish, G. L., and McLuckey, S. A. (1988) *Mass Spectrometry/Mass Spectrometry: Techniques and Applications of Tandem Mass Spectrometry.* VCH, New York.
16. Hogeland, K. E., Jr. and Deinzer, M. L. (1994) Mass spectrometric studies on the *N*-linked oligosaccharides of baculovirus-expressed mouse interleukin-3. *Biol. Mass Spectrom.* **23,** 218–224.

3

Charged Derivatives for Peptide Sequencing Using a Magnetic Sector Instrument

Joseph Zaia

1. Introduction

Peptides are sequenced using a magnetic sector tandem mass spectrometer by analyzing the product ions resulting from high-energy collisions between the peptide precursor ion and an inert gas *(1)*. These collisions result in product ion spectra with features specific to high-energy collision-induced dissociation (CID). Ideally, the product ion pattern will be sufficient to sequence the peptide analyte. For some peptides, however, the fragmentation pattern will not be amenable to complete interpretation, and further measures must be taken to improve the results. One such measure that has been found to improve the interpretability of CID spectra is derivatization of the N-terminus of the peptide with a fixed-charge bearing group *(2–4)*. Section 1.1. summarizes the salient features of the high-energy CID spectra of peptides and describes the rationale behind the best use of N-terminal charged derivatives. Several examples are shown in which charged derivatives improve the interpretability of CID spectra. Section 1.2. discusses synthetic schemes for attaching a charged group to a peptide N-terminus, and Sections 2. and 3. describe the synthesis of trimethylammoniumacetyl (TMAA)-peptide derivatives.

1.1. High-Energy CID

1.1.1. Collision Energy

Sector mass spectrometers work best in the high-collision-energy regime, which for practical purposes may be considered to be >2 keV in the laboratory frame of reference. Analyses at lower collision energies are accomplished only

From: *Methods in Molecular Biology, Vol. 61: Protein and Peptide Analysis by Mass Spectrometry*
Edited by: J. R. Chapman Humana Press Inc., Totowa, NJ

at the expense of ion transmission *(5,6)*. By contrast, low-energy collisions, in the range of 10–100 eV, are produced in triple-quadrupole *(7,8)* and hybrid i.e., sector-quadrupole *(9,10)*, mass spectrometers.

The ion series produced from high-energy collisions of peptides have been described *(3,11,12)*. These series are denoted a_n, b_n, c_n, and d_n for cleavages of the nth peptide bond, with charge retained by the N-terminal fragment. Ions produced with charge retention by the C-terminal fragment are denoted v_n, w_n, x_n, y_n, and z_n *(13)*. The ions denoted d_n, v_n, and w_n are produced by cleavage of the amino acid side chain, and such ions are found only in high-energy CID spectra *(3,11)*. The production of d_n and w_n series is considered to be favorable for the interpretation of CID spectra, because these ions allow leucine and isoleucine residues to be distinguished, whereas the v_n ion results from complete loss of the side chain and is therefore less useful *(4)*.

1.1.2. High-Energy CID Spectra

Generally, only a few of the described ion series are observed in each peptide CID spectrum. This is fortunate, because otherwise the pattern of product ions would be very complex, and the ion current would be divided between many product ions. The most interpretable spectra are those with one or two complete ion series, so that each peptide bond in the molecule is represented in the spectrum. In such a case, the sequence is determined by the mass difference between successive ions in a series. The production of more than two ion series occurs in high-energy spectra generally at the expense of the completeness of any given series, thus making interpretation more difficult. In the worst case, a sizeable region of the spectrum contains no product ions, and thus complete sequencing of the peptide is not possible.

It is worthwhile mentioning at this point that complete sequencing of a peptide is not necessary in cases where the sequence information is to be used in identifying the protein from which the peptide was produced. Recently, several computer methods have been described for matching partial CID data to a sequence data base for the purpose of identifying an unknown *(14)*.

1.1.3. Effects of Peptide Structure on High-Energy CID Fragmentation

1.1.3.1. Unmodified Peptides

The presence and location of arginine are a strong influence on the appearance of high-energy CID spectra of unmodified peptides. Arginine is by far the most basic amino acid residue in the gas phase *(15)*, and thus its side chain is the most favored site of protonation for singly charged ions *(3)*. Therefore, most of the ions produced in the CID spectrum of an arginine-containing peptide will contain that residue. For peptides containing an arginine at the C-terminus, it has been observed that C-terminal ions are produced and that these ions are predominantly v_n and w_n series *(3,11)*. For the corresponding peptides with an N-terminal arginine, a_n and d_n series predominate. Since the proton resides on the arginine side chain, the generation of product ions must occur by charge-remote fragmentation in these peptides *(16)*. A terminal arginine residue provides for optimal sequencing of a peptide by high-energy CID because only two ion series are produced, and these series are generally observed to be complete.

An arginine residue situated in the middle of a peptide results in a mixture of N-terminal and C-terminal product ions, depending on the site of backbone cleavage relative to the arginine. The result is a spectrum that is more difficult to interpret and may include gaps in the product ion pattern. Such a pattern can be seen clearly in Fig. 1A, which shows the CID spectrum of ACTH (SYSMEHFRWGKPVG).

Peptides lacking an arginine residue will generally produce a mixture of N-terminal and C-terminal ions resulting from CID. The completeness of the CID spectra and the presence of gaps in the fragmentation patterns vary greatly from peptide to peptide. Generally, of course, the interpretability of the CID spectra of these peptides will decrease with increasing mass. An example is shown in Fig. 2A for the peptide α-endorphin (YGGFMTSEKSQTPLVT). This spectrum, like the one shown in Fig. 1A, has sizable gaps in its product ion profile and would not be fully interpretable if the precursor ion were produced from an unknown peptide.

1.1.3.2. Modified Peptides

The presence of modifying groups strongly influences the appearance of high-energy CID spectra. Since cysteines are often alkylated in protein sequencing procedures, it is fortunate that the most common sulfhydryl-modifying groups (methyl, peracetyl, acetamidomethyl, vinylpyridyl) produce no detrimental effects on the CID profile of peptides. A peptide bearing a phosphate group on a serine, threonine, or tyrosine side chain will fragment during CID to form a moderately abundant ion, resulting from the loss of H_3PO_4 from the protonated molecular ion *(17,18)*. This ion serves as an identifying tag for a phosphorylated peptide, while not adversely influencing the peptide sequence fragmentation of the molecule. A stronger effect is observed for a peptide with a palmitylated cysteine. This modification has been shown to produce a very facile cleavage, resulting in the loss of the palmityl group from the precursor ion *(19)*. The resulting ion is the major product ion in the spectrum and renders interpretation of the spectrum troublesome.

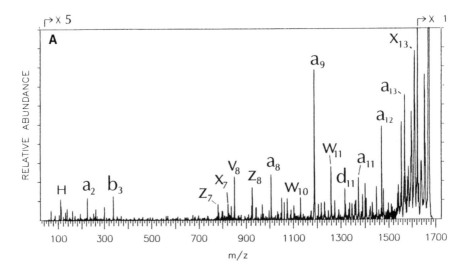

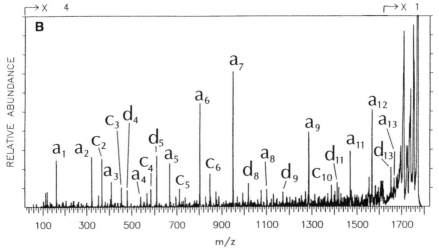

Fig. 1. CID spectra of **(A)** ACTH 1–14 (SYSMEHFRWGKPVG) $[M + H]^+ = 1680.3$ and **(B)** TMAA-SYSMEHFRWGKPVG $[M]^+ = 1780.0$.

Xenobiotic groups may have a strong influence on CID fragmentation, particularly as the size of the group increases. For example, diol-epoxide metabolites of polycyclic aromatic hydrocarbons have been detected covalently bound to peptides, and the resultant CID spectra are often dominated by ions produced from the adduct moiety *(20–22)*. An example is shown in Fig. 3A, for the peptide EGVYVHPV, which carries a benzo[*a*]-pyrene-triol moiety on the imidazole side chain of histidine. The abundant

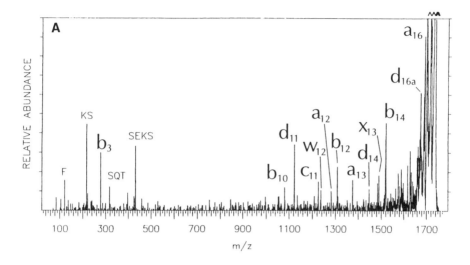

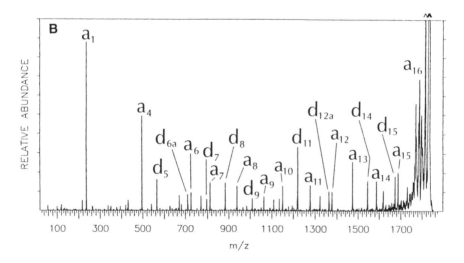

Fig. 2. CID spectra of **(A)** α-endorphin (YGGFMTSEKSQTPLVT) $[M + H]^+ =$ 1745.5 and **(B)** TMAA-YGGFMTSEKSQTPLVT $[M]^+ = 1844.4$.

ion at m/z 257 is produced from the triol moiety by cleavage of the bond between the triol and the imidazole nitrogen *(22)*. The peptide sequence ions are low in abundance relative to the ion at m/z 257. The nitrogen mustard drugs chlorambucil and melphalan are known to alkylate cysteine residues in metallothionein *(23,24)*. CID spectra of peptides derivatized with these drugs are also strongly influenced by fragmentation of the modifying moiety.

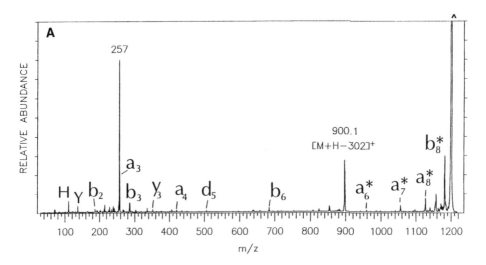

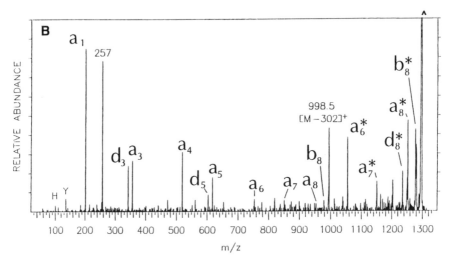

Fig. 3. CID spectra of B*a*P-triol modified **(A)** EGVYVHPV [M + H]$^+$ = 1201.9 and **(B)** TMAA-EGVYVHPV [M]$^+$ = 1300.9. The B*a*P-triol is attached at the histidine side chain. The ions labeled with an asterisk include the triol moiety in their structure.

1.2. Fixed-Charge Derivatives

1.2.1. Background

An effective approach for directing the production of complete series of a$_n$ and d$_n$ ions from peptides is the use of N-terminal fixed-charge-bearing derivatives *(3)*. The N-terminus is chosen because there are methods available for the

Table 1
Comparison of N-Terminal-Charged Derivatives
Used for High-Energy CID of Peptides

Class	Derivative[a]	Product ion pattern[b]	Ionization efficiency[c]	Synthesis[d]	References
Dimethylalkyl-	TMAA	++	0	+	*4,27,28*
ammonium	DMHAA	+	+	+	
	DMOAA	+	+	+	
Substituted	TMPB	–	ND	–	*3,28*
pyridinium	PyrA	0	–	+	*28*
	4-NMPA	–	ND	–	
	3-NMPA	–	ND	–	
Quaternary	TPPE	+	+	–	*29*
phosphonium					
Sulfonium	THTA	–	ND	+	*e*

[a]Abbreviations: TMAA, trimethylammoniumacetyl; DMHAA, dimethylhexylamineacetyl; DMOAA, dimethyloctylamineacetyl; TMPB, trimethylpyryliumbutyl; PyrA, pyridiniumacetyl; 4-NMPA, *N*-methyl-4-pyridiniumacetyl; 3-NMPA, *N*-methyl-3-pyridiniumacetyl; TPPE, triphenylphosphoniumethyl; THTA, tetrahydrothiopheniumacetyl.

[b]All derivatives direct the formation of a_n and d_n ions. ++, minimal unfavorable fragmentation of the derivative moiety is observed; +, derivative moiety fragments to produce moderately abundant ions *(28)*; 0, derivative moiety has been observed to produce abundant ions that may detract from the quality of the spectrum; –, derivative is generally unsuitable because of high-abundance ions produced from fragmentation of the derivative moiety.

[c]Ionization efficiency has been measured for TMAA, DMHAA, DMOAA, PyrA, and TPPE derivatives. +, positive effect; 0, neutral effect; –, negative effect of derivatization on ionization efficiency; ND, not determined.

[d]The ease of synthesis of derivatized peptides: +, facile synthesis; –, problematic synthesis. For the derivatives receiving a (–) rating in either of the other columns, the synthesis has not been optimized.

[e]Zaia and Biemann, unpublished results.

selective modification of the N-terminus in the presence of lysine *(25)*. The selective modification of the C-terminus, on the other hand, is not straightforward. To date, quaternary ammonium (dimethylalkylammoniumacetyl and pyridinium) *(3,4,26–28)*, triphenylphosphoniumethyl (TPPE) *(29)* and tetrahydrothiophenium (THTA) (Zaia and Biemann, unpublished results) groups have been used for the charged derivatization of the N-terminus for high-energy CID (*see* Table 1).

The precursor ion bearing a fixed charge carries an immobile charge rather than a potentially mobile proton charge. In theory, all product ions are produced by charge-remote fragmentation directed by the fixed-charge group *(16)*. In practice, the expected fragmentation is observed, although rearrangements that allow the charge to be transferred to other sites on the peptide occur *(28)*.

These charge transfer processes occur most often when an arginine is present in the peptide and is located near the N-terminus.

Three factors that should be considered when choosing a charged derivative are the effect on peptide product ion yield, the effect on ionization efficiency, and the ease of synthesis. These criteria are summarized in Table 1. All the listed derivatives are known to produce the expected a_n and d_n ions, although the substituted pyridines are prone to rearrangements that give rise to other types of product ions *(22)*. The dimethyloctylamineacetyl (DMOAA) *(22,27)*, dimethylhexylamineacetyl (DMHAA) *(27)*, and TPPE *(29)* derivatives are reported to have a positive effect on product ionization efficiency, whereas the effect of the TMAA derivative is neutral *(22,27)*. The dimethylalkylammonium derivatives may be synthesized by placing an iodoacetyl group on the N-terminus of the peptide and then reacting with the appropriate amine (for further discussion, *see* Section 1.2.3.1.). The substituted pyridinium derivatives are attached to the N-terminus by using carbodiimide chemistry. The TPPE derivative is synthesized by reacting the peptide with vinyltriphenylphosphonium bromide. The derivatives found to be least likely to rearrange to produce detrimental product ions, to have a positive or neutral effect on ionization efficiency, and to be easily synthesized are the dimethylalkylammonium derivatives *(28)*.

1.2.2. Examples of the Use of Charged Derivatives

Figures 1A and 2A show the CID spectra of two poorly fragmenting peptides, adrenocorticotropic hormone (ACTH) 1–14 and α-endorphin, respectively. The former peptide has a midchain arginine, and the latter has no arginine residue. TMAA derivative spectra of these peptides are shown in Figs. 1B and 2B, respectively. The effect of TMAA derivatization is to produce a complete series of a_n and d_n ions for both peptides, although the glycines do not produce charge-remote ions, as seen in Fig. 2B by the absence of the a_2 and a_3 ions. With the knowledge that a fixed positive charge at the N-terminus produces a_n and d_n ions, useful information can be gained by noting the mass difference between a_n and d_n ions for the same residue in an unknown peptide. This mass difference identifies the residue proximal to the backbone cleavage and serves as another tool to aid in the interpretation of the spectrum.

Figure 3A shows the CID spectrum of the peptide EGVYVHPV, which has been modified at histidine with a B*a*P-triol molecule. Figure 3B shows the TMAA derivative spectrum of the same peptide *(22)*. The peptide sequence product ions are more abundant relative to the *m/z* 257 ion in the latter spectrum than in the former. The improved peptide fragmentation allows much easier interpretation of the CID spectrum of the TMAA-derivatized peptide.

1.2.3. Synthesis of N-Terminal Fixed-Charge Derivatives

Discussion in this section is limited to the dimethylalkylammonium and the TPPE derivatives, which are the most widely used for peptide CID. Sections 2. and 3. then provide practical details for the specific preparation of trimethylammonium acetyl derivatives. Preparation of dimethyloctylammonium acetyl or dimethylhexylammonium acetyl derivatives may be accomplished by substitutuing the appropriate amine for trimethylamine.

1.2.3.1. GAS-PHASE SYNTHESIS OF DIMETHYLALKYLAMMONIUM DERIVATIVES

The dimethylalkylammonium peptide derivatives were first synthesized with a gas-phase derivatization procedure employing chloroacetylchloride followed by treatment with the appropriate amine *(4)*. This procedure is appropriate for subnanomolar quantities of peptide, although some specialized vacuum equipment is required. Despite the fact that the peptides are dried for the derivatization procedure, selective alkylation of the N-terminus is reported. Since pH control is necessary for selective alkylation, it would appear that drying the peptide from an acidic solution leaves the peptide in protonated form, thus providing the conditions for preferential alkylation of the N-terminus. A significant advantage of gas-phase synthesis is that no sample cleanup is required before acquiring mass spectra.

1.2.3.2. SOLUTION-PHASE ACETYLATION
IN DIMETHYLALKYLAMMONIUM DERIVATIVE PREPARATION

One procedure entails treating the peptide at pH 6.0 with excess iodoacetic anhydride, followed by a dimethylalkylamine to form the derivative *(25,27)*. The advantages of this procedure are that no specialized equipment is required and that the pH of the reaction solution can be easily controlled. Practically, it may be necessary to adjust the reaction conditions in order to obtain an acceptable yield of the derivatized peptide. In a separate study, it was necessary to lower the pH to 4.0 in order to obtain high selectivity and yield *(28)*. A purification step is necessary, which is best accomplished using a short C4 reversed-phase high-performance liquid chromatography (HPLC) column *(see* Section 3.).

The procedure described in this chapter employs carbodiimide chemistry to attach an iodoacetic acid (IAA) molecule to the N-terminus. Carbodiimide chemistry works well in the pH range of 4.0–5.5, which favors modification of the N-terminus in the presence of lysine. The method described in Section 3. will give a 70–90% yield, depending on the peptide.

1.2.3.3. PREPARATION OF TPPE DERIVATIVES

The published method uses bromoethyltriphenylphosphonium bromide to alkylate the peptide at pH 9.0 *(29)*. The commercially available vinyltriphenylphosphonium bromide may be substituted as an alkylating agent (Y.-S. Chang, personal communication). A disadvantage of this procedure is that at pH 9.0, it is not possible to alkylate the N-terminus of a peptide selectively in the presence of lysine. Purification of the alkylated product by HPLC is also a problem, because excess reagent has been found to coelute with many derivatized peptides *(28)*. These drawbacks make use of TPPE less attractive than use of the dimethylalkylammonium derivatives.

2. Materials

1. 1-(3-Dimethylaminopropyl)-3-ethylcarbodiimide (EDC) (Sigma, St. Louis, MO).
2. IAA (Sigma).
3. 4-Morpholinoethanesulfonic acid (MES) (Sigma).
4. HPLC-grade acetonitrile (ACN) (J.T. Baker, Phillipsburg, NJ).
5. 25% Aqueous trimethylamine (TMA) (Aldrich, Milwaukee, WI).
6. Glacial acetic acid (Aldrich).

3. Method for TMAA Derivatives (*see* Note 1)

1. Dissolve the peptide (see Notes 2 and 3) in MES/ACN (30% MES pH 4.0 [*see* Note 4]/70% ACN [v/v]).
2. Add 2% by volume IAA solution (*see* Note 5) (100 mg/mL neutralized IAA; see Note 6) and vortex.
3. Add 2% by volume EDC solution (*see* Note 7) (10 mg/mL in 50% ACN) and vortex.
4. Allow the reaction to proceed for 30 min at room temperature.
5. Add 25% by volume TMA solution (25% [w/w] aqueous).
6. Let the reaction proceed for 30 min at room temperature.
7. Acidify the reaction mixture by adding 10% by volume glacial acetic acid.
8. Remove the ACN by placing the mixture in a vacuum rotary evaporator for 15 min.
9. Desalt the mixture using a short reversed-phase column (*see* Note 8).

4. Notes

1. DMOAA or DMHAA derivatives may also be prepared by this procedure by substituting the appropriate amine for TMA.
2. The reagents are used in the proportions of EDC:IAA:TMA (1:10:1000). The concentration of peptide in the reaction is 100 pmol/µL for nanomolar quantities of peptide, although a lower concentration must be used for lower quantities of peptide.
3. Conical glass vials are recommended for this reaction.
4. MES is adjusted to pH 4.0 with HCl. This pH, below the normal buffering range for MES, is necessary to ensure that the N-terminus is selectively alkylated in the presence of lysine. MES is used because it has neither amino nor carboxyl groups, which interfere with the carbodiimide reaction.
5. IAA is added to the reaction before EDC to minimize reaction between EDC and peptide carboxyl groups.

6. The IAA must be adjusted to approx pH 4.0. This may be accomplished by adding 1 µL 10*N* NaOH to 25 µL of 100 mg/mL IAA solution. Alternatively, IAA may be purchased as its sodium salt and dissolved in MES, pH 4.0.

7. Care must be taken to store EDC under nitrogen or argon in a desiccated container at −20°C.

8. Desalting is best accomplished using a reversed-phase precolumn, such as an LC-304 (Supelco Inc., Bellefonte, PA).

Acknowledgments

This work was carried out in the laboratory of Klaus Biemann at the Massachusetts Institute of Technology. Grant support was provided by The National Institutes of Health (grant numbers NIEHS ES04675, NIH GM05472, and RR00317).

References

1. Sato, K., Asada, R., Ishihara, M., Kunihiro, F., Kammei, Y., Kubota, E., et al. (1987) High-performance tandem mass spectrometry: calibration and performance of linked scans of a four-sector instrument. *Anal. Chem.* **59,** 1652–1659.

2. Kidwell, D. A., Ross, M. A., and Colton, R. J. (1984) Sequencing of peptides by secondary ion mass spectrometry. *J. Am. Chem. Soc.* **106,** 2219,2220.

3. Johnson, R. S., Martin, S. A., and Biemann, K. (1988) Collision-induced fragmentation of $(M + H)^+$ ions of peptides. Side chain specific sequence ions. *Int. J. Mass Spectrom. Ion Processes* **86,** 137–154.

4. Vath, J. E. and Biemann, K. (1990) Microderivatization of peptides by placing a fixed positive charge at the N-terminus to modify high energy collision fragmentation. *Int. J. Mass Spectrom. Ion Processes* **100,** 287–299.

5. Yu, W. and Martin, S. A. (1994) Enhancement of ion transmission at low collision energies via modifications to the interface region of a four-sector tandem mass spectrometer. *J. Am. Soc. Mass Spectrom.* **5,** 460–469.

6. Cheng, X., Wu, Z., Fenselau, C., Ishihara, M., and Musselman, B. D. (1995) Interface for a four sector mass spectrometer with a dual-purpose collision cell: high transmission at low to intermediate energies. *J. Am. Soc. Mass Spectrom.* **6,** 175–186.

7. Yost, R. A. and Enke, C. G. (1978) Selected ion fragmentation with a tandem quadrupole mass spectrometer. *J. Am. Chem. Soc.* **100,** 2274,2275.

8. Hunt, D. F., Yates, J. R., Shabanowitz, J., Winston, S., and Hauer, C. R. (1986) Protein sequencing by tandem mass spectrometry. *Proc. Natl. Acad. Sci. USA* **83,** 6233–6237.

9. Schoen, A. E., Amy, J. W., Ciupek, J. D., Cooks, R. G., Dobberstein, P., and Jung, G. (1985) A hybrid BEQQ mass spectrometer. *Int. J. Mass Spectrom. Ion Processes* **65,** 124–140.

10. Bean, M. F., Carr, S. A., Thorne, G. C., Reilly, M. H., and Gaskell, S. J. (1991) Tandem mass spectrometry of peptides using hybrid and four-sector instruments: a comparative study. *Anal. Chem.* **63,** 1473–1481.

11. Johnson, R. S., Martin, S. A., Biemann, K., Stults, J. T., and Watson, J. T. (1987) Novel fragmentation process of peptides by collision-induced decomposition in a tandem mass spectrometer: differentiation of leucine and isoleucine. *Anal. Chem.* **59,** 2621.

12. Biemann, K. (1990) Sequencing of peptides by tandem mass spectrometry and high-energy collision-induced dissociation. *Methods Enzymol.* **193,** 455.

13. Biemann, K. (1988) Contributions of mass spectrometry to peptide and protein structure. *Biomed. Environ. Mass Spectrom.* **16,** 99–111. (This reference is a modification of that proposed by Roepstorff, P. and Fohlman, J. [1984] Proposal for a common nomenclature for sequence ions in mass spectra of peptides. *Biomed. Mass Spectrom.* **11,** 601.)

14. Mann, M. and Wilm, M. (1994) Error tolerant identification of peptides in sequence databases by peptide sequence tags. *Anal. Chem.* **66,** 4390–4399.

15. Wu, Z. and Fensleau, C. (1992) Proton affinity of arginine measured by the kinetic approach. *Rapid Commun. Mass Spectrom.* **6,** 403–405.

16. Tomer, K. B., Crow, F. W., and Gross, M. L. (1983) Location of double bond position in unsaturated fatty acids by negative ion MS/MS. *J. Am. Chem. Soc.* **105,** 5487,5488.

17. Biemann, K. and Scoble, H. A. (1987) Characterization by tandem mass spectrometry. *Science* **237,** 992–998.

18. Gibson, B. W. and Cohen, P. (1990) Liquid secondary ion mass spectrometry of phosphorylated and sulfated peptides and proteins. *Methods Enzymol.* **193,** 480–501.

19. Papac, D. I., Thornburg, K. R., Bullesbach, E. E., Crouch, R. K., and Knapp, D. R. (1992) Palmitylation of a G-protein coupled receptor. *J. Biol. Chem.* **267,** 16,889–16,894.

20. Hutchins, D. A., Skipper, P. L., Naylor, S., and Tannenbaum, S. R. (1988) Isolation and characterization of the major fluoranthene-hemoglobin adducts formed *in vivo* in the rat. *Cancer Res.* **48,** 4756–4761.

21. Day, B. W., Skipper, P. L., Zaia, J., and Tannenbaum, S. R. (1991) Enantiospecificity of covalent adduct formation by benzo[*a*]pyrene *anti*-diol epoxide with human serum albumin. *J. Am. Chem. Soc.* **113,** 8505–8509.

22. Zaia, J. and Biemann, K. (1994) Characteristics of high energy collision-induced dissociation tandem mass spectra of polycyclic aromatic hydrocarbon diolepoxide adducted peptides. *J. Am. Soc. Mass Spectrom.* **5,** 649–654.

23. Yu, X., Wu, Z., and Fenselau, C. (1995) Covalent sequestration of melphalan by metallothionein and selective alkylation of cysteines. *Biochemistry* **34,** 3377–3385.

24. Jiang, L. and Fenselau, C. (1994) Identification of Metallothionein and Anticancer Drug Interaction Sites by Mass Spectrometry. *Proceedings of the 42nd ASMS Conference on Mass Spectrometry Allied Topics,* p. 65.

25. Wetzel, R., Halualani, R., Stults, J. T., and Quan, C. (1990) A general method for highly selective cross-linking of unprotected polypeptides via pH-controlled modifications of N-terminal α-amino groups. *Bioconjugate Chem.* **2,** 114–122.

26. Vath, J. E., Zollinger, M., and Biemann, K. (1988) Method for the derivatization of organic compounds at the sub-nanomole level with reagent vapor. *Fresnius Z Anal. Chem.* **331,** 248–252.

27. Stults, J. T., Lai, J., McCune, S., and Wetzel, R. (1993) Simplification of high-energy collision spectra of peptides by amino-terminal derivatization. *Anal. Chem.* **65,** 1703–1708.

28. Zaia, J. and Biemann, K. (1995) Comparison of charged derivatives for high energy collision-induced dissociation tandem mass spectrometry. *J. Am. Soc. Mass Spectrom.* **6,** 428–436.

29. Wagner, D. S., Salari, A., Gage, D. A., Leykam, J., Fetter, J., Hollingsworth, R., and Watson, J. T. (1991) Derivatization of peptides to enhance ionization efficiency and control fragmentation during analysis by fast atom bombardment tandem mass spectrometry. *Biol. Mass Spectrom.* **20,** 419–425.

4

New Instrumental Approaches to Collision-Induced Dissociation Using a Time-of-Flight Instrument

Bernhard Spengler

1. Introduction

Described here is one of the most powerful methods of mass spectrometry (MS) for molecular structure analysis, introduced under the acronym "PSD-MALDI" (postsource decay with matrix-assisted laser desorption/ionization) MS *(1–6)*. It has found rapid and wide acceptance *(7–11)*, mainly because of its attractive features, such as high sensitivity, high tolerance of various sample conditions, speed of analysis, and limited instrumental requirements. All major mass spectrometer companies have now extended their product line to support the PSD-MALDI mode for bioorganic structure analysis.

The method is based on mass analysis of product ions (PSD ions) formed in the first field-free drift region of a reflectron time-of-flight (TOF) mass spectrometer. In contrast to mol-wt analysis by regular MALDI-MS (optimized precursor ion stability), the yield of fragment ions is enhanced by choosing an appropriate ion source design while product-ion mass analysis is performed by evaluation of characteristic flight times of PSD ions in the reflector.

The method is easy to use, but the success of an analysis most of the time depends on intelligent use of instrumental or methodological parameters. Knowledge of the basic mechanisms of PSD-MALDI is therefore rather important. Section 1.1. focuses on this topic. Section 1.1. is then followed by a description of the materials and instrumentation required and finally by two protocols for typical applications.

1.1. Mechanisms

The number of instrumental and methodological parameters that can be used for control of PSD-MALDI analysis is rather large, but the underlying mecha-

From: *Methods in Molecular Biology, Vol. 61: Protein and Peptide Analysis by Mass Spectrometry*
Edited by: J. R. Chapman Humana Press Inc., Totowa, NJ

nisms of fragmentation are quite simple. The detailed physical processes are not fully investigated yet, so that the following brief description just outlines a helpful working model.

1.1.1. Sample/Matrix Crystallization

The matrix works as a mediator for sensitive analyte molecules by performing three main tasks:

1. Separation and solvation of analyte molecules in the solid phase, avoiding agglomeration of analyte material;
2. Uptake of laser energy used for desorption and ionization of matrix and analyte molecules; and
3. Analyte ionization (protonation) via acid/base reactions in the gas phase.

The matrix in general has a high spectral absorption at the laser wavelength used, whereas the analyte usually (not necessarily) has a low absorption. Analyte solvation and codesorption require the formation, during preparation, of matrix crystals, which incorporate analyte molecules in the crystal lattice *(12,13)* (*see* Section 3.1.).

1.1.2. Desorption and Ionization

Matrix molecules are rapidly heated and excited by photon absorption, leading to a fast ejection of molecular material. Protonated matrix molecules (=matrix ions) are formed after a series of rapid photochemical reactions in a dense phase just above the surface *(14)*. Somewhat later, in a less dense phase (still within 1 mm of the surface), ionization of analyte molecules by proton transfer from matrix ions to neutral analyte molecules takes place, driven by the higher proton affinity of the analyte *(15)*.

1.1.3. In-Source Activation

Since the desorption plume is dominated by neutral matrix molecules, low-energy collisions of ions with these neutral target molecules during acceleration are rather frequent within the first microsecond after the laser pulse (Fig. 1). The density of the neutral cloud and the kinetic energy of colliding analyte ions are the two key factors for subsequent fragmentation *(16)*. Both a higher density and a higher collision energy result in more efficient activation and subsequent fragmentation (higher fragment ion signals). The neutral density is controlled mainly by the laser irradiance, the matrix sublimation temperature, and the focus diameter (*see* Note 1). The collision energy, on the other hand, is controlled mainly by the initial acceleration field strength.

A second mode of activation, driven by high-energy collisions of fully accelerated ions with residual gas molecules or with other target gas molecules *(1)*, is not discussed here, since its benefits have not been evaluated yet.

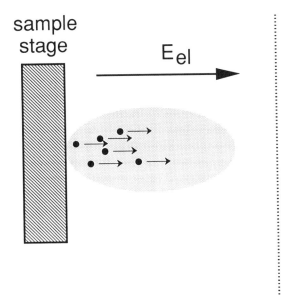

Fig. 1. In-source activation mechanism. Internal-energy uptake is a result of frequent low-energy collisions of analyte ions (black circles) with stationary matrix molecules (shaded area) during acceleration. The collision energy and the number of collisions are defined by the acceleration field strength and the particle density of the desorption plume.

1.1.4. Ion Decay

Collisionally activated ions of biomolecules undergo decay in the first field-free drift region of the mass spectrometer (Fig. 2) with unimolecular decay rates k typically in the range of 2000–30,000/s (1). Decay rates generally decrease with increasing ion mass. The long drift region of a TOF mass spectrometer is thus a good choice to achieve highly efficient fragmentation.

A considerable fraction of the fragment ions undergoes another or even a third or fourth fragmentation during the drift time. These secondary, tertiary, or quarternary products are usually called "internal fragments" if they contain neither end of the original analyte molecule.

1.1.5. Mass Analysis

Fragment ions (PSD ions) formed in the field-free drift region have the velocity of their precursor ions and, in a linear TOF instrument, would be detected at the same time. An electrostatic reflector, however, is able to disperse in time ions with different kinetic energies (*see* Fig. 2). The kinetic energy of PSD ions is a direct measure of their fraction of the precursor ion mass. Smaller frag-

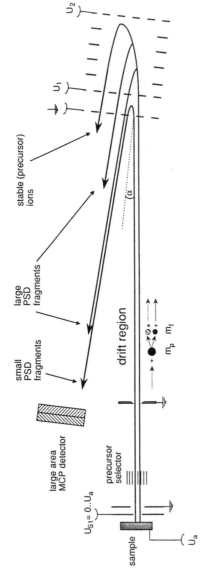

Fig. 2. Scheme of the main characteristics of a two-stage grid reflectron instrument for PSD analysis.

46

ments (with lower kinetic energy) are reflected earlier and thus take shorter pathways through the reflector than larger fragments. To image the complete range of PSD ions down to the lowest fragment masses, several spectra are acquired at different reflector potentials for different kinetic energy ranges. After passing the second field-free region of the TOF instrument, PSD ions are detected by a common ion detector, such as a microchannel plate or secondary electron multiplier.

1.1.6. Mass Calibration

The flight time dispersion function of PSD ions is not, as for stable ions, a simple square root function, but can only be fitted by a higher polynominal equation. In grid reflector instruments (using homogeneous electrical fields), flight times of PSD ions can additionally be calculated exactly, if the instrumental parameters (dimensions and voltages) are known precisely, and mass calibration can be performed on the basis of these calculations.

2. Instrumentation and Materials

2.1. Instrumentation

Commercial instruments designed for PSD-MALDI mass analysis are now available from all major mass spectrometer companies. Most of them, however, were originally developed as regular MALDI machines for mol-wt determination and were only upgraded later for the PSD mode. They cannot necessarily be regarded as optimized instruments for structure analysis yet. A skillful home-built modification of a MALDI instrument might therefore be a reasonable alternative to buying a commercial instrument. A brief description of the instrumentation necessary for PSD analysis is listed in the following:

1. Laser irradiation: All commercial instruments and most of the home-built machines use off-axis laser irradiation between 20 and almost 90° relative to the ion-beam axis. It has been found, however, that the mean angle of initial ion emission is not always directed along the surface normal, but can be tilted back into the direction of the laser beam, depending on the irradiance used and on the surface morphology *(15)*. The stability of the ion signal can thus deteriorate when using off-axis irradiation, especially under low-acceleration-field conditions. Using coaxial laser optics (irradiation of 0° relative to the ion beam) has been found to result in a considerable improvement in ion signal stability *(2,3)*. This configuration can be implemented by using a quartz focusing lens and a 45° mirror centrally bored for ion transmission.

2. Ion acceleration: Two-stage acceleration is mandatory for control of in-source activation (*see* Section 1.1.) at constant total kinetic energy. This can be done by using two grid electrodes in front of the sample. The initial field strength (first stage) should be variable between almost zero and at least 5×10^4 V/cm. If

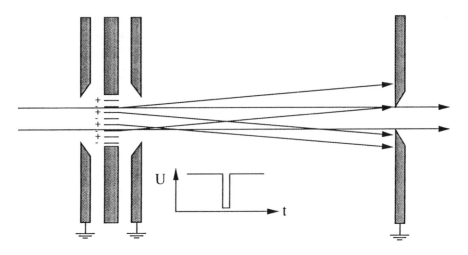

Fig. 3. Scheme of a precursor selection device using alternating deflection voltages.

gridless electrodes (ion lenses) are used, they should be designed to keep the ion transmission constant with variation of the initial field strength.

The optimal value for the total ion kinetic energy (the final acceleration potential) depends on the physical dimensions of the instrument (length of drift tube, deflection angle of ion beam, diameter of ion detector, detector type). The two parameters that determine the PSD ion detection yield are the decay rate and the kinetic energy released on unimolecular fragmentation. A longer drift time (obtained by using a lower total kinetic energy) results in a higher total yield of fragment ions, but on the other hand, leads to geometrical loss of ions (and thus to a lower detection yield) owing to radial velocities from the fragmentation event. Furthermore, a lower kinetic energy leads to peak broadening, since in that case, the distribution of additional energies released on fragmentation has a greater contribution to the total flight time distribution. In most instruments, a final acceleration potential of 10–20 kV has been found to be a good compromise.

3. Precursor ion selection: One of the most important features in PSD-MALDI analysis is its MS–MS capability, used for structure analysis of a certain precursor ion preselected from a set of stable ions. Use of this option is extremely helpful in the analysis of mixtures or unpurified samples, since it allows the omission of additional purification steps.

A precursor selector can, for example, be a set of wires or metal ribbons (Fig. 3), pulsed with potentials of alternate polarity for deflection of nonselected precursor ions. An aperture located at a distance of 100–300 mm is used to suppress deflected ions and to transmit selected ones. A typical selectivity of $M/\Delta M = 50$–100 (full width at half maximum) can be achieved with this setup.

The use of metal ribbons means that lower potentials are necessary for ion deflection, but wire systems are easier to build without considerable loss of ion

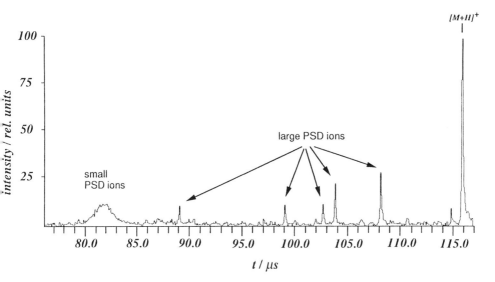

Fig. 4. Typical first section (full reflector potentials) of a peptide spectrum from a PSD mass analysis in raw data format.

transmission. In both cases, care has to be taken to avoid fringe fields influencing the total flight time of selected precursor ions and PSD ions. It has to be kept in mind that at the position of the precursor ion selector, a certain fraction of the precursor ions has already decayed into PSD ions, which are easily deflected or decelerated even by low fringe fields.

Deflection pulses should have short rise and fall times (≤50 ns) and should be controllable by the computer system. If the pulsing electronics are not home-built, a good (but rather expensive) alternative is to use a digital pulse/delay generator (Stanford Research Systems, Sunnyvale, CA, Model DG535) and a fast high-voltage switch (Behlke Electronic GmbH, Frankfurt, Germany, Model HTS 80-PGSM).

4. Ion reflector: Ion reflectors are generally used as flight time correction devices to improve the attainable mass resolving power in conventional TOF MS. In PSD analysis, they are used as energy filters since they result in different flight times for ions of identical velocity, but different kinetic energy. Different designs of ion reflectors have been used for PSD analysis, including single/dual-stage systems, on-axis/off-axis reflection, and grid/gridless electrodes. The question of which type of ion reflector is the best for PSD analysis cannot be answered in general. Two-stage grid reflectors have the advantage of a larger useful PSD mass range per reflector voltage setting. The broad peak of small ions reflected in the first stage can easily be used as an internal measure for the total ion current (Fig. 4). Since ion beams are not spatially refocused by grid reflectors, the rather strong divergence of PSD ion beams (as a result of kinetic energy release) requires

a large area detector to avoid ion signal loss. Microchannel plates with active diameters of 70 mm (Galileo Electro-Optics Corporation, Sturbridge, MA) have been used with great success *(2,3)*.

Gridless reflectors, owing to their inhomogeneous fields, work as ion lenses and are thus able to refocus divergent ion beams onto a relatively small detector. This, however, is true only for ions within a rather limited energy range. Smaller ions, which are reflected earlier, are not refocused, but are strongly defocused and are thus not detected even by a large area detector.

Single-stage grid reflectors have the advantage of a simple mass calibration function. The useful mass range is somewhat smaller compared to the two-stage grid reflector. Just as with gridless reflectors, no small-ion signal is obtained, so that the total ion current can only be monitored by simultaneously detecting neutral products in the linear mode. The large physical size is another disadvantage of the single-stage grid reflector, since the useful drift path available for ion decay is reduced.

On-axis reflectors using annular detector systems have some advantages over off-axis systems, since they are able to reduce the lateral dispersion of PSD ions of different mass. Secondary electrons and/or secondary ions formed on the grids of the reflector may, however, cause a considerable noise level in PSD analysis. In off-axis reflector systems, secondary particles can be eliminated geometrically.

2.2. Data Acquisition and Instrument Control

PSD mass analysis requires a much more sophisticated data-acquisition software compared with regular MALDI mass analysis. Instrumental parameters have to be controlled with high precision, and mass calibration has to be performed by complex fitting algorithms. The data systems of commercial PSD-MALDI instruments have implemented these functions with reasonable success. A generic software package for PSD mass analysis and instrument control, which can be adapted to almost any TOF system, is also available (Chips at Work Communication Systems GmbH, Bonn, Germany, "ULISSES 7.3").

High-voltage control has to be performed with high precision. For calibration fitting, only relative voltages have to be known, which is an easy task compared to the determination of absolute voltages needed for mass calculation procedures. In two-stage grid reflectrons, the required precision and stability of high voltages are roughly one and a half times the requested PSD mass accuracy. To obtain, for example, a mass accuracy of $\Delta m/m = \pm125$ ppm (=2000 $\pm$ 0.25 u), the reflecting potential has to be controlled with a precision of ±85 ppm. Digitally controlled high-voltage supplies with ±10 ppm stability are available, for example, from FuG Elektronik GmbH, Rosenheim, Germany (*see* Note 2).

The transient recorder and the computer should be fast enough to allow data acquisition and spectra summation with a rate of at least 10 Hz. This is possible with a Pentium PC and a transient recorder card (Precision Instruments Inc., Knoxville, TN, Model 9845) or with an averaging digital scope (e.g., LeCroy Corporation, Chestnut Ridge, NY, Model 9350).

2.3. Materials

1. 2,5-Dihydroxybenzoic acid.
2. α-Cyano-4-hydroxycinnamic acid.
3. Ethanol, methanol, acetonitrile.
4. Trifluoroacetic acid (TFA).

3. Methods

Two typical procedures for peptide structure analysis are described in the following.

3.1. Sequence Analysis of Unknown Peptides

1. Mix 0.5 μL of analyte solution (10^{-4}–$10^{-5}M$) in an Eppendorf cup with 0.5 μL of saturated matrix solution (e.g., 10 g/L 2,5-dihydroxybenzoic acid). The solvents used can be pure water or mixtures of water with ethanol, methanol, acetonitrile, TFA, or others, depending on the solubility of the analyte (*see* Notes 4 and 5).
2. Deposit the mixture or part of it on a polished preheated aluminum target (*see* Note 6) and dry it in a stream of warm air (using a hair dryer). For comments on the handling of small amounts of analyte, *see* Note 7.
3. When using α-cyano-4-hydroxycinnamic acid as a matrix, quickly wash the sample by addition of 2 μL of cold water. The water droplet must be removed after a few seconds.
4. Insert the sample into the instrument, and run a spectrum with the parent selector turned off and the reflector at full potential to make sure that the sample preparation is usable. Set the instrument to achieve optimized mass resolution for PSD ions. Check mass calibration for stable ions.
5. Set the precursor selector to the analyte ion, and acquire a few spectra. Set the acceleration field and the laser irradiance to obtain a high small-ion signal (the broad signal at about 70% of the precursor ion flight time in two-stage grid reflectrons) while still retaining a high precursor-ion signal (*see* Note 8).
6. Acquire the first averaged spectrum by summing 10–100 single-shot spectra. Monitor the small-ion signal to keep a fairly constant total fragment-ion intensity. If the sample spot is used up, move the target to focus on another spot. Changing the sample spot can be done without changes of mass calibration or spectra quality if the sample is flat and homogeneous. Store the averaged spectrum.
7. Switch reflector potentials for acquisition of the next section of the PSD spectrum. Acquire another 10–100 spectra for averaging, and store the spectrum. Repeat the procedure until the PSD ion mass range below $m/z = 70$ u has been reached.
8. Use the software concatenation procedures to form a recombined PSD spectrum (Fig. 5), or print out the acquired section spectra.

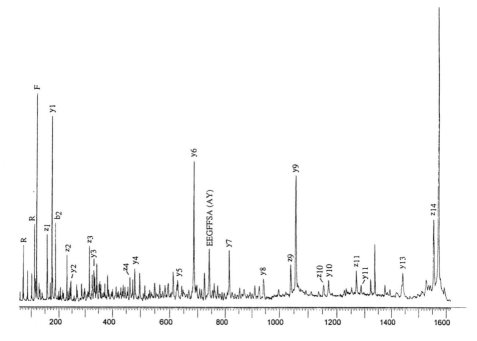

Fig. 5. Recombined PSD spectrum of (Glu^1)-fibrinopeptide B (EGVNDNEEGF-FSAROH). Monoisotopic molecular weight = 1569.7 u. Concatenation of section spectra includes mass scale linearization and vertical scaling to uniform relative sensitivities. No background subtraction or signal-to-noise enhancement routines were used.

9. Look for immonium ions and their smaller fragments in the mass range <200 u. Sort out amino acids (AAs) that are not part of the peptide.

10. Look for typical internal fragments in the mass range <400 u containing chains of two or three AAs.

11. List possible AA sum formulae resulting from the selected precursor mass on the basis of the detected immonium ions.

12. Try to verify AA chains starting at the N- or C-terminus by looking for a- and b-, or y- and z-type ions. Search for characteristic mass differences, such as 28 u (between a- and b-type ions) or 17 u (between y- and z- type ions). Expect to find (m − 17) ions (where m are fragment ion masses) resulting from loss of ammonia, especially if the peptide contains an arginine residue. If the peptide contains a basic residue, then the charge is usually located on the fragment that includes this residue rather than on the corresponding counterfragment.

13. Combine the information obtained to deduce the sequence information. If the information is not sufficient for a complete sequence, run on-target derivatization procedures (steps 14–18).

14. Perform on-target deuteration with the same sample *(17)*, and run another PSD spectrum.

15. Verify the AA composition, and rule out ambiguities by analyzing the immonium ion signals.
16. Determine the number of exchangeable hydrogens from the mass of the deuterated precursor ion. Use this information to eliminate incompatible AA sum formulae from the list in addition to impossible structural assumptions.
17. Compare the PSD spectra of the native and the deuterated sample, and rule out incompatible structural assumptions.
18. If the information is still insufficient or the number of possibilities is still too high, perform derivatization procedures to unequivocally distinguish between N-terminal and C-terminal ions, or to direct fragmentation to certain fragment ion classes.

3.2. Sequence Confirmation or Localization of Modifications in Known Peptides

1. Carry out PSD spectra acquisition as in Section 3.1., steps 1–8.
2. For modified peptides, detect the mass difference between the (expected) unmodified peptide and the recorded precursor ion mass.
3. Look for immonium ions and their smaller fragments in the mass range <200 u. Sort out AAs that are not part of the peptide. Verify the expected peptide composition.
4. Look for typical internal fragments in the mass range <400 u containing chains of two or three AAs. Verify the expected peptide structure, and try to find internal fragments that additionally contain the modification.
5. Try to verify the expected AA sequence starting at the N- or C-terminus by looking for a- and b-, or y- and z-type ions. Search for characteristic mass differences, such as 28 u (between a- and b-type ions) or 17 u (between y- and z-type ions). Expect to find $(m - 17)$ ions resulting from loss of ammonia, especially if the peptide contains an arginine residue. If the peptide contains a basic residue, then the charge is usually located on the fragment that includes this residue rather than on the corresponding counterfragment.
6. Determine whether the N- or the C-terminal ions contain the additional modification. Try to find the location of the modification by detecting the largest fragment ion of a series (N- or C-terminal) that still contains (or does not yet contain) the modification.
7. If information is insufficient, perform on-target deuteration with the same sample *(17)*, and run another PSD spectrum.
8. Verify the results from the undeuterated spectrum, and rule out ambiguities.
9. Determine the number of exchangeable hydrogens of the complete modified peptide and of the modification to aid in elucidation of its composition.
10. If the problem is still not solved, try other derivatization procedures or regard the sample as an unknown (*see* Section 3.1.).

4. Notes

1. It is a common misunderstanding that increased laser irradiance will lead to increased decay owing to direct photon-induced internal-energy uptake of analyte molecules/ions rather than conventional low-energy collisions of ions. It has been

shown that even very high laser irradiances lead to almost no fragmentation if the initial acceleration field is turned off.

2. Accuracy of PSD mass calibration is mainly limited by the accuracy of high-voltage control. Limiting parameters of high-voltage supplies are long-term and short-term stability, the temperature coefficient, and high-frequency ripple.

3. Different analyte classes require very different strategies for structure determination. Branched oligosaccharides usually form sets of PSD fragment ions of identical mass. In this case, evaluation of internal fragments (secondary, tertiary, quarternary ions) is much more helpful than evaluation of only the regular primary PSD ions, since internal fragments tend to image the branching structure *(18)*. PSD analysis of oligonucleotides is possible, but requires more fundamental studies of fragmentation behavior *(19)*.

4. Sample preparation plays an important role in the success of PSD analysis, although it is no more critical than for general MALDI analysis. Preparational artifacts can be obtained, especially in oligonucleotide analysis, but also in the analysis of peptides and oligosaccharides. Among the possible side effects are hydrolysis, chemical degradation, and adduct formation, often catalyzed by the acidic matrix.

5. MALDI is known to be rather tolerant with respect to sample impurities and physicochemical properties (acidity, hydrophobicity). About 100 fmol–1 pmol of sample is usually sufficient for a complete structural analysis of a peptide or glycopeptide. Oligosaccharides and oligonucleotides can be analyzed with about the same sensitivity. Buffer concentrations of up to 1 mmol/L are usually tolerated without loss of analytical information. Alkali impurities can be a problem if the precursor selector is unable to separate the protonated ion from the alkali-attached ion, since mixed PSD spectra are more complicated. It is therefore worth trying to have the analyte substance as alkali-free as possible. Doubly distilled water should be used for preparation. Oligonucleotides can be purified from alkali ions by means of ammonium ion exchange *(20)*.

6. The choice of the sample substrate is mainly determined by the analyte and by the chosen preparation procedure. For general purposes, polished aluminum targets are a good choice. Oligonucleotides, however, tend to form adducts with aluminum ions and show much clearer spectra with gold targets. Silver targets can be used with similar success, but cause adduct formation if the surface is visibly oxidized.

7. For the analysis of smaller amounts of analyte or more dilute analyte solutions (down to $10^{-6}M$), analyte and matrix solutions can be mixed on the target, or the analyte solution can be dropped onto a predried matrix spot (especially with α-cyano-4-hydroxycinnamic acid matrix diluted in acetonitrile *[21]*).

8. PSD ion analysis in general is possible only if the ion intensity of the precursor ion is rather high. The intensity of PSD ions can be expected to be not more than a small percent of the precursor ion intensity. It is, however, possible in cases of extremely low analyte precursor ion stability that the precursor ion is not detected at all, whereas PSD ions show up in normal intensities.

References

1. Spengler, B., Kirsch, D., and Kaufmann, R. (1992) Fundamental aspects of postsource decay in matrix-assisted laser desorption mass spectrometry. *J. Phys. Chem.* **96,** 9678–9684.
2. Kaufmann, R., Spengler, B., and Lützenkirchen, F. (1993) Mass spectrometric sequencing of linear peptides by product-ion analysis in a reflectron time-of-flight mass spectrometer using matrix-assisted laser desorption ionization. *Rapid Commun. Mass Spectrom.* **7,** 902–910.
3. Kaufmann, R., Kirsch, D., and Spengler, B. (1994) Sequencing of peptides in a time-of-flight mass spectrometer: evaluation of postsource decay following matrix-assisted laser desorption ionization (MALDI). *Int. J. Mass Spectrom. Ion Proc.* **131,** 355–385.
4. Spengler, B., Kirsch, D., and Kaufmann, R. (1991) Metastable decay of peptides and proteins in matrix assisted laser desorption mass spectrometry. *Rapid Commun. Mass Spectrom.* **5,** 198–202.
5. Spengler, B. and Kaufmann, R. (1992) Gentle probe for tough molecules: matrix-assisted laser desorption mass spectrometry. *Analusis* **20,** 91–101.
6. Spengler, B., Kirsch, D., Kaufmann, R., and Jaeger, E. (1992) Peptide sequencing by matrix assisted laser desorption mass spectrometry. *Rapid Commun. Mass Spectrom.* **6,** 105–108.
7. Yu, W., Vath, J. E., Huberty, M. C., and Martin, S. A. (1993) Identification of the facile gas-phase cleavage of the Asp-Pro and Asp-Xxx peptide bonds in matrix assisted laser desorption time-of-flight mass spectrometry. *Anal. Chem.* **65,** 3015–3023.
8. Huberty, M. C., Vath, J. E., Yu, W., and Martin, S. A. (1993) Site-specific carbohydrate identification in recombinant proteins using MALD-TOF MS. *Anal. Chem.* **65,** 2791–2800.
9. Critchley, W. G., Hoyes, J. B., Malsey, A. E., Wolter, M. A., Engels, J. W., and Muth, J. (1994) Structural Analysis of Mixed Base Oligonucleotides Using Negative Ion Matrix Assisted Laser Desorption Ionisation Time of Flight (MALDI-TOF) Mass Spectrometry Metastable Analysis. *Proceedings of the 42nd ASMS Conference on Mass Spectrometry and Allied Topics,* Chicago, IL, May 29–June 3, p. 140.
10. Hoyes, J. B., Curbishley, S. G., Doorbar, P., Tatterton, P., and Bateman, R. H. (1994) A New High Performance MALDI-TOF Instrument for PSD and CID Metastable Ion Analysis. *Proceedings of the 42nd ASMS Conference on Mass Spectrometry and Allied Topics,* Chicago, IL, May 29–June 3, p. 682.
11. Talbo, G. and Mann, M. (1994) Localization of Phosphorylation Sites in Peptides by MALDI Metastable Decay at the Subpicomole Level. *13th International Mass Spectrometry Conference,* Budapest, Hungary, August 29–September 2.
12. Beavis, R. C. and Bridson, J. N. (1993) Epitaxial protein inclusion in sinapic acid crystals. *J. Phys. D: Appl. Phys.* **26,** 442–447.
13. Strupat, K., Karas, M., and Hillenkamp, F. (1991) 2,5–Dihydroxybenzoic acid: a new matrix for laser desorption-ionization mass spectrometry. *Int. J. Mass Spectrom. Ion Proc.* **111,** 89–102.

14. Ehring, H., Karas, M., and Hillenkamp, F. (1992) Role of photoionization and photochemistry in ionization processes of organic molecules and relevance for matrix-assisted laser desorption ionization mass spectrometry. *Org. Mass Spectrom.* **27,** 472–480.

15. Bökelmann, V., Spengler, B., and Kaufmann, R. (1995) Dynamical parameters of ion ejection and ion formation in matrix-assisted laser desorption/ionization. *Eur. Mass Spectrom.* **1,** 81–93.

16. Spengler, B., Lützenkirchen, F., Kirsch, D., Deppe, A., and Kaufmann, R. (1994) Structure Analysis of Biopolymers by PSD-MALDI in a Reflectron Time-of-Flight Mass Spectrometer. *Proceedings of the 42nd ASMS Conference on Mass Spectrometry and Allied Topics*, Chicago, IL, May 29–June 3, p. 885.

17. Spengler, B., Lützenkirchen, F., and Kaufmann, R. (1993) On-target deuteration for peptide sequencing by laser mass spectrometry. *Org. Mass Spectrom.* **28,** 1482–1490.

18. Spengler, B., Kirsch, D., Kaufmann, R., and Lemoine, J. (1995) Structure analysis of branched oligosaccharides using post-source decay in matrix-assisted laser desorption ionization mass spectrometry. *J. Mass Spectrom.* **30,** 782–787.

19. Kirsch, D., Rood, H.-A., Kaufmann, R., and Spengler, B. (1994) Ion Stability of Deoxyoligonucleotides and its Implication for PSD-MALDI-MS. *Proceedings of the 42nd ASMS Conference on Mass Spectrometry and Allied Topics*, Chicago, IL, May 29–June 3, p. 821.

20. Spengler, B., Pan, Y., Cotter, R. J., and Kan, L.-S. (1990) Molecular weight determination of underivatized oligodeoxyribonucleotides by positive-ion matrix-assisted ultraviolet laser-desorption mass spectrometry. *Rapid Commun. Mass Spectrom.* **4,** 99–102.

21. Vorm, O., Roepstorff, P., and Mann, M. (1994) Improved resolution and very high sensitivity in MALDI TOF of matrix surfaces made by fast evaporation. *Anal. Chem.* **66,** 3281–3287.

5

Mass Spectrometric Quantification of Neuropeptides

Dominic M. Desiderio

1. Introduction

The molecular specificity that is attached to the quantification of an endogenous neuropeptide is an important experimental parameter *(1)*, because the other analytical methods that are commonly used, such as radioimmunoassay (RIA), provide a very high level of detection sensitivity, but only limited molecular specificity *(2)*. The purpose of this chapter is to describe the use of tandem mass spectrometry (MS/MS) methods to increase the specificity of the qualitative and quantitative analysis of endogenous neuropeptides in human tissue extracts.

The theory behind this experimental protocol for quantitative analytical MS is straightforward. The amino acid sequence of a neuropeptide is the only experimental datum that provides the highest level of molecular specificity *(3)*. An MS/MS measurement approaches that high level of specificity, because, following the desorption and ionization of the peptide to produce the protonated molecule ion, $(M + H)^+$, of the peptide, the first mass spectrometer (MS1) then selects only that $(M + H)^+$ ion, and the second mass spectrometer (MS2) collects only the product ions that derive from that $(M + H)^+$ ion. MS/MS therefore provides the needed structural link between a precursor ion, $(M + H)^+$, and its corresponding product ions, which are the amino acid sequence-determining fragment ions. An internal standard (IS), using a stable isotope-incorporated amino acid residue in the peptide, optimizes the molecular specificity *(4)*.

In this chapter, methionine enkephalin (ME = YGGFM), which derives from the protein preproenkephalin A, and β-endorphin (BE), which derives from proopiomelanocortin (POMC), are measured by MS/MS in extracts of human pituitary tissue (controls and tumors) *(4–8)* (*see* Note 1).

From: *Methods in Molecular Biology, Vol. 61: Protein and Peptide Analysis by Mass Spectrometry*
Edited by: J. R. Chapman Humana Press Inc., Totowa, NJ

2. Materials

1. Human pituitary tumor tissue: This is obtained in the surgery operating suite, and is placed into liquid nitrogen as soon as possible to inhibit any peptide and protein metabolism *(9)* *(see* Note 2).
2. Stable isotope-incorporated synthetic peptide internal standards; for example, [^{2}H$_5$-Phe^4]-ME and [^{2}H$_4$- Ile^{22}]-BE. All of the ^{2}H-atoms are in nonexchangeable positions *(see* Note 3).
3. A high-performance liquid chromatography (HPLC) instrument, including analytical columns, UV detector (200 nm), and recorder *(10–12)*.
4. An octadecylsilyl (ODS) analytical column, which is used for the gradient reversed-phase (RP)-HPLC purification of selected endogenous peptides in a biologic extract, using acetonitrile as the organic modifier and a volatile ion-pairing buffer, such as triethylamine-formic acid (TEAF).
5. A polystyrene-divinylbenzene analytical column for the isocratic purification of the ODS-purified peptide fraction, and for the removal of the ODS "bleed" resulting from the gradient elution. Bleed usually interferes with the liquid secondary ion mass spectrometry (LSIMS) desorption/ionization of the peptide.
6. A tandem mass spectrometer (MS/MS instrument), consisting of MS1, a collision-induced decomposition (CID) cell, and MS2. Various types of MS/MS instrumental configuration are commercially available, and include triple-sector quadrupole (TSQ), three- and four-sector, and hybrid (for example, E_1BE_2qQ) instruments *(see* Notes 4 and 5).
7. A sample desorption/ionization system, such as LSIMS *(see* Note 6), fast-atom bombardment (FAB), or alternatively, matrix-assisted laser desorption/ionization (MALDI) to provide a sufficient amount of $(M + H)^+$ ion current from the peptide for MS/MS analysis. Electrospray ionization (ESI) could also be used.
8. A computer for data acquisition and processing.

3. Methods
3.1. Sample Preparation

1. Tissue: The frozen pituitary tissue (postsurgical; postmortem) *(see* Note 1) is homogenized in acetic acid (1*M*; 10:1 [w:v]) with a homogenizer (Brinkmann, 13,000 rpm, 30 s) after adding the appropriate amount of internal standard *(see* Note 7). The ratio of the endogenous peptide:internal standard is usually within the range 1:4–4:1.
2. A SepPak ODS cartridge *(see* Note 8) is prepared for use by washing with water, methanol, and water; the sample is placed onto the cartridge; the cartridge is washed with water; and the peptide-rich fraction is eluted with a bolus of acetonitrile *(13)*.
3. A gradient of organic modifier, such as acetonitrile (generally 1–2%/min), and volatile ion-pairing buffer is used to separate the endogenous peptides. The UV absorption of the amide bond is monitored at 200 nm. The retention time of each peptide of interest is determined in a separate gradient, and the analytical HPLC

column is cleaned extensively to remove all traces of synthetic peptide; RIA is used to verify that the column is clean. Fractions are collected at the retention time of each native peptide of interest. A window for the peptide collection (usually, the calibrated retention time ± 1 min) ensures that all of the native peptide is collected and compensates for any slight shift (usually only a fraction of a minute to an earlier elution time) in retention time for endogenous peptides extracted from biological extracts.

4. The ODS bleed (*see* Note 9) is removed from the gradient RP-HPLC gradient fraction with a synthetic hydrocarbon column (e.g., as polystyrene-divinyl-benzene) (*see* Notes 10 and 11), and the corresponding signal-to-noise (S/N) ratio of the LSIMS-produced $(M + H)^+$ ion of the peptide is improved *(4)*.

3.2. Mass Spectrometry

1. The mass spectra of the endogenous peptide and of the internal standard are recorded using the first stage (MS1) of the tandem mass spectrometer. Although many of the expected amino acid sequence-determining fragment ions may be observed in this MS spectrum (which is usually a normal magnet scan spectrum), the precursor–product relationship between $(M + H)^+$ and its corresponding amino acid sequence-determining fragment ions cannot be established unequivocally. The $(M + H)^+$ ions of the endogenous peptide and of the internal standard are selected.

2. The operating parameters for CID of the $(M + H)^+$ ion are optimized. CID of the $(M + H)^+$ precursor ion will either produce, or increase, the intensity of the fragment ions. Thus, the nature of the collision gas, the collision gas pressure, and the collision energy can be optimized to produce the maximum amount of ion current of the particular amino acid sequence-determining fragment ions that will be used for peptide corroboration and quantification. In particular, the pressure of the collision gas within the CID cell is adjusted to reduce the intensity of the $(M + H)^+$ ion to ~50% of its original value and to produce an appropriate ion current of the fragment ions (*see* Notes 12 and 13).

3. The fragment-ion spectrum of the $(M + H)^+$ ion of the endogenous peptide is recorded using the second stage (MS2) of the tandem mass spectrometer (*see* Note 14).

4. Before quantification of a peptide, qualitative analysis of the peptide is performed to determine the presence of the appropriate amino acid sequence-determining fragment ions (*see* Note 15). The product-ion spectrum of the $(M + H)^+$ ion of the native peptide is compared to the product-ion spectrum of the $(M + H)^+$ ion of the appropriate synthetic peptide. If the fragment ions in these two spectra are equivalent, then the amino acid sequence of the native peptide is established unequivocally *(7)*. The optimum product-ion spectrum would consist either of all of the N-terminus- or of all of the C-terminus-containing amino acid sequence-determining fragment ions. Such an optimum spectrum is usually not obtained (*see* Note 16).

5. Data analysis: The quantity of the native peptide is obtained readily by comparing the ion current from the selected precursor → product-ion transition of the endogenous peptide to the corresponding current from the internal standard (*see*

Table 1
m/z Values for the Precursor and Product
Ions for ME and Its Internal Standard (IS)

	m/z		
	$(M + H)^+$	$\rightarrow$	YGGF-$^+$
ME	574	$\rightarrow$	425
IS	579	$\rightarrow$	430

Table 2
Neuropeptide Measurements in Postmortem
Human Pituitary Controls[a]

Control	ME	BE
1	59.4[b]	250.7[b]
2	33.9	379.3
3	50.7	503.2
4	43.9	419.3
5	151.7	180.9
6	28.0	88.5
7	66.5	71.8
8	251.8	181.5
9	30.8	88.7
10	25.2	134.6
11	19.9	241.8
12	27.6	71.9
Avg.[c]	65.8 ± 19.8	217.7 ± 44.2

[a]Controls: nine males and three females.
[b]MS/MS measurements of ME and BE are given as pmol/mg protein.
[c]Mean ± SEM.

Notes 17–19). For example, for the pentapeptide ME, Table 1 contains the precursor $\rightarrow$ product transitions that are monitored at the indicated m/z values.

3.3. Representative Quantitative Data

Quantitative analytical data are presented here, briefly, to indicate the type of data that can be obtained with MS/MS. For ME, the transitions shown in Table 1 were used. For BE, the tryptic pentapeptide BE_{20-24} (NAIIK, mol wt = 557 Da) and the NAI^+ fragment ion were used. Table 2 contains the measurements of ME and BE in 12 human postmortem pituitary controls, and Table 3 the

Table 3
Neuropeptide Measurements in Postsurgical
Human Nonsecreting Pituitary Tumors

Tumor	ME	BE
1	13.3[a]	4.5[a]
2	12.8	115.6
3	33.1	1.1
4	23.2	13.5
5	18.8	0.3
6	48.9	38.0
Avg.[b]	25.0 ± 5.7	28.8 ± 18.3

[a]MS/MS measurements of ME and BE are given as pmol/mg protein.
[b]Mean ± SEM.

measurements of ME and BE in six human postsurgical, nonsecreting pituitary tumors (5). These measurements were made to test whether metabolic defects in these neuropeptidergic systems may be a contributing factor to human anterior pituitary tumor formation, because, in addition to the hypersecretion of hormones (ACTH, GH, FSH, LH), hyposecretion of an opioid neuropeptide may be a new factor.

The decreases observed for the preproenkephalin A neuropeptide ME (controls 65.8 ± 19.8; tumors 25.0 ± 5.7 pmol ME/mg protein) and for the POMC peptide BE (controls 217.7 ± 44.2; tumors 28.8 ± 18.3 pmol BE/mg protein) corroborate the downregulation of these two opioid neuropeptidergic systems in the "nonsecreting" (or, "neuroendocrinologically silent") pituitary tumors vs controls (5; see also 14).

3.4. Summary

The molecular specificity of MS/MS methods for the accurate quantification of endogenous neuropeptides, such as ME and BE, exceeds that of all other methods because:

1. MS/MS qualitative analysis first establishes the amino acid sequence of the endogenous peptide;
2. MS/MS establishes and maintains the structural link between the $(M + H)^+$ ion and its corresponding amino acid sequence-determining ion(s) during quantification; and
3. A stable isotope-incorporated synthetic peptide internal standard is used.

Endogenous ME and BE have been measured in human pituitary tissue by this quantitative analytical MS method.

4. Notes

1. Intact endogenous preproenkephalin A and POMC OPCP are also being analyzed in pituitaries *(15,16)*.
2. The human pituitary tissue must be collected as fresh as possible and placed immediately into liquid nitrogen. The acquisition of human postmortem control tissue will have some unavoidable delays. The postmortem stability of several endogenous neuropeptides has been studied *(9)*, and it has been determined that the combined use of an acid and a rapid temperature lowering suffices to inactivate peptide-processing enzymes. The postsurgical tissue is collected into liquid nitrogen within seconds after excision. The objective during this rapid cooling step is to avoid any peptide biosynthesis and/or degradation. For example, the biosynthesis of an opioid peptide-containing protein (OPCP; *15,16*) in that tissue may continue within the immediate postmortem or postsurgical time frame, POMC may be degraded to BE, and BE may be metabolically degraded.

    During storage of tissue at $-70°C$, spontaneous air oxidation may occur of the sulfur atoms in the methionine residues. Thus, biologic tissue or fluid samples should not be stored for long periods of time.
3. The stable isotope (usually ^2H) in each amino acid must be in a nonexchangeable position during peptide synthesis, sample incubation, and peptide desorption/ionization.
4. A variety of MS/MS instruments is available including sector instruments, such as BEEB, hybrid instruments, such as EBEqQ, and triple sector quadrupoles (QqQ). Each type of instrument has its own inherent advantages and disadvantages (such as high/low ion energy, cost, and the mass resolution of the precursor ion and of the product ions), and thus an MS/MS instrument must be chosen rationally (*see* Note 16).
5. An important philosophy exists within this analytical scheme. Namely, the need for a very high level of chromatographic resolution decreases commensurately with the increase in the molecular specificity of the detection system. In this particular analytical scheme, because the MS/MS specificity is so high, chromatographic resolution could be intentionally decreased, but only within certain well-defined, specified, and experimentally tested limits. This point is important, because peptide losses generally increase with the number of manipulations, and it is always advantageous to minimize the number of sample-handling steps (transfers, lyophilizations, etc.). As the amount of sample decreases, the amount of those losses increases.
6. LSIMS is a very effective method for the desorption/ionization of peptides, and the high level of $(M + H)^+$ ion current produced is sufficient for MS/MS analysis.
7. The stable isotope-incorporated synthetic peptide internal standard should be added to the tissue homogenate as early as feasible in the separation scheme so that peptide losses, enzymatic degradation, oxidation, and so forth, can be compensated for accurately *(4)*. The objective in this step is not necessarily to achieve a very high recovery of peptide; that optimization study should be done separately *(13)*. Also, some sample incubation time is needed before separating the

peptides so that the exogenous peptide experiences the same cellular microenvironment as the native peptide before both of those peptides are extracted from the tissue homogenate, because the endogenous and exogenous peptides derive from two very different sources. In some studies, artificially higher levels of peptide recovery may be calculated when the internal standard is added to, and extracted quickly from, the tissue homogenate late in the separation scheme.

8. The SepPak cartridge removes cell debris, enzymes, proteins, lipids, and saccharides, and preferentially enriches the neuropeptide:buffer ion-pair before RP-HPLC. The rates of sample deposition, column washing, and sample elution are important experimental parameters, and they must be optimized *(13)*.

9. When it is necessary to collect several different neuropeptides *(6)*, such as ME, substance P (SP), and BE, that elute across a wide range of hydrophobicity, an RP-HPLC gradient is usually used. However, the ODS material from the analytical column will usually bleed and interfere with LSIMS, which is a surface-active ionization phenomenon.

10. Isocratic RP-HPLC (16:84 [v:v], CH_3CN:TEAF, 40 mM, pH 3.1) with a polystyrene-divinylbenzene analytical column effectively removes any ODS bleed, and improves the S/N ratio of the $(M + H)^+$-to-background ion current during LSIMS.

11. One could use exclusively an isocratic separation method of peptides from a tissue extract on a synthetic hydrocarbon column, but over time, background material would accumulate and the column must be cleaned on a regular basis. In that case, it is important to monitor the absolute amount of UV absorbance.

    It is important to be aware of these processes, because in general, one simply "electrically zeros" the HPLC recorder or computer, and a deceptive measurement of "peak purity" is obtained. However, a constant HPLC column bleed will always produce a background level of organic compound(s), and the mass spectrometer will detect all types of compounds in an HPLC fraction.

12. The experimental parameters of CID (e.g., the percent reduction of the precursor ion intensity, the various types of collision gases that could be used, and the optimization of the collision energy) must be studied carefully, optimized experimentally, and used appropriately for the MS/MS production of the product-ion spectrum from the precursor ion $(M + H)^+$. In particular, the ion yield of the unique amino acid sequence-determining fragment ion(s) used for MS/MS quantification must be optimized by CID.

13. It may be difficult for some mass spectrometrists to "throw away" such a high percentage of the $(M + H)^+$ ion current when using MS/MS for the quantification of an endogenous peptide. However, the noise level decreases even more rapidly in MS/MS, and the analytically more important S/N ratio increases.

14. MS2 is used to collect the fragment-ion spectrum that derives from either the unimolecular decomposition or the CID of the $(M + H)^+$ precursor ion *(17,18)*.

15. Differences exist in the sensitivity of qualitative analytical MS, which uses a full-scan analysis for the amino acid sequence determination of a peptide, and quantitative analytical MS, which uses MS/MS methods that have an increased sensitivity.

16. The ideal case for amino acid sequence determination would be the presence of all N-terminus-containing, all C-terminus-containing, or both sets of amino acid sequence-determining ions. However, usually only a few N- and C-terminus-containing ions, which overlap, are found.

17. The selection of the unique amino acid sequence-determining fragment ion(s) is important, and it must take into account several unique features of the experiment and of peptide neurochemistry. For example, the ion current of the selected $(M + H)^+ \rightarrow$ product-ion transition should be abundant; it should contain an amino acid residue that can have a stable isotope readily incorporated; and it must represent accurately the target peptide.

As an example of that last point, all opioid peptides contain the N-terminal tetrapeptide sequence YGGF-, and thus they usually produce an abundant product ion at m/z 425. That point is important, because even though a C-terminus-containing fragment ion would confer a higher level of molecular specificity to the quantitative measurement of the endogenous peptide, not all C-terminus-containing product-ions are present, nor do they have a sufficient level of intensity for MS/MS analysis and quantification.

18. Even though HPLC is employed in this analytical scheme, one must always question the purity of an HPLC-purified endogenous peptide. For this reason, the maximum molecular specificity that is provided by MS/MS as an HPLC detector is a significant advantage over all of the other analytical methods.

19. The use of several different product-ion transitions for the quantification of an endogenous peptide would increase even further the molecular specificity of this quantitative analytical MS procedure.

20. The need for mass resolution, which has been the goal of mass spectrometry for many years, is now decreased in this analytical procedure. Rather, what is now needed for peptide quantification is optimal "structural resolution," which is readily provided by the link that is established instrumentally by MS/MS between the $(M + H)^+$ of a peptide and the corresponding amino acid sequence-determining fragment ions *(1)*.

Acknowledgment

The author gratefully acknowledges the financial assistance from NIH (GM26666).

References

1. Desiderio, D. M. (1992) Quantitative analytical mass spectrometry of endogenous neuropeptides in human pituitaries, in *Mass Spectrometry: Clinical and Biomedical Applications,* vol. 1 (Desiderio, D. M., ed.), Plenum, New York, pp. 133–165.
2. Lovelace, J. L., Kusmierz, J. J., and Desiderio, D. M. (1991) Analysis of methionine enkephalin in human pituitary by multi-dimensional RP-HPLC, radioreceptorassay, radioimmunoassay, FAB-MS, and FAB-MS-MS. *J. Chromatogr. Biomed. Appl.* **562,** 573–584.
3. Desiderio, D. M. (1992) Minireview: quantitative analytical mass spectrometry of endogenous neuropeptides in human pituitaries. *Life Sci.* **51,** 169–176.

4. Kusmierz, J. J., Sumrada, R., and Desiderio, D. M. (1990) Fast atom bombardment mass spectrometric quantitative analysis of methionine enkephalin in human pituitary tissues. *Anal. Chem.* **62**, 2395–2400.

5. Desiderio, D. M., Kusmierz, J. J., Zhu, X., Dass, C., Hilton, D., Robertson, J. T., and Sacks, H. S. (1993) Opioid and tachykinin neuropeptides in non-secreting and ACTH-secreting adenomas. *Biol. Mass Spectrom.* **22**, 89–97.

6. Kusmierz, J., Dass, C., Robertson, J. T., and Desiderio, D. M. (1991) Mass spectrometric measurement of β-endorphin and methionine enkephalin in human pituitaries: tumors and post-mortem controls. *Int. J. Mass Spectrom. Ion Proc.* **111**, 247–262.

7. Dass, C., Kusmierz, J. J., and Desiderio, D. M. (1991) Mass spectrometric quantification of endogenous β-endorphin. *Biol. Mass Spectrom.* **20**, 130–138.

8. Dass, C., Kusmierz, J. J., Desiderio, D. M., Jarvis, S. A., and Green, B. N. (1991) Electrospray mass spectrometry for the analysis of opioid peptides and for the quantification of endogenous methionine enkephalin and β-endorphin. *J. Am. Soc. Mass Spectrom.* **2**, 149–156.

9. Zhu, X. and Desiderio, D. M. (1993) Post-mortem stability study of methionine enkephalin, substance P, and beta-endorphin in rat pituitary. *J. Chromatogr. Biomed. Appl.* **616**, 175–187.

10. Fridland, G. H. and Desiderio, D. M. (1987) Measurement of opioid peptides with combinations of reversed phase high performance liquid chromatography, RIA, RRA, and MS. *Life Sci.* **41**, 809–812.

11. Desiderio, D. M. and Fridland, G. (1984) A review of combined liquid chromatography and mass spectrometry. *J. Liquid Chromatogr.* **7**, 317–351.

12. Desiderio, D. M. and Fridland, G. (1986) Mass spectrometric measurement of neuropeptides, in *Mass Spectrometry in Biomedical Research* (Gaskell, S., ed.), Wiley, New York, p. 443.

13. Higa, T. and Desiderio, D. M. (1989) Optimizing recovery of peptides from an octadecylsilyl (ODS) cartridge. *Int. J. Pept. Prot. Res.* **33**, 250–255.

14. Zhu, X., Robertson, J. T., Sacks, H. S., Dohan, F. C., Jr., Tseng, J.-L., and Desiderio, D. M. (1995) Opioid and tachykinin neuropeptides in prolactin-secreting human pituitary adenomas. *Peptides* **16**, 1097–1107.

15. Yan, L., Fridland, G., Tseng, J., and Desiderio, D. M. (1993) Evidence of an intermediate-sized β-endorphin-containing precursor molecule in bovine pituitary by mass spectrometry. *Biochem. Biophys. Res. Commun.* **196**, 521–526.

16. Yan, L., Tseng, J. L., Fridland, G. H., and Desiderio, D. M. (1994) Characterization of an opioid peptide-containing protein and of bovine α-lactalbumin by electrospray ionization and liquid secondary ion mass spectrometry. *J. Am. Soc. Mass Spectrom.* **5**, 377–386.

17. Dass, C. and Desiderio, D. M. (1989) Characterization of neuropeptides by fast atom bombardment and B/E linked-field scan techniques. *Int. J. Mass Spectrom. Ion Proc.* **92**, 267–287.

18. Desiderio, D. M. (1986) FAB-MS/MS study of two neuropeptides, dynorphins 1-7 and 1-13. *Int. J. Mass Spectrom. Ion Proc.* **74**, 217–233.

6

The Identification
of Electrophoretically Separated Proteins
by Peptide Mass Fingerprinting

John S. Cottrell and Chris W. Sutton

1. Introduction
1.1. Overview

A mass spectrum of the peptide mixture resulting from the digestion of a protein by an enzyme can provide a fingerprint that is specific enough to identify the protein uniquely. The general approach is to take a small sample of the protein of interest and digest it with a proteolytic enzyme, such as trypsin. The resulting digest mixture is analyzed by mass spectrometry (MS), the ionization techniques of choice being matrix-assisted laser desorption/ionization (MALDI) or electrospray ionization. The experimental mass values are then compared with a database of peptide mass values, calculated by applying the enzyme cleavage rules to the entries in one of the major collections of sequence data, such as SwissProt or PIR. By using an appropriate scoring algorithm, the closest match or matches are identified. If the "unknown" protein was present in the sequence database, then the goal is to retrieve that entry. If the sequence database does not contain the unknown protein, then a successful search will identify those entries that exhibit the closest sequence homology, often equivalent proteins from related species.

This method of identification is considerably faster and more sensitive than automated Edman sequencing, and much more specific than fingerprints based on polyacrylamide gel electrophoresis (PAGE) migration patterns or high-performance liquid chromatography (HPLC) retention times. However, peptide mass fingerprinting is limited to the identification of proteins for which sequences are already known. It is not a method of structural elucidation.

From: *Methods in Molecular Biology, Vol. 61: Protein and Peptide Analysis by Mass Spectrometry*
Edited by: J. R. Chapman Humana Press Inc., Totowa, NJ

Evaluation exercises in a number of laboratories have shown that a typical protein can be uniquely identified from a database, such as SwissProt, using just 4 or 5 peptide mass values, while mass measurement errors of 1 or 2 Da have little effect *(1)*. Sample throughput is determined by the time taken in isolating the sample and preparing the digest. Analysis by MS on an automated system might take 5–10 min/sample, whereas the search itself takes 2 min at most. Under favorable conditions, substantially less than a picomole of sample is required for mass spectrometric analysis, allowing the remainder of the digest to be reserved for subsequent Edman sequencing, should it prove necessary.

The first publications describing the technique of peptide mass fingerprinting appeared as recently as 1993 (for a review, *see* ref. *2*), but several substantial applications can already be found in the literature. Examples include the identification of proteins from human myocardial tissue *(3)*, mouse brain tissue *(4)*, cancerous human colonic crypts *(5)*, *Escherichia coli* cells expressing human growth hormone *(6)*, human keratinocytes *(7)*, and channel catfish virus *(8)*.

1.2. Choice of Instrumentation

For peptide and protein work, two types of mass spectrometer predominate: MALDI coupled to a time-of-flight mass analyzer and electrospray ionization (ESI) coupled to a quadrupole mass analyzer. Detailed descriptions of these instruments can be found in Chapter 2.

Two characteristics of MALDI make it particularly well suited to the direct analysis of intact digest mixtures: It is relatively tolerant of low levels of contaminants, such as buffer salts, and discrimination between components of mixtures is relatively low compared with alternative MS ionization methods. On the other hand, in comparison with many chromatographic and electrophoretic methods, the tolerance to contaminants is poor and the quantitation accuracy is poor. If, however, the aim is to obtain a mass fingerprint from a picomole or less of a digest mixture, in the presence of buffer salts and stain residue, MALDI is the only practical choice. Mass measurement error with a simple, linear instrument is generally <0.1%, and can approach 0.01% if the spectrum contains a peak of known mass that can be used as an internal calibrant. This equates to an uncertainty of 1 or 2 Da for a typical tryptic peptide.

1.3. Protein Isolation and Purification

Peptide mass fingerprinting does not work well on mixtures of proteins. This makes gel electrophoresis a natural choice for sample isolation. Many of the methods that have been developed for obtaining internal sequence data from proteins separated on gels are relevant to this application *(9,10)*. Specific difficulties in analyzing proteins from gels by MS are that the associated reagents can cause severe signal suppression and may also result in covalent modification of the sample, altering its molecular weight. Some of the worst offenders are sodium dodecyl sulfate (SDS), which can only be tolerated in trace amounts, acrylamide monomer, and many common stains.

Practical methods of producing proteolytic peptides from a band on a gel fall into three broad categories:

1. The intact protein is eluted from the gel for subsequent digestion;
2. The protein is digested within the gel, or the peptides eluted from the gel *(11)*; and
3. The protein is electroblotted onto a membrane and then digested *in situ* on the membrane, under conditions that release the peptides into solution *(12)*.

Blotting onto a membrane has the great advantage that most of the SDS can be rinsed away prior to digestion. Otherwise, removal of the SDS by selective precipitation of the protein *(13)* or HPLC purification of the peptides is likely to be necessary. Polyvinylidene fluoride (PVDF) is preferred to nitrocellulose membrane for *in situ* digests that are to be analyzed by MALDI because of the cleaner chemical background, (D. J. C. Pappin, personal communication). A cationic PVDF membrane (Immobilon-CD, Millipore, Bedford, MA), designed specifically for *in situ* digestion, is available *(14)*.

In the case of MALDI, it is possible to eliminate HPLC cleanup entirely by choosing reagents that are reasonably compatible with the ionization process. Stains, such as sulforhodamine B for PVDF *(15)* or zinc imidazole for gels *(16)*, cause less signal suppression in MALDI than Coomassie brilliant blue and, hence, do not require destaining. Volatile buffers, such as ammonium bicarbonate, are readily removed by lyophilization. Octyl glucoside is better tolerated in both MALDI and ESI than SDS *(17)*.

1.4. Choice of Enzyme

An enzyme of low specificity, which digests a protein to a mixture of free amino acids and very short peptides, is not a good choice for several reasons. At the practical level, a mixture containing very large numbers of components of similar mass will result in a spectrum containing many coincident and overlapping peaks. More fundamentally, large peptides have greater discriminating power than small ones because they are fewer in number, and because the number of possible amino acid permutations increases geometrically with length.

1.5. Database Considerations

1.5.1. Enzyme Cleavage Rules

A reference database of proteolytic peptide mass values is created by simulating the observed behavior of an enzyme on the entries in a collection of sequence data, such as SwissProt or PIR. The cleavage rules must be chosen to reflect the actual behavior of the enzyme as closely as possible. If the experimental conditions are such that the digest fails to go to completion, then the calculated database should also include partial peptides spanning potential cleavage sites. In practice, at the low picomole level, it is difficult to ensure that digests go to completion, so that much of the art of database building lies in the accurate simulation of incomplete digestion. However, defining inaccurate rules for partial cleavage can be worse than having none at all. Although the precise behavior of an enzyme depends on the specific digest conditions as well as the nature of the substrate, cleavage probabilities for trypsin and chymotrypsin derived from a large pool of experimental data are tabulated in Keil's book *(18)*.

1.5.2. Molecular Mass Calculations

In calculating the molecular weights of the proteolytic peptides, it is essential that the mass values for individual amino acid residues correspond to those found in the samples. The most common ambiguity is the state of the cysteine residues. It is generally desirable to cleave disulfide bridges prior to digestion so as to make the protein more accessible to the protease. This may involve nothing more than the addition of a reducing agent, in which case the appropriate mass value to use for cysteine is that for its reduced state. Alternatively, if the preferred protocol calls for cysteines to be carboxymethylated or pyridylethylated, then the assumed molecular weight of cysteine should be modified accordingly.

Posttranslational modifications present a more difficult problem. A significant proportion of database entries are derived from conceptual translations of nucleic acid sequences, so they contain no information on posttranslational modifications. For sequence entries based on protein characterization, annotations may include information on posttranslational processing, such as N-terminal acetylation, the loss of a signal peptide, conflicting or ambiguous regions of the sequence, phosphorylation, glycosylation, and so forth, all of which would help improve the accuracy of matching the experimental data.

Writing code to parse sequence annotations in general presents a formidable task, and has not been implemented in any published studies. The current release of MassMap, which was used in the work described here, includes options for just two common modifications: deletion of N-terminal methionine and N-terminal acetylation. Fortunately, in most cases, the mass spectrum of a digest contains such extensive redundancy that failure to obtain positive matches for peptides that incorporate other posttranslational modifications does not jeopardize the overall chances of success.

Molecular weights are generally calculated from chemical atomic masses, which are the isotopically averaged values for naturally occurring material *(19)*. It would only be appropriate to use monoisotopic atomic weights if the data were acquired at unit mass resolution, possibly from a magnetic sector instrument.

1.5.3. Sequence Databases

The major sequence databases are maintained by various independent groups around the world. Although there is extensive interchange of information between these groups and one database may incorporate updates from one or more of the others, there is no rigorous system to ensure that any one database contains all known sequences.

This presents a difficult choice: whether to choose just one of the databases and accept that a fraction of known sequence data are missing, or whether to merge multiple databases into a composite database, and suffer the resulting redundancy. The ideal solution is to construct a comprehensive, composite database from which duplicate entries are excluded. The best documented example of a nonredundant, composite database is OWL, available via anonymous file transfer protocol from s-ind2.dl.ac.uk or ncbi.nlm.nih.gov. OWL contains entries from SwissProt, PIR, PDB (Brookhaven), and GenBank conceptual translations *(20)*.

For users without access to the Internet, the Entrez Sequences CD-ROM distributed by the National Center for Biotechnology Information (Bethesda, MD) provides a convenient source for all the major sequence databases.

1.5.4. Scoring Schemes

A number of different search strategies and scoring schemes have been described in the literature *(2)*. In the simplest approaches, the calculated peptide masses for each entry in the sequence database are compared with the set of experimental masses. Each calculated value that falls within a given mass tolerance of an experimental value counts as a match. The score is just the number of matched peptides per entry.

The problem with simply counting the number of matching peptides is that the highest scores go to very large proteins because they show a large number of spurious, random matches. To obtain a meaningful result, it is generally necessary to restrict the mol-wt range for the intact protein to a fairly narrow window. More sophisticated schemes avoid this problem by assigning a statistical weighting factor to each matched peptide mass *(1,21)*.

2. Materials

2.1. Reagents

2.1.1. Protease Digest

1. Autolysis-resistant trypsin (Promega, Southampton, UK).
2. Ammonium bicarbonate (Merck, Poole, UK).
3. *n*-Octyl glucoside (Sigma-Aldrich, Poole, UK).
4. Hydrochloric acid (Merck).

2.1.2. Electroblotting and Visualization

1. Boric acid (Merck).
2. Coomassie brilliant blue R-250 (Bio-Rad, Hemel Hempstead, UK).
3. Tris buffer (Bio-Rad).
4. FluoroTrans PVDF membrane (Pall, Havant, UK).
5. Sulforhodamine B (Eastman Chemicals, Phase Separations, Clwyd, UK).
6. Acetic acid (Merck).

2.1.3. MS

1. Trifluoroacetic acid (TFA) (Pierce & Warriner, Chester, UK).
2. Substance P (Sigma-Aldrich).
3. α-Cyano-4-hydroxycinnamic acid (4HCCA) (Sigma-Aldrich).

2.1.4. Other Reagents

1. Solvents (Merck).
2. Dithiothreitol (DTT) (Sigma-Aldrich).
3. SDS (Bio-Rad).

2.2. Instrumentation

1. Following 2D-gel electrophoresis, protein spots were electroblotted onto PVDF membrane using a Hoefer Iso-Dalt system (Hoefer, Newcastle-under-Lyme, UK).
2. The MALDI mass spectrometer used for analyzing the protein digests was a Lasermat 2000 (Finnigan MAT Ltd., Hemel Hempstead, UK).

2.3. Software

1. The C language source code for Fragfit is available on request from Henzel et al. *(22)*.
2. The software described by Mann et al. *(23)* has been implemented as a commercially available Apple Macintosh program, PeptideSearch (Bruker Franzen GmbH, Bremen, Germany).
3. The MassSearch module of the Darwin system at ETH Zürich can be searched by e-mail *(21)*. For instructions, send a message to cbrg@inf.ethz.ch containing just the phrase *help mass search* in the body of the message. There is also a World Wide Web server at location http://vinci.inf.ethz.ch.
4. A Mowse search of the nonredundant OWL sequence database can be conducted by e-mail *(1)*. For instructions, send a message to mowse@dl.ac.uk containing just the word *help* in the body of the message. The Mowse algorithm has also been incorporated into a commercially available Microsoft Windows program, MassMap (Finnigan MAT), which uses the Entrez Sequences composite database on CD-ROM (National Center for Biotechnology Information).

3. Methods and Discussion

The experimental strategy is illustrated in Fig. 1. We take as our starting point a 230 × 200 × 1.5 mm SDS-PAGE gel on which 1 mg of a mixture of

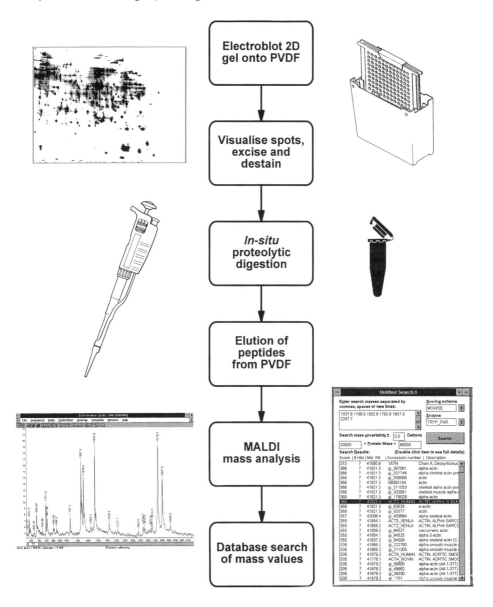

Fig. 1. A schematic illustration of the protocol for protein identification by peptide-mass fingerprinting.

proteins from human myocardial tissue has been separated by 2D electrophoresis. A digitized image of one of these gels is shown in Fig. 2. Details of the sample isolation and electrophoresis procedures can be found elsewhere *(24)*.

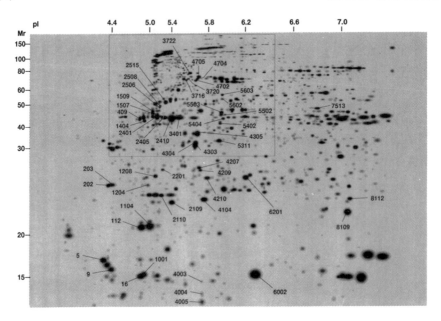

Fig. 2. A digitized image of a silver-stained two-dimensional gel on which a mixture of human myocardial proteins has been separated. The numbered spots are those that have been identified by peptide-mass fingerprinting *(3)*.

Such details are not important to what follows, although differences in the size and composition of the gel may dictate changes to the electroblotting conditions.

3.1. Electroblotting and Visualization

1. Equilibrate the gel for 30 min in the blotting buffer, 50 mM Tris/boric acid, pH 8.5.
2. Electroblot onto FluoroTrans membrane by applying 500 mA for 6 h at 10°C in a Hoefer Iso-Dalt system *(24)*.
3. After the transfer is complete, rinse the membrane several times in deionized water to remove the bulk of the buffer salts.
   Proteins may be visualized using either Coomassie brilliant blue R-250 or sulforhodamine B:
4. Immerse the membrane in a 0.1% solution of Coomassie brilliant blue in 50% methanol for 5 min.
5. Rinse with several changes of deionized water.
6. Dry the membrane *in vacuo* for 20 min.
7. Immerse the membrane in a solution of 30% aqueous methanol containing 0.2% (v/v) acetic acid and 0.005% (w/v) sulforhodamine B, agitating mildly for 60 s.
8. Rinse the membrane in several changes of deionized water to remove excess dye and enable the bands to be visualized in UV light *(15)*.

After staining, the rinsed blots are stored in sealed plastic bags. Selected protein spots may be excised using a scalpel.

3.2. Protease Digests

1. Prepare a stock solution of 1 mg/mL autolysis-resistant trypsin in 2 mM HCl and store at −20°C.
2. Prepare a fresh working solution of 20 μg/mL, each day, in a buffer composed of 25 mM ammonium bicarbonate, 1% *n*-octyl glucoside, and 10% methanol (*see* Note 1).
3. Destain Coomassie-stained spots with two changes of 70% acetonitrile. Sulforhodamine B-stained spots do not require destaining.
4. Dry each spot and, if necessary, cut into pieces no more than 2-mm square and place in the bottom of a 0.5-mL Eppendorf tube.
5. Add an aliquot of between 2 and 4 μL of trypsin solution, just enough to wet the membrane.
6. Cap the tube, and incubate at room temperature for 5–6 h followed by 27–28°C overnight.
7. Extract the peptides from the membrane by adding 10 μL of a 1:1 mixture of formic acid and ethanol, and incubating at 27–28°C for 2 h.
8. Remove the extract by pipet and repeat the process with a second 10-μL aliquot of formic acid and ethanol.
9. Pool the two extracts and lyophilize.
10. Resuspend the residue in 10 μL of 10% methanol.

3.3. MALDI-MS

1. Mix an aliquot of the digest solution (0.5 μL) with an equal volume of matrix solution (10 mg/mL 4HCCA in 70% acetonitrile) on a stainless-steel sample target, and allow the mixture to dry naturally.
2. Introduce the sample target into the vacuum system of the mass spectrometer, a Lasermat 2000.
3. Adjust the laser power and aim empirically to give an optimal balance between signal-to-noise and peak width. Depending on signal strength, data from 10–50 laser shots are averaged to produce each final spectrum.
4. For mass calibration, a second sample target can be prepared with the addition of a spike of a peptide of known mass, usually Substance P (MH$^+$ = 1348.6 Da).

3.4. Selecting Experimental Masses for Searching

1. For digestion by trypsin, roughly 10% of peptides will be longer than 20 residues. Because of their relative scarcity, it is these larger peptides that provide the greatest discriminating power in a search. On the other hand, it is advisable to avoid picking very large mass values, because these are likely to be extended partials that are not represented in the calculated peptide mass database. A good mass range for tryptic peptides is 1000–4000 Da.

2. Most searching schemes allow for limits to be placed on the molecular weight of the intact protein. In setting such limits, it is well to remember that many sequence database entries are for the least processed form of a protein. For example, the SwissProt entry for bovine insulin, INS_BOVIN, is actually the sequence of the precursor protein, including signal and connecting peptides. This adds up to a mol wt of 11,394 Da, so that a search based too tightly around an experimental measurement of the molecular weight of this protein (5734 Da) would fail to find a correct match. For this reason, even though the gel provides an indication of the molecular weight of the intact protein, it is best to be fairly generous with these limits, say −50 and +100% of the measured value as a first guess.
3. It is advisable to screen the experimental data for enzyme autolysis fragments, which are particularly troublesome when performing digests at the low-picomole level.

3.4.1. An Example of Mass Search Data (see Note 2)

Figure 3 shows a typical spectrum, obtained from the digestion of spot 3401 (Fig. 2). It had not been possible to identify this protein by Edman sequencing because it appeared to have a blocked N-terminus. The peak at m/z 1348.8 is a spike of 100 fmol of Substance P, which was added to provide an internal standard for mass calibration. The mass values selected for searching were 1037.6, 1199.9, 1502.6, 1792.8, 1957.9, and 2247.7. These peaks were chosen according to the considerations mentioned in Section 3.4., and also because the peaks were sharp and symmetrical. From its migration distance on the gel, the intact molecular weight of the protein appeared to be around 43 kDa. The error on the experimental peptide mass values was estimated to be <±3 Da based on the observed peak widths and the reproducibility of values from multiple determinations.

The result of searching these mass values against a database built from the complete release 15.0 of Entrez Sequences (>217,000 sequences) is shown in Fig. 4. It can be seen that the list of highest scoring proteins is dominated by actins. (The first entry, 1ATN, is a complex of actin with deoxyribonuclease I from the Brookhaven structural database). The next 10 entries are all the same sequence, human skeletal α actin. The sequence annotations for ACTS_HUMAN state that this protein is known to be acetylated at the N-terminus, which explains why Edman sequencing failed.

3.5. Outlook

Database search software can be expected to increase in sophistication, particularly with respect to improved support for posttranslational modifications.

Fig. 3. *(opposite page)* The MALDI mass spectrum of the tryptic digest mixture from spot 3401 taken off the gel shown in Fig. 2. The peak at m/z 1348.8 is Substance P, added as an internal mass calibrant.

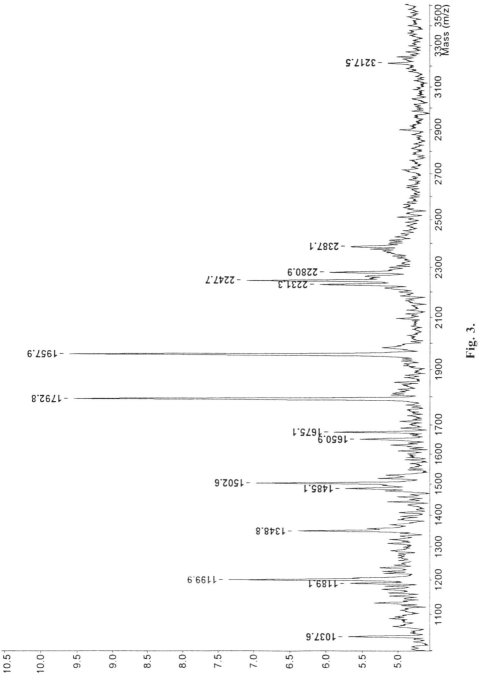

Fig. 3.

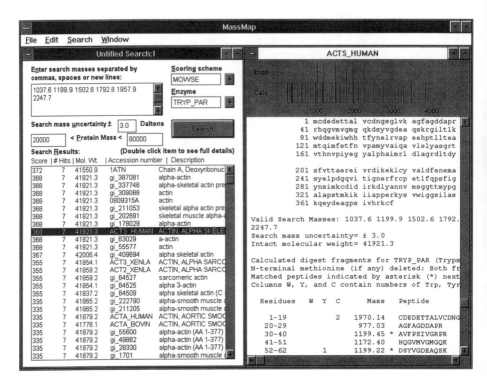

Fig. 4. A screen dump of the result of searching mass values from the spectrum in Fig. 3 against a database built from Entrez Sequences release 15.0 using the MassMap software. The protein was clearly identified as skeletal α actin.

It would also make sense for the software, rather than the user, to refine the initial search results by trying to assign all of the experimental peptide masses to extended partials, products of nonspecific cleavage, enzyme autolysis fragments, mass spectral artifacts, and so on (*see* Note 3).

Another trend is to combine structurally significant information from a number of sources. For example, partial sequence information from either a cycle or two of manual Edman degradation or from tandem mass spectrometry (MS/MS) can be combined with peptide mass fingerprints to improve search specificity quite dramatically *(25)*. The mass shifts observed on deuteration, which correspond to the number of exchangeable hydrogens in a peptide, provide additional search constraints that can also improve the accuracy of identification *(26)*. In fact, any chemical modification that produces a structurally related mass shift can be used in this way. The database search concept has also been adapted to the interpretation of incomplete or ambiguous MS/MS sequence data *(27,28)*.

Finally, as data become available from the systematic sequencing of genomic DNA, the sequence databases will rapidly come to contain the full complement of expressed protein sequences for certain organisms. This will enhance the utility of peptide fingerprinting as the primary technique for protein identification.

4. Notes

1. The *n*-octyl glucoside performs a dual function. It wets the membrane so that absolute minimum volumes of solution can be used, and it also eliminates the need to block the membrane with polyvinylpyrrolidone or similar material. Unlike many other detergents, *n*-octyl glucoside is reasonably compatible with MALDI and does not introduce a large number of background peaks into the mass range of interest. Ammonium bicarbonate was chosen for the buffer because much of it can be removed by lyophilization.

2. Interpretation of search results: In general terms, there are four possible outcomes to a search *(23)*:
   a. The sample protein was present in the database and achieved a high score, which differentiated it from unrelated entries (true positive).
   b. The sample protein was not in the database, but an unrelated protein achieved a well-differentiated high score (false positive).
   c. The sample protein was not in the database, and either related proteins achieved the highest scores or there was little to choose between any of the entries (true negative).
   d. The sample protein was present in the database, but did not achieve a high score which differentiated it from unrelated entries (false negative).

   Given a good scoring scheme and some knowledge of the origin of the sample, the incidence of false positives should be comparatively low. It is the negative results that can present ambiguities.

   It is inadvisable to accept the highest scoring match when there is little to choose between any of the scores. After all, something has to be at the top of the list. When faced with this situation, it is worth rerunning the search with a greater mass uncertainty and wider limits on the protein molecular weight. Possibly the mass calibration was incorrect, or maybe the experimental molecular weight was not for the intact protein, but actually for a fragment or a conjugate.

   In most cases, there will be information other than mass spectrometric data concerning the sample that can assist in deciding whether a result is true or false. Possibly the most readily available is the species. A second, valuable piece of information is the experimentally determined molecular weight of the intact protein. As was mentioned earlier, it is best not to constrain a search to this parameter too tightly. On the other hand, it is quite in order to refine ambiguous results by postsearch comparison of the experimental molecular weight with the calculated weights of suggested matches, after taking into account any posttranslational processing and modifications listed in the sequence report annotations.

Another good technique for dealing with an ambiguous result is to try to assign all of the experimental peptide masses. For example, can mass values be interpreted as extended partials, posttranslationally modified peptides, enzyme autolysis fragments, or mass spectral artifacts? The latter category would include multimers, multiply charged species, sodium or potassium adducts, matrix peaks, and calibration errors.

If several of the experimental masses resist all attempts at assignment, then the match is unlikely to be a good one. However, the converse, looking for all the calculated peptides in the experimental data, is a less reliable indicator. One reason is that it is possible for a protein to have a core region that is resistant to digestion. A second reason is that all desorption/ionization methods show some degree of discrimination between the components of mixtures. That is, some species ionize strongly, whereas others produce only weak peaks, or none at all. It is, therefore, unrealistic to expect to see all of the components of a complex digest in a single mass spectrum.

If the ambiguity from a single search cannot be resolved, then a possible course of action is to do a second digest using a different enzyme, search both sets of data, and pull out the results that are common to both lists *(26)*. The chances of a protein obtaining a high score in both searches through spurious matches are extremely small.

3. Artifacts: Any inadvertent chemical modification to the sample that alters the molecular weight of a peptide can jeopardize the chances of a successful match. Some of the most common modifications that have been observed to occur during electrophoresis are the formation of cysteine-acrylamide, +71 Da, cysteine-β-mercaptoethanol, +76 Da, and methionine sulfoxide, +16 Da *(10)*. If it is suspected that the modified species predominates, then the best solution is to use the mass for the modified residue when building the peptide mass database.

References

1. Pappin, D. J. C., Højrup, P., and Bleasby, A. J. (1993) Rapid identification of proteins by peptide-mass fingerprinting. *Curr. Biol.* **3,** 327–332.
2. Cottrell, J. S. (1994) Protein identification by peptide mass fingerprinting. *Pept. Res.* **7,** 115–118.
3. Sutton, C. E., Pemberton, K. S., Cottrell, J. S., Corbett, J. M., Wheeler, C. H., Dunn, M. J., and Pappin, D. J. (1995) Identification of myocardial proteins from two-dimensional gels by peptide mass fingerprinting. *Electrophoresis* **16,** 308–316.
4. Pappin, D. J. C., Rahman, D., Hansen, H. F., Jeffery, W., and Sutton, C. W. (1995) Peptide mass fingerprinting as a tool for the rapid identification and mapping of cellular proteins, in *Methods in Protein Sequence Analysis* (Proc. Int. Conf.), 10th, Meeting Date 1994 (Atassi, M. Z. and Appella, E., eds.), Plenum, New York, in press.
5. Ji, H., Whitehead, R. H., Reid, G. E., Moritz, R. L., Ward, L. D., and Simpson, R. J. (1994) Two dimensional electrophoretic analysis of proteins expressed by normal and cancerous human crypts: application of mass spectrometry to peptide-mass fingerprinting. *Electrophoresis* **15,** 391–405.

6. Henzel, W. J., Grimley, C., Bourell, J. H., Billeci, T. M., Wong, S. C., and Stults, J. T. (1994) Analysis of two-dimensional gel proteins by mass spectrometry and microsequencing. *Methods (San Diego)* **6**, 239–247.

7. Rasmussen, H. H., Mortz, E., Mann, M., Roepstorff, P., and Celis, J. E. (1994) Identification of transformation sensitive proteins recorded in human two dimensional gel protein databases by mass spectrometric peptide mapping alone and in combination with microsequencing. *Electrophoresis* **15**, 406–416.

8. Davison, A. J. and Davison, M. D. (1995) Identification of structural proteins of channel catfish virus by mass spectrometry. *Virology* **206**, 1035–1043.

9. Tempst, P., Link, A. J., Riviere, L. R., Fleming, M., and Elicone, C. (1990) Internal sequence analysis of proteins separated on polyacrylamide gels at the submicrogram level: improved methods, applications and gene cloning strategies. *Electrophoresis* **11**, 537–553.

10. Patterson, S. D. (1994) From electrophoretically separated protein to identification: strategies for sequence and mass analysis. *Anal. Biochem.* **221**, 1–15.

11. Rosenfeld, J., Capdevielle, J., Guillemot, J. C., and Ferrara, P. (1992) In-gel digestion of proteins for internal sequence analysis after one- or two-dimensional gel electrophoresis. *Anal. Biochem.* **203**, 173–179.

12. Aebersold, R. H., Leavitt, J., Saavedra, R. A., Hood, L. E., and Kent, S. B. H. (1987) Internal amino acid sequence analysis of proteins separated by one- or two-dimensional gel electrophoresis after *in situ* protease digestion on nitrocellulose. *Proc. Natl. Acad. Sci. USA* **84**, 6970–6974.

13. Henderson, L. E., Oroszlan, S., and Konigsberg, W. (1979) A micromethod for complete removal of dodecyl sulfate from proteins by ion-pair extraction. *Anal. Biochem.* **93**, 153–157.

14. Patterson, S. D., Hess, D., Yungwirth, T., and Aebersold, R. (1992) High-yield recovery of electroblotted proteins and cleavage fragments from a cationic polyvinylidene fluoride-based membrane. *Anal. Biochem.* **202**, 193–203.

15. Coull, J. M. and Pappin, D. J. C. (1990) A rapid fluorescent staining procedure for proteins electroblotted onto PVDF membranes. *J. Protein Chem.* **9**, 259,260.

16. Ortiz, M. L., Calero, M., Patron, C. F., Castellanos, L., and Mendez, E. (1992) Imidazole-SDS-zinc reverse staining of proteins in gels containing or not SDS and microsequence of individual unmodified electroblotted proteins. *FEBS Lett.* **296**, 300–304.

17. Vorm, O., Chait, B. T., and Roepstorff, P. (1993) Mass Spectrometry of Protein Samples Containing Detergents. *Proceedings of the 41st American Society for Mass Spectrometry (ASMS) Conference on Mass Spectrometry and Allied Topics.* ASMS, San Francisco CA, pp. 621.

18. Keil, B. (1992) *Specificity of Proteolysis.* Springer-Verlag, Berlin.

19. De Laeter, J. R. (1988) Atomic weights of the elements 1987. *J. Phys. Chem. Ref. Data* **17**, 1791–1793.

20. Bleasby, A. J. and Wootton, J. C. (1990) Construction of validated, non-redundant composite protein sequence databases. *Protein Eng.* **3**, 153–159.

21. James, P., Quadroni, M., Carafoli, E., and Gonnet, G. (1993) Protein identification by mass profile fingerprinting. *Biochem. Biophys. Res. Commun.* **195**, 58–64.

22. Henzel, W. J., Billeci, T. M., Stults, J. T., Wong, S. C., Grimley, C., and Watanabe, C. (1993) Identifying proteins from two-dimensional gels by molecular mass searching of peptide fragments in protein sequence databases. *Proc. Natl. Acad. Sci. USA* **90,** 5011–5015.
23. Mann, M., Højrup, P., and Roepstorff, P. (1993) Use of mass spectrometric molecular weight information to identify proteins in sequence databases. *Biol. Mass Spectrom.* **22,** 338–345.
24. Baker, C. S., Corbett, J. M., May, A. J., Yacoub, M. H., and Dunn, M. J. (1992) A human myocardial two-dimensional electrophoresis database: protein characterisation by microsequencing and immunoblotting. *Electrophoresis* **13,** 723–736.
25. Mann, M. (1995) Role of mass accuracy in the identification of proteins by their mass spectrometric peptide maps, in *Methods in Protein Sequence Analysis* (Proc. Int. Conf.), 10th, Meeting Date 1994 (Atassi, M. Z. and Appella, E., eds.), Plenum, New York, in press.
26. James, P., Quadroni, M., Carafoli, E., and Gonnet, G. (1994) Protein identification in DNA databases by peptide mass fingerprinting. *Protein Sci.* **3,** 1347–1350.
27. Mann, M. and Wilm, M. (1994) Error-tolerant identification of peptides in sequence databases by peptide sequence tags. *Anal. Chem.* **66,** 4390–4399.
28. Eng, J. K., McCormack, A. L., and Yates, J. R. (1994) An approach to correlate tandem mass spectral data of peptides with amino acid sequences in a protein database. *J. Am. Soc. Mass Spectrom.* **5,** 976–989.

7

Liquid Chromatography/Mass Spectrometry for the Analysis of Protein Digests

Thomas Covey

1. Introduction

Proteolytic digestion, for example with trypsin, in combination with reversed-phase high-pressure liquid chromatographic (HPLC) separation/electrospray ionization (ESI) mass spectrometric detection has become an essential tool in protein analysis. This technique, known as peptide mapping, separates and provides mol-wt information on the peptides resulting from digestion of the protein. In addition, using the instrumentation described in this chapter, sequence information also may be made available (*see* Sections 3.3.2. and 3.3.5.). Thus, peptide mapping is a highly effective approach to *inter alia* characterization of protein primary structure and the elucidation of sites of posttranslational modifications. It is the purpose of this chapter to describe the experimental details regarding the chromatographic systems most effectively employed with mass spectrometric detection of peptides, as well as details concerning the mass spectrometer tuning, data acquisition, and data interpretation.

The effects of HPLC column diameter (and thus mobile-phase flow rate) on the performance of ESI-mass spectrometry (MS) is the single most important consideration when designing peptide mapping protocols. ESI is commonly viewed as a liquid-phase ionization technique that requires very low liquid flow rates and, thus, very small diameter HPLC columns. Embedded in this view is an element of truth, but mostly misperception. It is true that the highest absolute sensitivity (greatest signal for a fixed sample amount injected) is obtained with the lowest possible flows and column diameters. This is a consequence of both the increased concentration in peaks eluting from smaller column diameters and an increased efficiency of the ionization process at the lower liquid flows.

Increasing sensitivity with decreasing column diameter implies that the electrospray mass spectrometer is behaving as a concentration-sensitive detec-

From: *Methods in Molecular Biology, Vol. 61: Protein and Peptide Analysis by Mass Spectrometry*
Edited by: J. R. Chapman Humana Press Inc., Totowa, NJ

tor, much like a UV detector. One important analytical implication of this is that postcolumn splitting of the mobile-phase stream, in order to collect a large percentage of the purified peptides simultaneously with mass spectrometric detection, does not reduce detection sensitivity. Postcolumn stream splitting thus becomes an important tool to consider *(1)*.

The decreasing concentration in peaks eluting from larger diameter columns is an unalterable reality that will decrease sensitivity. In the world of protein chemistry, situations requiring high absolute sensitivity (<1 pmol) are therefore gravitating toward the use of packed capillary columns *(2–4)*. When both high sensitivity (1–100 pmol) and on-line sample fractionation for further characterization are desired, 1-mm id columns are more practical *(1,5)*. For intermediate sensitivity (1–100 pmol), but routine chromatography, 2-mm id columns are being widely used *(6)*. Low sensitivity (>1 nmol) is generally observed with 4.6-mm id columns, but very routine and reproducible chromatography is the advantage. A good example of the use of these columns is the quality control of recombinant proteins *(7)*.

2. Materials

2.1. Enzymatic Digestion

1. TPCK trypsin (Pierce, Rockford, IL).
2. Ammonium bicarbonate, dithiothreitol (DTT), iodoacetic acid (Sigma, St. Louis, MO).
3. PD-10 G-25 columns (Pharmacia, Piscataway, NJ).

2.2. Chromatography

1. Columns (*see* Note 1) can be obtained from LC Packings (San Francisco, CA), Keystone Scientific (Bellefonte, PA), Applied Biosystems (Foster City, CA), or prepared according to the procedures described in Note 2.
2. Fused-silica tubing can be obtained from Polymicro Technologies (Phoenix, AZ), 150-μ od × 50, 75, and 100 μ id.
3. Fused-silica fittings should be made with standard 1/16-in. chromatography fittings by sleeving the fused silica with 1/16-in. od × 150–200 μ id Teflon™.
4. Mobile phases should be prepared using Milli-Q purified water and UV-grade acetonitrile.

2.3. Mass Spectrometry

1. Ionization can be acomplished with a pneumatically assisted electrospray (Ion Spray) system.
2. A mass analyzer of the triple quadrupole design.

3. Methods

3.1. Enzymatic Digestion

No special considerations come into play regarding procedures for enzymatic digestion prior to liquid chromatography (LC)/MS, and standard procedures can generally be employed. Ionic buffers (e.g., Tris) can be used with the caveat that a sufficient gradient is applied during the chromatographic run to separate the peptides from the buffer; otherwise severe ionization supression will occur. Nonionic surfactants should be avoided primarily because it is frequently difficult to separate them chromatographically from the region of peptide elution. When possible, volatile buffers are substituted for nonvolatile buffers. As an example, a typical tryptic digestion method for a disulfide-bridged protein for LC/MS is outlined in the following:

1. Reduce an aliquot of the protein solution in 8M urea for 30 min at 37°C with 10 mM DTT.
2. S-Carboxymethylate the product from step 1 with 25 mM iodoacetic acid (*see* Chapter 14, Section 3.1., step 3 for detailed instructions).
3. Buffer exchange the solution into 100 mM ammonium bicarbonate using a disposable G-25 column (Pharmacia).
4. Add TPCK-trypsin to the solution from step 3 to give a final enzyme-to-substrate ratio of 1:50 (w/w).
5. Allow the digestion to proceed at room temperature for 8 h, and then add a second aliquot of enzyme.
6. Stop the digestion by freezing after 24 h.

3.2. Chromatography

3.2.1. Stationary Phases

There are also no special considerations regarding the LC stationary phases utilized for LC/MS. The standard reversed-phase column packings found to provide superior resolution for UV detection are used equally effectively for LC/MS. Occasionally a particular lot of stationary phase will be observed to produce excessive chemical background, which dramatically increases during the course of a solvent gradient, but is transparent to the UV detector. Frequently, this can be attributed to the bleeding of hydrolysis byproducts of dimethyl-dichlorosilane, used as a capping agent in silica deactivation from the stationary phase. Protonated molecular ions at *m/z* 297 (octamethlycyclotetrasiloxane), 446 (dodecamethylcyclohexasiloxane), and 668 (octadecylmethylcyclononasiloxane) are seen to persist for the life time of the column. The only solution is to obtain columns made from a different lot of stationary phase.

3.2.2. Column Length

Because of the extra specificity provided by the mass spectrometer, columns of shorter length are used to decrease run times. Columns as short as 5 cm and less are routinely employed with excellent results. The shorter columns have the added advantage of reduced adsorptive losses.

3.2.3. Mobile Phases

The popular mobile-phase gradient system for UV detection, which consists of acetonitrile, water, and 0.1% trifluoroacetic acid (TFA), continues to to be used for LC/MS. Typical gradient rates of 1%/min from 0 to 80% acetonitrile are employed. TFA has been noted to have some sensitivity suppression effects (twofold lower signal than with other organic acids) and, for this reason, is sometimes replaced with formic acid. Generally, however, the chromatographic peak widths are significantly broadened with the formic acid modifier, resulting in no net sensitivity gain.

3.2.4. Degassing

The issue of solvent degassing continues to be important, although the disruptive effect of bubble formation at higher liquid flow rates (>50 μL/min) is hardly noticeable when pneumatically assisted electrospray is used. With packed capillary columns, however, thorough degassing is imperative for successful operation. Periodic helium degassing during the day is highly recommended.

3.2.5. Packed Capillary Columns

Packed capillary columns ranging in internal diameter from 50–500 μ are commercially available and can also be prepared in the laboratory (*see* Note 2). The 250- and 320-μ id columns, whose optimum flow rates of 1–5 μL/min nicely match the lower flow rate specification of the most commonly used pneumatically assisted electrospray (ion spray) ion sources, have been particularly successful. Flow rates below this range require either an additional sheath make-up flow or a restricted sprayer aperture, which adds to operational complexity.

Currently there are no commercial LC systems that will deliver solvent gradients in the required flow range of 100 nL/min to 5 μL/min for packed capillary chromatography. A practical approach to low flow rate gradient chromatography with conventional HPLC pumps involves the use of a flow splitter prior to the injector (Fig. 1) *(8,9)*. The flow from the HPLC pump at 50–500 μL/min is split at a T-fitting between back-pressures generated by either an analytical column or a length of 50-μ id fused-silica capillary. Although the majority of the eluent flows through the lower resistance analytical column or fused-silica capillary, a small portion will flow through the small-diameter column. The final flow is dictated by empirically choosing the appropriate analytical column or length of fused silica, and then varying the initial flow from the pump to control the pressure. Commercial back-pressure regulators based on this principle are available *(9)*. Flow meters can

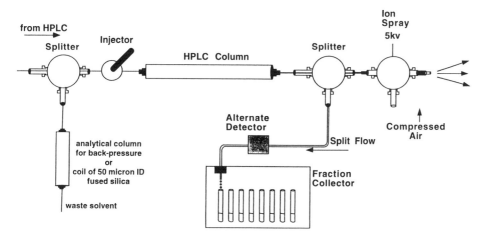

Fig. 1. Diagram of an LC/MS inlet system.

be easily assembled in the lab (*see* Note 3) to measure the split flow. Further elaboration on the details of setting-up packed capillary systems may be found in refs. *10* and *11*.

3.2.6. Columns of 0.5–1 mm id

For routine separations in the 1–100 pmol range, the 0.5–1-mm id column size is the most practical choice *(1)*. Flow rates are in the range of 15–25 μL/ min for the 0.5-mm columns and 40–50 μL/min for the 1-mm id columns. Postcolumn flow splitting can be utilized for simultaneous sample purification and mass spectrometric detection. Flow reduction is, however, not required from the point of view of the mass spectrometer *(5)*. Columns of 1-mm id are commercially available, and several pump manufacturers offer systems that will deliver reliable gradients. Disposable 1-mm columns, which have surprisingly high efficiencies, can be quickly and inexpensively dry packed in the lab *(12)*, although the high-pressure slurry-packed columns continue to give the highest efficiencies.

3.2.7. Columns of 2.1 mm id

Columns of 2.1 mm id are generally recognized as having a distinct advantage over 1-mm columns with regard to reliability, ruggedness, and routine application. In this respect, they perform in a very similar manner to 4.6-mm columns. Operation in the 200 μL/min range is relatively routine for most pumping systems. Postcolumn splitting is useful for sample purification, but is not an absolute requirement from the point of view of the mass spectrometer ion source.

3.2.8. Columns of 4.6 mm id

In general, columns of this size are seldom used for LC/MS of protein digests, except in quality-control (QC) situations where significant amounts of sample are available *(7)*. Methods can be transferred directly to the mass spectrometer from the QC labs, which have historically standardized on the use of 4.6-mm columns. Postcolumn stream splitting is invariably used to reduce the solvent load to the electrospray nebulizer and increase the ionization efficiency *(5)*. Postcolumn stream splitters are similar in principle to the preinjector splitter described earlier (Fig. 1). A low dead volume "tee" is plumbed into the eluant path between the column exit and the sprayer. Typically, a 50–100-µ id fused-silica capillary will be connected from the splitter to the sprayer. A second, larger id, but longer length fused-silica line is connected to regulate the split ratio. Flow into the mass spectrometer should be between 5 and 100 µL/min. The flow out the split line is measured (*see* Note 3), and the split ratio can be adjusted by simply cutting successive pieces off or coupling longer pieces on with connectors. A splitter, once it has been calibrated, has been proven to regulate the flow rate reliably without constant attention.

3.3. Mass Spectrometry

3.3.1. General Considerations

The ion source should be tuned for maximum sensitivity at the flow rate to be used for the separation.

1. Remove the column and plumb the injector directly in-line to the sprayer.
2. Flow inject 1-µL plugs of a known concentration of a peptide at the flow rate to be used in practice. A comprimise solvent composition (typically 20% acetonitrile) is used for this optimization procedure.
3. Quickly optimize the sprayer position, nebulizer gas flows, and source temperatures for maximum signal and minimum background while monitoring the molecular ion of the known peptide.

The other important considerations are to set up the mass spectrometer with a proper mass calibration and with unit resolution. Good resolution is especially important for the interpretation of the mass spectra of proteolytic digests because of ambiguities regarding the identity of doubly charged molecular ions, singly charged molecular ions, and singly and multiply charged multimers (*see* Section 3.3.2.).

Scan rates are typically 2–4 s/scan over a range of *m/z* ca. 300–2400. Some exceptions to this exist and are highlighted in the sections on glycopeptide and phosphopeptide mapping (Sections 3.3.3. and 3.3.4.).

3.3.2. Spectrum Interpretation

Spectra from the LC/MS analysis of enzyme digests of proteins are generally simple to interpret, but there are situations where there is some ambiguity in the assignment of the molecular weights of the peptides. For example, it is well known that multiply charged ions are frequently produced from peptides when using ESI *(13)*. In this case, situations can arise where interpretation of the spectra can be ambiguous unless the instrument was tuned with sufficient resolution to distinguish doubly from singly charged molecular ions on the basis of the isotope spacing. Figure 2 illustrates a case in point.

The spectrum in Fig. 2A is of three coeluting peptides from an LC/MS analysis of a tryptic digest of the Rho protein. On first inspection of the spectrum in Fig. 2A, it appears that there are three peptides with $(M + H)^+$ ions at m/z 1436.8, 1508.8, and 1625.8 and molecular weights of 1435.8, 1507.8, and 1624.8 Da, respectively. Closer inspection of the data reveals that only the third mass assignment is correct. The ions at m/z 813.2 and 1625.8 were correctly identified as the doubly and singly charged ions from a single peptide. The spacings of the isotopes of m/z 813.2 (Fig. 2C) and 1625.8 (Fig. 2E) are clearly 0.5 and 1.0 u, respectively, thus verifying their charge states. The other apparent doubly/singly charged pair at m/z 754.9 and 1508.8 is an artifact of the ionization process. The ion at m/z 1508.8 is a dimer of the structure $(2M + H)^+$ that can form during the ionization process, and that is especially prominent when larger amounts of sample are injected. The expansion of the spectrum in Fig. 2B clearly identifies the ion at m/z 754.9 as singly charged, indicating a mol wt of 753.9 Da. The same situation exists for the 718.9/1436.8 pair.

Incorrect mass assignments from spectra of peptides sometimes result from confusion regarding the differences between monoisotopic mass, exact mass, and average mass, and from issues regarding the relative abundance of the ^{13}C isotope in nature. At a molecular weight of around 1800 u, the first ^{13}C isotope begins to dominate the isotope cluster. The exact mass (within the ±0.2 u error window of a quadrupole mass spectrometer) of the $(M + H)^+$ of the peptide is one mass unit lower than that of the dominant peak in the isotope cluster, and frequently the mass is misassigned because of this. Figure 3 shows the molecular ion region of a spectrum of a tryptic peptide from an LC/MS run. The shift in relative abundance of the isotopes is apparent. For an excellent discussion of these issues, *see* ref. *14*.

In addition to mol-wt information, some peptide sequence information can be made available from an LC/MS run. Electrospray ion-source voltages can be adjusted to induce some collision-induced dissociation (CID) of the molecular ion into fragments (*see* Note 4). This is frequently referred to as CID-MS to distinguish it from the CID experiment in tandem mass spectrometry (MS/MS) (*see* Note 5). The result is an MS spectrum that includes ion peaks from the

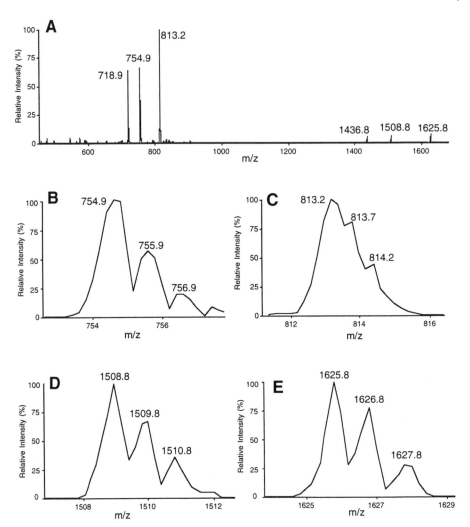

Fig. 2. (A) A typical spectrum of three coeluting peptides from the LC/MS analysis of a tryptic digest. (B) Expansion of m/z 754.9 from the spectrum in (A). (C) Expansion of m/z 813.2 from the spectrum in (A). (D) Expansion of m/z 1508.8 from the spectrum in (A). (E) Expansion of m/z 1625.8 from the spectrum in (A).

molecular ion as well as some sequence information. The spectrum will, however, also contain ion peaks from background chemical noise and any other coeluting components and these can complicate interpretation and make full-sequence determination difficult. On the other hand, partial sequence information can usually be extracted from data of this type, and this can be extremely useful for searching protein databases (15). Figure 4A–C shows an example of

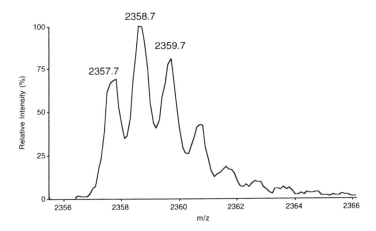

Fig. 3. Molecular ion region [(M + H)⁺] of a tryptic peptide of mol-wt 2356.7.

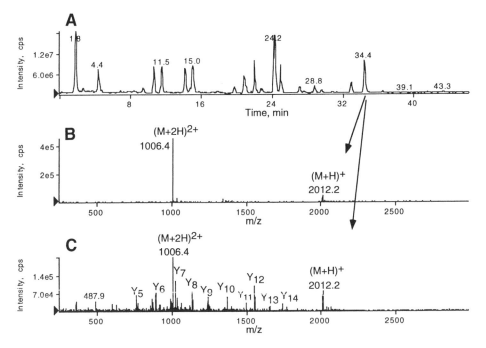

Fig. 4. (A) The total-ion current chromatogram from the LC/MS analysis of a tryptic digest of bovine cytochrome c using alternating low/high ion source fragmentation potentials. A 1-mm id column was used in this example. **(B)** Spectrum from retention time 34.4 min at a low source fragmentation voltage. **(C)** The next spectrum following the one displayed in (B), but recorded at a high source fragmentation voltage.

the LC/MS analysis of a tryptic digest of bovine cytochrome c using this technique. The instrument is set to provide alternate low and high ion-source fragmentation voltages with each scan. Figure 4B shows an unambiguous mol-wt assignment for the T12 peptide. During the following scan (Fig. 4C), the fragmentation potential was raised, and clear sequence information was obtained. For further comments on the interpretation of CID spectra, *see* Section 3.3.5. and Note 5.

3.3.3. Glycopeptide Mapping

The location of glycosylated peptides in an HPLC map of a protein digest is a task that is frequently encountered in protein analysis. An elegant method for localizing glycopeptides in a peptide map produced by LC/MS analysis has been developed by Carr and coworkers *(16,17)*, and software improvements in electrospray systems have further extended the utility of this technique. The method is based on the observation that all glycopeptides contain hexose, *N*-acetylhexosamine, and/or neuraminic acid isomers, which produce diagnostic fragment ions when subjected to high collision voltages in the CID-MS experiment described at the end of the preceding section. The most useful diagnostic ions (positive-ion mode) are: m/z 163 (hexose), m/z 204 (*N*-acetylhexosamine), m/z 292 (neuraminic acid), and m/z 366 (hexose-*N*-acetylhexosamine).

1. Set the mass spectrometer to scan by sequentially monitoring the four ions for 100 ms each while holding a high CID-MS fragmentation voltage to ensure efficient production of fragment ions.
2. After monitoring the four marker ions (400 ms), use the mass spectrometer to scan the full mass range (2–4 s) at a low CID fragmentation voltage to obtain mol-wt information from the same peptides in the digest.
3. Begin the cycle again with monitoring of the four marker ions.

Points in the chromatogram where a signal is recorded for the marker ions indicate the elution of a glycopeptide and, because of the low CID potential used in the full scan mode, the molecular weight of the corresponding peptide is immediately available. Figure 5A–D demonstrates an example of LC/MS

Fig. 5. *(opposite page)* LC/MS map and glycopeptide screen of a tryptic digest of interleukin-4. Chromatography on a 1-mm id column was used during this run with approx 50 pmol injected. **(A)** The extracted-ion profile for the carbohydrate marker ion at m/z 204 (*N*-acetylhexosamine) recorded at a high source fragmentation potential. **(B)** The extracted-ion profile for the carbohydrate marker ion at m/z 292 (neuraminic acid) recorded at a high source fragmentation potential. **(C)** The extracted-ion profile for the carbohydrate marker ion at m/z 366 (hexose-*N*-acetylhexosamine) recorded at a high source fragmentation potential. **(D)** The total-ion current trace recorded simultaneously with the traces in A, B, and C at a low ion source fragmentation voltage.

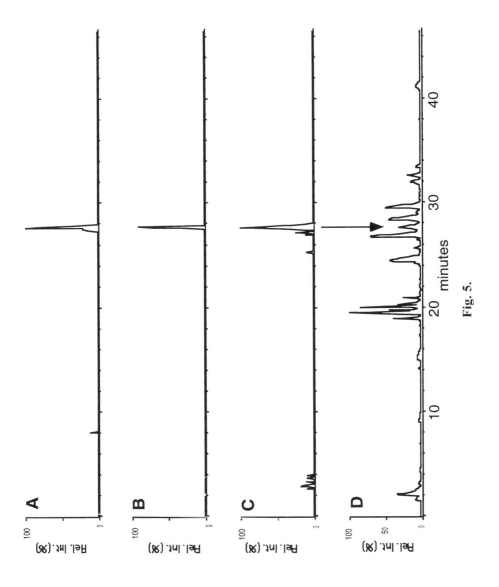

Fig. 5.

93

mapping and glycopeptide screening on recombinant interleukin-4. The strong signal for the three marker ions at retention time 27 min indicates which of the peaks in the total-ion current (Fig. 5D) corresponds to the glycopeptide. Bovine fetuin *(16)* is a good test protein for glycopeptide screening.

3.3.4. Phosphopeptide Mapping

A general MS-based screen for selectively identifying phosphopeptides containing phosphate linked to Ser, Thr, and Tyr in complex mixtures similar to the glycopeptide mapping approach has been developed *(18–20)*. This method uses CID-MS to produce negative-ion fragments at low masses that are sensitive, specific markers for all three common phosphate-to-peptide linkage types. The marker ions for Ser- and Thr-linked phosphate are m/z 97 ($H_2PO_4^-$), m/z 79 (PO_3^-), and m/z 63 (PO_2^-), whereas only m/z 79 and 63 are formed from Tyr-linked phosphate.

1. Set the mass spectrometer to scan by sequentially monitoring the three ions in negative mode for 100 ms each while holding a high CID-MS fragmentation voltage to ensure efficient production of fragment ions.
2. After monitoring the three marker ions (300 ms), use the mass spectrometer to scan the full mass range (2–4 s) at a low CID fragmentation voltage in the positive-ion mode to obtain mol-wt information from the same peptides in the digest.
3. Begin in the cycle again with monitoring of the three marker ions.

Points in the chromatogram where a signal is recorded for the marker ions indicate the elution of a phosphopeptide and, because of the low CID potential used in the full scan mode, the molecular weight of the corresponding peptide is immediately available. Casein *(18,21)* is a good test protein for phosphopeptide screening.

3.3.5. Tandem Mass Spectrometry

Tandem mass spectrometry (MS/MS) is considered an important alternative means to Edman sequencing for obtaining sequence information from peptides. One of the noted drawbacks of the technique is the relatively complex and often ambiguous process for the interpretation of CID spectra. The ideal spectrum has all the cleavages located at the amide bonds and proceeds along the entire length of the molecule. This seldom occurs, and internal clevages, secondary fragmentation, side-chain clevages, and other peptide backbone series usually complicate the spectrum. A trained eye is required to decipher the spectrum and frequently a full-sequence determination is not possible.

One of the main determinants of the fragmentation pathways during the CID process is the location of the charges on the molecular ion. The ideal case is observed when the charges are localized at the amino-terminus and carboxy-

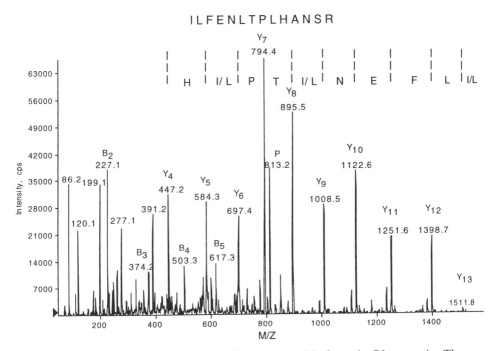

Fig. 6. Product-ion spectrum of the T17 tryptic peptide from the Rho protein. The precursor ion chosen for collision was the doubly charged molecular ion at m/z 813.2.

terminus of the peptide (a doubly charged ion). With this situation, charge-remote fragmentations occur, and these are dominated by a series of amide bond cleavages, which give a clear indication of a large portion of the peptide sequence. In this respect, a fortuitously suitable peptide structure is produced when tryptic digestion of a protein is performed. Because enzymatic cleavage is at the C-terminus of arginine or lysine residues, the resulting tryptic peptides have highly basic functional groups at each end of the molecule, which is where the charges reside (amino-terminus and C-terminal arginine or lysine). Thus, the ideal peptide structure for sequencing by MS/MS is generated.

The CID fragmentation pattern of the doubly charged ions of tryptic peptides is very consistent from peptide to peptide, and obtaining sequence information from such spectra is relatively simple *(5)*. Cleavages predominate at the amide bonds allowing the user to obtain sequence information by equating the mass differences between peaks with the mass of an amino acid. Figure 6 shows an example of a product-ion spectrum from an LC/MS/MS analysis of a tryptic digest of the Rho protein (peptide T17). Simple inspection of the Y- series ions in the spectrum yields 10 consecutive residues. The isobaric leucine and isoleucine are indistinguishable. The Y-ion series (C-terminal series) can always be easily identified in the spectra of tryptic peptides. They are the predominant series of ions with some m/z values greater than that of the precursor ion (the doubly charged molecular ion chosen as the precursor ion for fragmentation).

4. Notes

1. Chromatography columns: Stationary phases are Vydac, ABI Aquapore, and Nucleosil, all 300 Å pore-size, C18-, C8-, and C4-phase material. Column lengths vary from 5–15 cm.
2. Preparation of packed capillary columns of 320-μ id: The following describes a method developed to prepare, in the laboratory, packed capillary HPLC columns of 320-μ id, which operate well at flow rates between 3 and 6 μL/min.

   Materials:
   a. Column: 320 μ id × 430 μ od fused-silica capillary.
   b. Transfer line: 50 μ id × 186 μ od fused-silica capillary.
   c. Injector end sealing sleeve: 0.020 in. id × 1/16 in. od PEEK tubing.
   d. Frit: glass-wool fiber filter paper (Whatman, Clifton, NJ).
   e. Sealant for transfer line to column: epoxy glue.
   f. Packing reservoir: A 2 × 100-mm emptied HPLC column plumbed in-line between the pumps and the column to be packed.

   Packing procedure for a 5-cm long column (the same procedure is used for longer columns):
   a. Cut a 6-cm length of fused-silica capillary for the column.
   b. Using the end of the column as a punch, push out a disk of glass-fiber filter paper. Push the disk approx 1 cm into the column using the transfer line fused-silica capillary.
   c. Cut a transfer line with a length of 30–45 cm. Insert the transfer line approx 3 mm into the other end of the column from the frit, and dab some glue at the joint. Glue will then wick in by capillary action. If a microscope is available, it is helpful to observe the advance of the glue up the column. It is not necessary for the glue to wick all the way to the end of the transfer line. However, it is important that it does not plug the end of the line. Allow the assembly to dry overnight.
   d. Using a piece of the transfer line as a ramrod, push the glass fiber frit down to the transfer line from the injector end of the empty column.
   e. Cut a 3-cm piece of the PEEK tubing to be used as the sleeve for the fitting on the injector end of the column. Put on a nut and ferrule that is compatible with the injection valve to be used to inject the sample, and seat the assembly of PEEK tubing and column in the injector. The author's experience is with the Rheodyne 8125 injection valve (low dead volume and variable loop size) using the long Rheodyne nuts. The fitting reach is very short in this valve. This allows the user to insert this swaged fitting into the end of the packing reservoir (step f) without altering the reach set for the injector valve. Equiva-

lent low dead volume injectors are available from other manufactures of injector valves.

f. After seating the ferrule, remove the column assembly from the injector, and tighten it into the bottom of the slurry reservoir.

g. Add acetonitrile to the top of the reservoir, and pump it through the empty column to remove air bubbles.

h. Open the top of the reservoir, and remove approx half of the column volume with a pipet. Top off the reservoir with the slurried packing (0.34 g/mL in acetonitrile). Preferred packing materials for peptide separations are C18 reversed-phase with 5–7 µ average particle size and 300-Å pore size.

i. Close the reservoir, and turn on the HPLC pumps. Either syringe pumps or reciprocating piston pumps will suffice, although the reciprocating pumps have the advantage of coming up to pressure more quickly. Allow the liquid to flow at a sufficient rate to generate 5000–6000 psi to pack the column thoroughly. A typical flow for this length of column is approx 450 µL/min to achieve 5000 psi. Once the column is filled, disconnect from the reservoir (allow pressure to drop slowly) and then reconnect to pump. Adjust the flow rate to generate approx 5000 psi and allow the liquid to flow for 1 h.

The column is now ready to use, and is complete with injector end fitting and transfer line to the mass spectrometer. Although no frit has been placed in the injector end, as long as rapid depressurization of the column is avoided, any back-flushing of the packing into the injector can be avoided.

3. Split-flow measurement: A flow meter can easily be assembled for the purpose of calibrating split flows. Couple an HPLC injector syringe to the split-flow line with a standard 1/16-in. HPLC fitting. With the syringe barrel removed, the split flow will rise up the barrel and can be measured accurately. The flow to the column (in the case of preinjector splitting) or to the mass spectrometer ion source (in the case of postcolumn splitting) is calculated knowing the flow delivered by the pumps.

4. CID-MS: In electrospray ion sources, ions are created at atmospheric pressure and then drawn into the vacuum system of the mass analyzer through a small aperture. In the region downstream of this aperture, the gas density is high and ions can be accelerated through this region by means of applied voltages, which cause them to collide energetically with the neutral gas molecules and consequently fragment. The primary difference between this experiment and the CID experiment with a tandem mass spectrometer (*see* Note 5) is that CID spectra due to mixtures can be recorded because there is no mass selection of a precursor ion before the collision process.

5. MS/MS: In the tandem mass spectrometer, the CID process occurs in a region between two mass analyzers (collision cell) into which a target gas has been introduced. With such a configuration, the first mass analyzer can selectively filter the molecular ion of the targeted peptide out of a complex mixture of many components. That precursor ion is then fragmented in the gas collision cell, and the product-ion masses determined by the second mass analyzer. The main difference between this experiment and the CID-MS experiment is the knowledge that

all the product ions in the spectrum are derived from the targeted molecular ion. On this basis, interpretation of the spectrum is much more straightforward. Chemical background is also reduced substantially in MS/MS experiments, which leads to lower limits of determination.

References

1. Hess, D., Covey, T. R., Winz, R., Brownsey, R. W., and Abersold, R. (1993) Analytical and micropreparative peptide mapping by high performance liquid chromatography/electrospray mass spectrometry of proteins purified by gel electrophoresis. *Protein Sci.* **2**, 1342–1351.
2. Huang, E. C. and Henion, J. D. (1991) Packed capillary liquid chromatography/ ion spray-tandem mass spectrometry determination of biomolecules. *Anal. Chem.* **63**, 732–739.
3. Hunt, D. F., Henderson, R. A., Shabanowitz, J., Sakaguchi, K., Michel, H., Sevilir, N., et al. (1992) Characterization of peptides bound to the class I MHC molecule HLA-A2.1 by mass spectrometry. *Science* **255**, 1261–1266.
4. Kassel, D. B., Shushan, B., Sakuma, S., and Salzman, J. P. (1994) Evaluation of packed capillary perfusion column HPLC/MS/MS for the rapid mapping and sequencing of enzymatic digests. *Anal. Chem.* **66**, 236–243.
5. Covey, T. R., Huang, E. C., and Henion, J. D. (1991) Structural characterization of protein tryptic peptides via liquid chromatography/mass spectrometry and collision-induced dissociation of their doubly charged molecular ions. *Anal. Chem.* **63**, 1193–1200.
6. Lewis, D. A., Guzzetta, A., and Hancock, B. (1994) Characterization of humanized anti-TAC, an antibody directed gainst the interleukin 2 receptor, using electrospray ionization mass spectrometry by direct infusion, LC/MS, and LC/ MS/MS. *Anal. Chem.* **66**, 585–595.
7. Ling, V., Guzzetta, A. W., Canova-Davis, E., Stults, J. T., Hancock, W. S., Covey, T. R., and Shushan, B. I. (1991) Characterization of the tryptic map of recombinant DNA derived tissue plasminogen activator by high-performance liquid-chromatography-electrospray ionization mass spectrometry. *Anal. Chem.* **63**, 2909.
8. Balogh, M. A. and Stacey, C. C. (1990) Chromatographic Developments for Low Flow Rate LC-MS Interfaces. *36th Conference on Mass Spectrometry and Allied Topics*, Tucson, AZ, p. 1209.
9. Chervet, J. P., Meijvogel, C. J., Ursem, M., and Salzmann, J. P. (1992) Development of reliable splitter systems for packed capillary HPLC. *LC/GC* **10**, 140–150.
10. Davis, M. T. and Lee, T. D. (1992) Analysis of peptide mixtures by capillary high performance liquid chromatography: a practical guide to small-scale separations. *Protein Sci.* **1**, 935–944.
11. Davis, M. T., Stahl, D. C., and Lee, T. D. (1995) Low flow high-performance liquid chromatography solvent delivery system designed for tandem capillary liquid chromatography-mass spectrometry. *J. Am. Soc. Mass Spectrom.* **6**, 571–580.

12. Southan, C. The use of glass capillary tubes as disposable microbore columns for RP-HPLC of proteins and peptides, in *Techniques in Protein Chemistry* (Hugli, T., ed.), Academic, New York, pp. 392–398.

13. Covey, T. R., Bonner, R. F., Shushan, B. I., and Henion, J. D. (1988) The determination of protein, oligonucleotide, and peptide molecular weights by ion-spray mass spectrometry. *Rapid Commun. Mass Spectrom.* **2,** 249–255.

14. Yergey, J., Heller, D., Hansen, G., Cotter, R. J., and Fenselau, F. (1983) Isotopic distributions in mass spectra of large molecules. *Anal. Chem.* **55,** 353–356.

15. Wilm, M. and Mann, M. (1995) Error-tolerant identification of peptides in sequence databases by peptide sequence tags. *Anal. Chem.* **136,** 167–180.

16. Carr, S. A., Huddleston, M. J., and Bean, M. (1993) Selective identification and differentiation of *N*- and *O*-linked oligosaccharides in glycoproteins by liquid chromatography-mass spectrometry. *Protein Sci.* **2,** 183–196.

17. Huddleston, M. J., Bean, M. F., and Carr, S. A. (1993) Collisional fragmentation of glycopeptides by electrospray ionization LC/MS and LC/MS/MS: methods for selective detection of glycopeptides in protein digests. *Anal. Chem.* **65,** 877–882.

18. Huddleston, M. J., Annan, R. S., Bean, M. F., and Carr, S. A. (1993) Selective detection of thr-, ser-, and tyr-phosphopeptides in complex digests by electrospray LC-MS. *J. Am. Soc. Mass Spectrom.* **4,** 710–717.

19. Ding, J., Burkhary, W., and Kassel, D. B. (1994) Identification of phosphorylated peptides from complex mixtures using negative-ion orifice-potential stepping and capillary liquid chromatography/electrospray ionization mass spectrometry. *Rapid Commun. Mass Spectrom.* **8,** 94–98.

20. Covey, T. R., Shushan, B., Bonner, R., Schroder, W., and Hucho, F. (1991) LC/MS and LC/MS/MS screening for the sites of post-translational modification in proteins, in *Methods in Protein Sequence Analysis* (Jornvall, Hoog, Gustavsson, eds.), Birkhauser, Verlag, Basel, Switzerland, pp. 249–256.

21. Nuwaysir, L. M. and Stultz, J. T. (1994) Electrospray ionization mass spectrometry of phosphopeptides isolated by on-line immobilized metal-ion affinity chromatography. *J. Am. Soc. Mass Spectrom.* **4,** 662–669.

8

Structural Analysis of Protein Variants

Yoshinao Wada

1. Introduction

A significant proportion of genetic disorders are caused by point mutations in proteins. Structural studies of these mutated proteins are now mostly carried out by genetic methods. Even so, mass spectrometry (MS) is a rapid and excellent method of mutation analysis, especially when sufficient amounts of the gene product are obtained. It is also useful for verifying the structure of recombinant proteins.

Mutation-based variants may be analyzed by comparing them with normal reference proteins, which is in contrast to more traditional methods of elucidating the primary amino acid sequence of structurally unknown proteins. A general strategy for the analysis of protein variants consists of two steps: detection and characterization.

For the detection of protein mutations, a chemical approach using electrophoresis and chromatography was often the most convenient method prior to the application of MS. In addition, some functional proteins could be detected using bioassays that tested for specific actions. In 1981, MS was first used for the analysis of a complex mixture of peptides from hemoglobin variants *(1)*. A few years later fast-atom bombardment (FAB) replaced field desorption as an ionization method for the same purpose *(2)*. These mass spectrometric approaches were peptide mapping in their principle, and aimed at both detection and characterization of mutations *(3,4)*.

In the last few years, MS has joined the pool of detection methods dealing with whole-protein molecules *(5)*. This was mainly brought about by the advent of electrospray ionization (ESI), which allows proteins with molecular masses of many kilodaltons that were previously considered to be well beyond the mass range of MS, to be analyzed by using multiply charged species that bring

From: *Methods in Molecular Biology, Vol. 61: Protein and Peptide Analysis by Mass Spectrometry*
Edited by: J. R. Chapman Humana Press Inc., Totowa, NJ

the mass-to-charge ratio (*m/z*) down to within the limits of the available mass analyzers *(6,7)*. Although none of these detection methods are absolutely perfect, MS seems to be very effective in the study of proteins of molecular mass <20,000–30,000.

For the characterization of protein mutations, a chemical approach consists of two steps: peptide mapping and peptide sequencing. As described, these successive steps may be replaced by a mass spectrometric analysis of a complex mixture of peptides, followed by tandem MS (MS/MS) of the species of interest. More general methods for determining the primary structure of peptides by MS/MS will be discussed in Chapter 3.

The mass spectrometric approach for structural analysis of protein variants is summarized as follows:

1. Procedures prior to MS (if necessary): Purification, alkylation, and desalting by reversed-phase liquid chromatography (RP-LC).
2. Detection of variants through mass measurement of the intact protein molecule with ESI-MS.
3. Chemical or enzymatic cleavage.
4. Identification of mutated peptides through "peptide mapping" of peptide mixtures using FAB-MS.
5. Characterization of the mutation.

2. Materials

2.1. Desalting and Purification

1. LC column (C4, 300 Å; diameter 4.6 mm; length 250 mm) (Vydac, Hestperia, CA).
2. Trifluoroacetic acid (TFA) (Sigma).
3. Acetonitrile (high-performance liquid chromatography [HPLC]-grade) (Sigma).

2.2. Reduction and Alkylation (Carboxymethylation)

1. Stock solution A: 1*M* Tris-HCl containing 4 m*M* EDTA, adjusted to pH 8.5 with HCl.
2. Stock solution B: 8*M* guanidium chloride.
3. Mix stock solutions A and B at a 1:3 ratio to make a solution of 6*M* guanidium chloride, 0.25*M* Tris-HCl, and 1 m*M* EDTA just prior to use.
4. Dithiothreitol (Sigma).
5. Iodoacetic acid, free acid (Sigma).

2.3. Cleavage

1. Cyanogen bromide (CNBr) (Sigma) **(toxic, handle with gloves in hood)**.
2. Trypsin (TPCK-treated) (Sigma type XIII).
3. Lysylendopeptidase (Wako, Osaka, Japan) or endoproteinase Lys-C (Boehringer).
4. *Staphylococcus aureus* V8 protease (Boehringer).
5. Endoproteinase Asp-N (Boehringer).
6. 50 m*M* Ammonium bicarbonate, pH 7.8 (pH adjustment is not required).

7. 50 m*M* Ammonium carbonate, pH 8.3 (pH adjustment is not required).
8. 50 m*M* Tris-HCl, pH 7.4.

2.4. Mass Spectrometry

1. Acetonitrile:water:acetic acid (49:49:2 [v/v/v]).
2. Glycerol (Sigma).
3. Trichloroacetic acid (Sigma).

3. Methods

3.1. Procedures Prior to MS (if Necessary): Alkylation and Desalting by RP-LC

In general, MS requires no special procedures for sample preparation. However, removing salts (desalting) is preferred because the formation of adduct ions with alkali metals, such as sodium and potassium, will reduce the abundance of the protonated molecular ion species, which are the most informative ions for the mass measurement of proteins and peptides. Adducts also unnecessarily complicate mass spectra. In order to prevent this, RP-LC is recommended as the final step of sample preparation prior to MS, and volatile buffers are preferably chosen for the enzymatic cleavage prior to mass spectrometric peptide mapping. RP-LC is also a convenient method for estimating the amount of the sample prepared.

3.1.1. RP-LC for Purification or Desalting

1. Separation conditions: flow rate 1 mL/min; temperature ambient.
2. Buffer system and gradient condition: 0.1% TFA to 70% acetonitrile in 0.08% TFA using a linear gradient for 60 min.
3. Detection: UV absorbance at 214 nm.
4. Collection in Eppendorf tubes and recovery by lyophilization.
5. Sample loading: <1 mg for 4.6-mm diameter column.
6. For the purposes of desalting, stepwise elution from 0–70% acetonitrile on a short (<50 mm) column is recommended.

3.1.2. Reduction and Carboxymethylation (Optional)

Cysteine residues are often subject to auto-oxidation during preparation procedures, leading to a variety of products and random formation of disulfide bridges during preparation procedures. To avoid these unexpected modifications, conversion to stable derivatives by alkylation with iodoacetic acid is often performed after reduction. Such a modification destroys the higher order structure of the protein by cleaving intra- or intermolecular bridges, thus facilitating the action of proteolytic enzymes by allowing them to access cleavage sites.

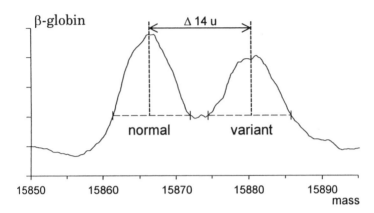

Fig. 1. Transformed ESI mass spectrum of a mixture of β-globin chains differing by 14 u.

1. Dissolve the protein at 1–20 μg/mL (<500 μL in volume) in a solution containing 6M guanidium chloride, 0.25M Tris-HCl, and 1 mM EDTA, pH 8.5.
2. Add a 100-fold molar excess, over cysteine residues, of dithiothreitol directly and incubate under argon gas at room temperature for 2 h.
3. Add the same amount (in moles) of iodoacetic acid (X mg) directly to the solution, quickly add X μL of 10N NaOH to maintain the pH, and then incubate at room temperature for 30 min in the dark.
4. Desalt by RP-LC.

3.2. Detection of Variants Through Mass Measurement of the Intact Protein Molecule with ESI-MS

Dissolve the sample in acetonitrile:water:acetic acid (49:49:2 [v/v/v]) at a concentration of 10–100 μM, and introduce it into the electrospray ion source. Any sample remaining in the delivery tube after analysis is recovered and lyophilized for further analysis, if necessary.

Figure 1 shows the transformed ESI mass spectrum of a mixture of β-globin chains differing by 14 u. Two different β-globin species are clearly differentiated in the mass spectrum taken at a resolution of 3000, and the mass separation was determined from this analysis to be 14 u (*see* Notes 1 and 2).

Among the naturally occurring amino acids, Leu and Ile have the same elemental composition and, thus, the same exact mass. Lys and Gln are different in composition, but have the same nominal mass. All other amino acids differ from each other in their molecular mass by ≤129 Da. Variants having substitutions of these amino acids are different in molecular mass from the corresponding normal proteins and are, theoretically, detectable by measuring the molecular mass of the intact protein. The smallest difference in molecular masses detectable by MS increases with the increasing size of the proteins. A

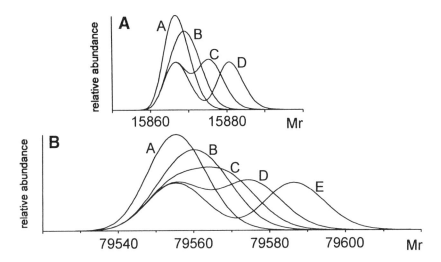

Fig. 2. Theoretical distributions of isotopic clusters corresponding to an equimolar mixture of **(A)** normal and variant β-globins. (A) Pure normal protein; (B) normal and +4 mass unit mutant; (C) normal and +9 mutant; (D) normal and +14 mutant. **(B)** Normal and variant transferrin. (A) Pure normal protein; (B) normal and +9 mass unit mutant; (C) normal and +14 mutant; (D) normal and +20 mutant; (E) normal and +31 mutant. The curves have been drawn by connecting the peak top of each isotopic ion from a display at sufficient resolution to achieve complete separation of ^{13}C isotopic peaks.

theoretical analysis can be considered using two model proteins: hemoglobin β-chain and transferrin (M_r 79,556) (Fig. 2). A separation of 14 u in transferrin is not clear (C in Fig. 2B), whereas the shapes of the clusters for 20 and 31 u differences, corresponding to D and E, respectively, in Fig. 3B, are similar to those for 9 and 14 u differences, corresponding to C and D, respectively, of the β-globin chain in Fig. 3A. Identification of a variant species differing from the normal one by 14 u in an actual analysis, corresponding to the theoretical pattern D in Fig. 3A, has been presented in Fig. 1. The minimum difference in molecular mass detectable as pattern D in Fig. 3B for an equimolar mixture with normal proteins of M_r over 100,000 is about 20 u. This means that about half of the possible amino acid substitutions in a protein of this size will escape detection. A strategy to detect these substitutions should start with ESI-MS analysis of cyanogen bromide-cleaved peptides (Section 3.3.).

3.3. Chemical or Enzymatic Cleavage

3.3.1. Cleavage with CNBr Followed by ESI-MS

CNBr cleaves only at methionine, a relatively rare amino acid in proteins. Therefore, it will generate large peptide fragments compared with enzymatic

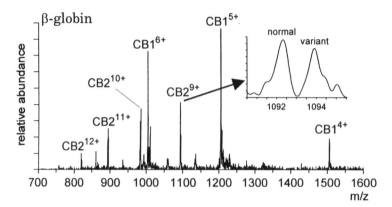

Fig. 3. ESI mass spectrum of CNBr peptides of the β-globin isolated from the same patient shown in Fig. 1. The region of the nine-charged ion for peptide CB-2 is depicted in an inset. The peptide has two Cys residues. Peptide CB2 is detected at the expected mass, although unalkylated in this case. The resolution used for measurement was 2000.

cleavage, but their molecular size will be <20,000–30,000, which is well within the range that can be measured to the nearest integer mass. Most of the resulting C-terminal homoserine residues of the peptides from this type of cleavage are in lactone form.

The choice of ionization method in this instance is ESI. The analysis of peptide mixtures with ESI-MS, however, has an associated problem derived from the fact that the most suitable ionization condition differs from peptide to peptide. Even so, this method is very useful for confining a suspected site of mutation to a small part of the protein molecule and for determining the size of the molecular shift precisely.

1. Dissolve the protein at 10 mg/mL in 70% (v/v) TFA (*see* Note 3).
2. Add a 100-fold molar excess, over Met residues, of CNBr dissolved in 70% TFA solution, and incubate in the dark under argon at room temperature for 2 h.
3. Add 10 vol of water, rotary evaporate to remove excess CNBr, freeze, and lyophilize. Alternatively, the cleavage mixture is directly subjected to RP-LC similar to the desalting procedure described in Section 3.1.1..
4. ESI-MS, as described in Section 3.2..

In the example shown in Fig. 3, a mutation causing a difference of 14 u is identified in peptide CB2 (M_r 9821.4 for the normal species) of β-globin. The separation of the normal and the variant species is apparently better than that of the intact proteins shown in Fig. 1.

3.3.2. Enzymatic Digestion Followed by FAB-MS

The following enzymes are amino acid-specific in their cleavage, and are employed for peptide mapping: trypsin (TPCK-trypsin), lysylendopeptidase

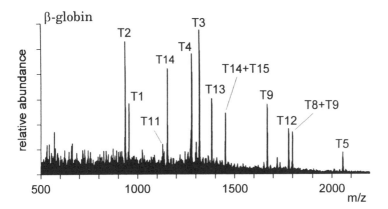

Fig. 4. FAB mass spectrum of the tryptic digest of carboxymethylated β-globin. The number above each peak is that of the predicted tryptic peptide starting from the N-terminus.

or endoproteinase Lys-C, *Staphylococcus aureus* V8 protease, endoproteinase Asp-N.

1. Dissolve protein in the digestion buffer at about 1 mg/mL.
2. Optimized buffers: for trypsin and V8 protease, 50 mM ammonium hydrogencarbonate (pH 7.8); for lysylendopeptidase (or Lys-C protease), 50 mM ammonium carbonate (pH 8.3); for Asp-N protease, 50 mM Tris-HCl (pH 7.4) (*see* Note 4).
3. Add protease at 1% (w/w) ratio and digest at 37°C for 6 h.
4. Lyophilize.
5. FAB-MS: The matrix used is acidified glycerol, containing 20 mg trichloroacetic acid/mL glycerol. In the case of the hemoglobin variants, approx 100 µg of lyophilized digest were mixed with 50 µL of the matrix, and 2 µL of the resultant mixture were placed on the FAB probe. The loaded sample was, therefore, about 5 µg (300 pmol) (*see* Note 5).

3.4. Peptide Mapping by FAB-MS

A typical tryptic peptide map of β-globin by FAB-MS is shown in Fig. 4. Most of the sequence of β-globin, except for the portions corresponding to some small peptide fragments, is covered by the mass spectrum. Low-mass peptides, such as peptides T6, T7, T8, and T15 in this map, are difficult to identify owing to the chemical noise in the low-mass region of the FAB mass spectrum. This low-mass noise is particularly prevalent in the FAB-MS of peptide mixtures. Fortunately, this will not often be a problem, since the probability of a point mutation arising in these smaller peptides will be low. Furthermore, this may be overcome by using multiple analyses with different

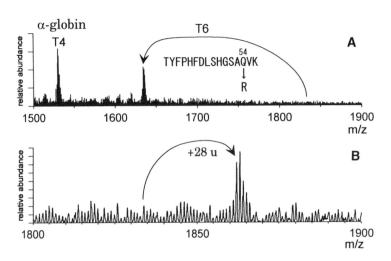

Fig. 5. FAB mass spectra of **(A)** tryptic and **(B)** lysylendopeptidase digests of an α-globin variant.

enzymes. LC-MS improves signal-to-noise ratio and facilitates the detection of small peptides. This technique is described in Chapter 7.

3.5. Characterization of the Mutation

Categorized strategies are described here using examples of variant hemoglobins (*see* Note 6).

3.5.1. Simple Cases: Mutation at the Cleavage Site of a Specific Enzyme

1. New cleavage site: The example shown in Fig. 5 is a hemoglobin variant with a mutation in α-globin. Tryptic peptide 6 (T6), which should be detected at *m/z* 1833.9, is missing, and a new peptide appears at *m/z* 1634.8 in the mass spectrum. The substitution of Arg for Gln at position 54 is the only possible substitution that fits the molecular mass change. The substitution can be verified by a digestion with Lys-C, which does not cleave arginyl peptide bonds; the peptide ion is shifted from the normal *m/z* 1833.9 to *m/z* 1862.0, with the increase of 28 u confirming the Gln→Arg mutation.
2. Missing cleavage site: In this example, a β-globin variant (Fig. 6), tryptic peptides 13 (T13) and 14 (T14) are missing in the digest, indicating a mutation at the Lys residue at position 132. The new peptide is observed at *m/z* 2495.3, indicating the replacement of Lys by Asn.
3. Unique replacement: In the case shown in Fig. 7, tryptic peptide 11 (T11) at *m/z* 1126.5 is missing and a new peptide appears at *m/z* 1152.5 in the digest of purified β-globin of the variant hemoglobin. This increase of 26 u suggests five pos-

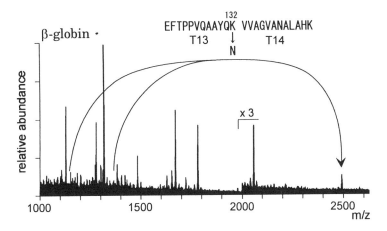

Fig. 6. FAB mass spectrum of the tryptic digest of a β-globin variant.

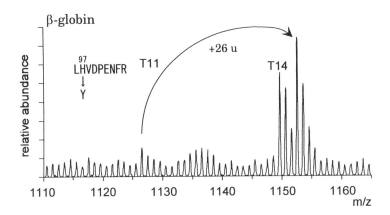

Fig. 7. FAB mass spectrum of the tryptic digest of a β-globin variant. The peak at *m/z* 1149.6 is the normal peptide-14.

sible mutations: His→ Tyr, Ala → Pro, Cys → Glu, Ser → Leu, and Ser→ Ile. Since the normal peptide has His, but not Ala, Cys, or Ser residues, and has a single His residue at position 97 in the sequence, it follows that the latter has been replaced by Tyr.

3.5.2. Complex Cases

1. Characterization based on additional information: In cases where the molecular weight of the mutated peptide does not permit the assignment of a unique type and/or position of substitution, chemical data, such as the charge carried on the variant, may provide an answer. In the FAB mass spectrum of a tryptic digest of

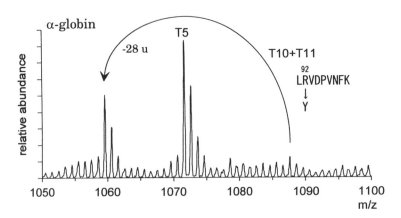

Fig. 8. FAB mass spectrum of the tryptic digest of an α-globin variant. Peptide-10 + 11 is normally identified at *m/z* 1087.6 owing to incomplete cleavage of arginyl bond at position 90. In this case, the peptide is missing and a new peak appears at *m/z* 1059.6. The peak at *m/z* 1071.6 is normal peptide-5.

a purified variant of the α-globin shown in Fig. 8, a combined peptide T10 + T11, which is normally present, at *m/z* 1087.6 owing to incomplete cleavage, is missing and a new peptide is observed at *m/z* 1059.6. The decrease of 28 u suggests six possible mutations: Arg → Lys, Arg → Gln, Val→ Ala, Asp→ Ser, Glu → Thr, and Met→ Cys. The first four substitutions are possible on the basis of the normal sequence of the peptide. The variant was determined to be acidic, relative to the normal globin, on electrophoresis under denaturing conditions, and therefore only the Arg → Gln mutation at position 92 is plausible (*see* Note 7).

2. Combination of different enzymes: In the case shown in Fig. 9, the structure could not be determined directly from the mass spectrum of the peptide mixture. The FAB mass spectrum of the tryptic digest of β-globin from the patient discloses new peaks near the normal peptide 9 (T9), suggesting that the mutation must be in T9 (Fig. 9). The difference of 14 u suggests seven possible substitutions: Gly → Ala, Ser→ Thr, Val → Leu, Val → Ile, Asn→ Gln, Asp → Glu, Asn → Lys, and Thr → Asp. However, even on the basis of both the normal sequence and the electrophoretic data, six different substitutions at seven possible positions are left as candidates. In this instance, the substitution of Glu for Asp at position 79 is determined, when the tryptic digest is then hydrolyzed with V8 protease. Otherwise, the mutation can be characterized by MS/MS with collision-induced dissociation (CID) *(8)*.

3. CID: CID may be used to differentiate among various possibilities, as shown in the last case, or for verifying a proposed structure. MS/MS with CID is a technique that should be tried before Edman sequencing. The technique is described in Chapters 2 and 3.

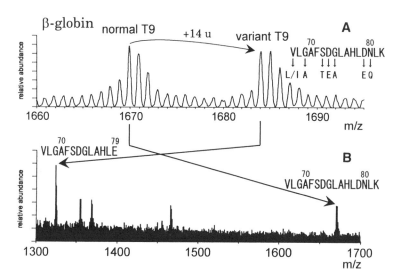

Fig. 9. FAB mass spectra of **(A)** tryptic and **(B)** V8-protease digests of a β-globin variant. The variant cannot be separated from the normal counterpart in sample preparation. The peptides at *m/z* 1669.9 and at *m/z* 1683.9 are derived from the normal and variant proteins, respectively. The increase of 14 u for this peptide gives eight possible types of mutations even with supporting chemical data. The substitution Asp79-Glu is determined by V8-protease digestion.

4. Notes

1. The theoretical isotopic distributions of human β-globin (146 residues, M_r 15,867) at resolutions of 15,000 (A) and 2000 (B) are shown in Fig. 10. In actual measurements of proteins of molecular mass over several kilodaltons, it is usually impossible to obtain the high ion current necessary for unit resolution as (A), nor is it possible to identify the ^{12}C species corresponding to the exact mass in the molecular ion cluster. Consequently, the average mass of the unresolved isotopic cluster or "chemical mass" is usually used.

2. The minimum amount of the sample required for ESI-MS will be a few hundred picomoles for proteins with molecular masses in the 20,000–30,000 range, although this is very much dependent on the size and the structure of the protein. However, most of the sample can be recovered after measurement.

3. Cleavage in formic acid is not recommended, because it has a tendency to formylate amino and hydroxyl groups, which creates a mixture consisting of products differing in molecular mass by multiples of 28 Da.

4. Lysylendopeptidase is active in urea solution concentration up to 4*M*. The digestion is often carried out in a solution of 4*M* urea, 0.1*M* Tris-HCl, pH 9.0, in which substrates are denatured and often become more susceptible to the cleavage. Although the presence of Tris or urea does not drastically hamper FAB ionization or affect the mass spectrum, it is better to remove them by RP-LC.

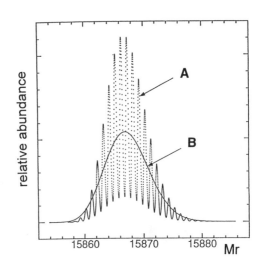

Fig. 10. Theoretical isotopic distributions of hemoglobin β-chain at resolutions of 15,000 **(A)** and 2000 **(B)** (10% valley definition).

5. The sample can be recovered from the probe after measurement and subjected to RP-LC to isolate it from the matrix. The amount consumed in MS, therefore, would be very small, about one-tenth of that loaded, when recovery is performed.
6. Variant hemoglobins are heterozygous in most cases, and therefore both normal and mutant proteins are present in the patients as shown in Fig. 1. In this chapter, variant hemoglobins, isolated from their normal counterpart, are presented, except for the variant shown in Figs. 1, 3, and 9.
7. Most of the protein variants are caused by a single nucleotide base mutation. If the normal DNA sequence of the protein is known, the number of nucleotide base changes leading to the proposed type of amino acid substitution may aid selection. However, this should not be considered a final answer because polymorphism may be present in the "normal" DNA sequence.

References

1. Wada, Y., Hayashi, A., Fujita, T., Matsuo, T., Katakuse, I., and Matsuda, H. (1981) Structural analysis of human hemoglobin variants with field desorption mass spectrometry. *Biochim. Biophys. Acta* **667,** 233–241.
2. Wada, Y., Hayashi, A., Masanori, F., Katakuse, I., Ichihara, T., Nakabushi, H., Matsuo, T., Sakurai, T., and Matsuda, H. (1983) Characterization of a new fetal hemoglobin variant, Hb F Izumi ^{A}γ6Glu → Gly, by molecular secondary ion mass spectrometry. *Biochim. Biophys. Acta* **749,** 244–248.
3. Wada, Y., Matsuo, T., and Sakurai, T. (1989) Structure elucidation of hemoglobin variants and other proteins by digit-printing method. *Mass Spectrom. Rev.* **8,** 379–434.
4. Wada, Y. and Matsuo, T. (1994) Structure determination of aberrant proteins, in *Biological Mass Spectrometry* (Matsuo, T., Caprioli, R. M., Gross, M. L., and Seyama, Y., eds.) Wiley, Chichester, UK, pp. 369–399.

5. Green, B. N., Oliver, R. W. A., Falick, A. M., Shackleton, C. H. L., Roitmann, E., and Witkowska, H. E. (1990) Electrospray MS, LSIMS, and MS/MS for the rapid detection and characterization of variant hemoglobins, in *Biological Mass Spectrometry* (Burlingame, A. L. and McCloskey, J. A., eds.), Elsevier, Amsterdam, pp. 129–146.
6. Fenn, J. B., Mann, M., Meng, C. K., Wong, S. F., and Whitehouse, C. M. (1989) Electrospray ionization for mass spectrometry of large biomolecules. *Science* **24,** 64–70.
7. Smith, R. D., Loo, J. A., Edmonds, C. G., Barinaga, C. J., and Udseth, H. R. (1990) New developments in biochemical mass spectrometry: electrospray ionization. *Anal. Chem.* **62,** 882–899.
8. Wada, Y., Matsuo, T., Papayannopoulos, I. A., Costello, C., and Biemann, K. (1992) Fast atom bombardment and tandem mass spectrometry for the characterization of hemoglobin variants including a new variant. *Int. J. Mass Spectrom. Ion Processes* **122,** 219–229.

9

Noncovalent Interactions Observed Using Electrospray Ionization

Brenda L. Schwartz, David C. Gale, and Richard D. Smith

1. Introduction

Molecules that form structurally specific noncovalent associations in solution are of fundamental biological importance. The weak bonds that constitute these liquid-phase noncovalent complexes determine their strength of interaction, which is indicated by the dissociation constant (K_d). Electrospray ionization mass spectrometry (ESI-MS) is rapidly becoming a useful technique for the detection of specific noncovalent complexes, and potentially for probing relative stabilities for these complexes. Several types of noncovalent association in solution, including enzyme–substrate, receptor–ligand, host–guest, protein–nucleic acid, and oligonucleotide interactions have been demonstrated to survive transfer into the gas phase using ESI (1–4). Through the careful control of instrumental variables in the electrospray atmosphere–vacuum interface region, these complexes can be detected intact.

There are several important instrumental variables in the interface region that can be carefully adjusted to either preserve noncovalent complexes or cause their dissociation into individual components (3,4). The interface conditions used to preserve such complexes are inherently more gentle than ones conventionally employed for ESI-MS analysis. Figure 1 shows the various regions in the electrospray process for ion desolvation/activation, and some of the variables that can be adjusted to affect observation of noncovalent complexes. In instruments that use a glass or metal inlet capillary for ion sampling, adjusting the temperature of this inlet capillary can be used to provide thermal desolvation of electrosprayed ions, such that noncovalent complexes can be detected intact. It should be noted that some instruments utilize a sampling aperture that effectively eliminates this heating region. Although at least some

From: Methods in Molecular Biology, Vol. 61: Protein and Peptide Analysis by Mass Spectrometry
Edited by: J. R. Chapman Humana Press Inc., Totowa, NJ

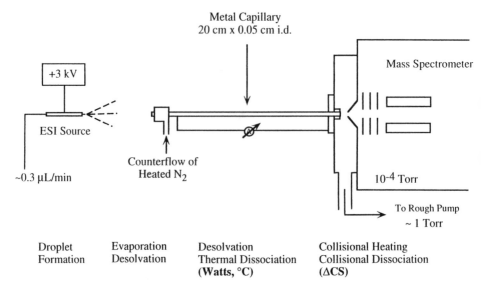

Fig. 1. Schematic diagram of the atmosphere–vacuum interface region showing the regions of droplet formation to ion formation together with the variables important for observation of noncovalent complexes.

noncovalent complexes can be detected using an aperture, it remains to be determined whether this difference imposes any fundamental limitation on the detection of noncovalent complexes by such instruments. Another variable that can be manipulated, and that induces collisional activation in the interface region, is the capillary-skimmer voltage difference (ΔCS), or orifice-skimmer voltage bias for those interfaces incorporating a sampling orifice rather than a capillary inlet. Typically, relatively low capillary temperatures and low ΔCS voltages are employed to provide gentle interface conditions, such that intact noncovalent associations can be observed. These variables can be adjusted to provide harsher interface conditions, which will generally increase removal of adduction/solvation to the molecular ions, but will also result in greater dissociation of the noncovalent complex. The use of a heated countercurrent drying gas at atmospheric pressure in the spray region has been shown to provide additional ion desolvation for high *m/z* ions, reducing the amount of heating necessary in subsequent stages *(5)*.

Generally, adjustment of the instrumental variables in the electrospray interface region in order to observe noncovalent associations involves a compromise between providing adequate heating/activation for ion desolvation, while ensuring that the intact complex is preserved. The extent to which the complex can be heated is expected to depend on the nature of its binding, and the relative stabilities of the relevant nonspecific complexes (with solvent, buffer components, and so forth) compared to the specific complex. At present, there is insufficient knowledge of the relevant kinetics and thermodynamics to predict whether these methods will be successful for any particular application.

In addition to instrumental considerations, proper solution conditions must obviously be maintained in order to detect ions corresponding to intact complexes. For conventional ESI-MS experiments, solution conditions are generally acidic (for detection of positive ions), often to cause sufficient gas-phase charging of biomolecules to allow detection at the relatively low mass-to-charge ratio (m/z) ranges (<2000) provided by most earlier quadrupole mass spectrometers. However, many noncovalent associations of physiological relevance are unstable under such solution conditions. For example, the physiologically active forms of many proteins are multimeric, and the destruction of the higher order quaternary structures generally occurs in acidic solution. In these instances, as with other noncovalent interactions, it is desirable to place the protein in aqueous solution conditions near neutral pH, such that the specific subunit associations are preserved. The use of "volatile" buffers, such as ammonium acetate, is generally preferred for this purpose, since the use of nonvolatile salts can lead to excessive adduction. However, utilizing buffers at physiological pH generally results in a reduced amount of gas-phase charging on the biomolecule, presumably because of the retention of a relatively compact higher order structure and the resulting Coulombic constraints on maximum charging. Therefore, for certain large proteins, mass spectrometers capable of extended m/z (>3000) operation are required in order to detect ions corresponding to the active form of the protein.

The fact that mass spectra showing the intact quaternary structure of proteins (such as the hemoglobin tetramer) and non self-complementary oligonucleotide duplexes can be produced without contributions owing to nonspecific or random aggregation is utterly convincing evidence of the feasibility of studying such interactions. However, in any particular application, there is a need for the ability to distinguish between structurally specific noncovalent interactions in the ESI mass spectra and nonspecific aggregation, possibly resulting from the electrospray process itself. There are several methods that can be used to provide evidence that the observed gas-phase interactions are structurally specific and reflect those associations existing in solution *(3)*. Manipulation of the electrospray interface conditions should permit the predominance of ions corresponding to the noncovalent complex of preferred stoichiometry (A · A, B · B) over species from complex dissociation or random aggregation (A, B, A · B_2, A_2 · B, B_3, and so on). In general, the complexes can easily be distinguished from any covalent species, since the noncovalent complexes will readily dissociate under harsher interface conditions. Tandem mass spectrometric (MS/MS) methods can also be used to readily dissociate the intact complex into its respective components. Further supportive control experiments can be provided by altering the solution conditions (e.g., pH, temperature) to affect the interactions present in solution, which should induce a corresponding change in the ESI mass spectra. Finally, and most ideally, chemical modifications of a component of the complex can be explored, since these may alter the complex formation in solution and generate a subsequent change in the mass spectra.

This chapter presents the application of an extended *m/z* range single quadrupole mass spectrometer for detection of ions at high *m/z* corresponding to large protein noncovalent complexes. Additionally, solution and instrumental conditions required to observe intact oligonucleotide noncovalent complexes (e.g., duplex formation from either self-complementary or sequence-specific oligonucleotides), and also drug–DNA complexes, at a more conventional *m/z* range (<2000 *m/z)* on another single-quadrupole mass spectrometer, are illustrated. Finally, the use of high-performance Fourier transform ion cyclotron resonance (FTICR) instrumentation for the study of such complexes is demonstrated.

2. Materials

2.1. Chemicals

1. Avidin (Sigma, St. Louis, MO).
2. Distamycin A (Sigma).
3. The sequence-specific 14-bp oligonucleotides, 5'-dCCCCAAATTTCCCC-3' and 5'-dGGGGAAATTTGGGG-3', were synthesized using phosphoramidite chemistry and purified by reversed-phase high-performance liquid chromatography (RP-HPLC), followed by Centricon-3 filtration.
4. Duplex formation was achieved by placing the two purified oligonucleotides, 5'-dCCCCAAATTTCCCC-3' (strand C, M_r = 4104 Da) and 5'-dGGGGAAA TTTGGGG-3' (strand G, M_r = 4424 Da), at concentrations of 75 μ*M* each, in a 10-m*M* ammonium acetate/10-m*M* ammonium citrate (pH 8.3) annealing buffer, heating the solution in a water bath to 95°C for 5 min, and then cooling it at room temperature for over 3 h.
5. The self-complementary 12-bp oligonucleotide, 5'-dCGCAAATTTGCG-3', was synthesized using phosphoramidite chemistry and purified by RP-HPLC followed by Centricon-3 filtration.
6. Preparation of the noncovalent complex formed between the minor groove binder, distamycin A, and the duplex formed by the oligonucleotide (5'-dCGC AAATTT GCG-3', M_r = 3645.5 Da) used annealed solutions containing concentrations of 50 μ*M* distamycin and 100 μ*M* oligonucleotide duplex in 30 m*M* ammonium acetate buffer (pH 8.3). The annealing procedure was similar to that in item 4.

2.2. Instrumentation

1. Protein complex positive-ion mass spectra were acquired on an extended m/z range (~45,000) single quadrupole mass spectrometer operated at low frequency, built in our laboratory and previously described *(5)*. Characteristics of this instrument include low resolution (<100) and discrimination against low m/z ions. The atmosphere–vacuum interface (*see* Fig. 1) consists of a resistively heated metal inlet capillary (1.59 mm od, 0.5 mm id, 20 cm long) held at a potential ΔCS, relative to that of the skimmer. A heated countercurrent N_2 gas flow in the spray region was utilized to assist in desolvation of the large ions.
2. Oligonucleotide complex negative-ion mass spectra were acquired on both a single quadrupole mass spectrometer, with typical experimental conditions previously described *(6)*, and a FTICR mass spectrometer, which already has been utilized for the analysis of other types of noncovalent complexes *(7)*.
3. The electrospray source (emitter) used for all studies utilizes a sheathless design, and is capable of low flow rates (0.3 μL/min) as previously described *(8)*. The source potential was 3.0 kV and SF_6 was used for electrospray stabilization.

3. Methods (Solution and Instrumental Conditions) and Discussion

3.1. Large Protein Noncovalent Complexes: High m/z Ions

Positive-ion spectra of the tetrameric protein avidin were acquired as follows:

1. Place the protein in 10 mM NH_4OAc pH 6.7 at 1 μg/μL (*see* Note 1).
2. Set the metal capillary heating to 15 W (*see* Note 2), corresponding to a temperature of approx 174°C as measured on the external capillary surface.
3. Set the capillary-skimmer voltage (ΔCS) to 80 V.
4. Acquire positive-ion spectra on the extended m/z range single-quadrupole instrument.

Figure 2 shows predominantly (Q^{18+} to Q^{16+}) ions in the 3550–4000 m/z range indicative of the intact tetrameric form of the protein (Note 3). The relatively narrow charge distribution and limited amount of charging observed for the protein reflect the retention of portions of higher order structure in the gas phase. The ESI-MS spectrum also indicates a small octamer contribution at higher m/z (~5000) from aggregation of the tetramer. The absence of any nonspecific aggregate species (i.e., trimer, pentamer) indicates that the ESI-MS spectrum reflects only the stoichiometrically specific protein subunit association into a tetramer, consistent with expected solution behavior.

There are several ways to provide further support that the observed noncovalent species in the ESI mass spectrum are structurally specific, exhib-

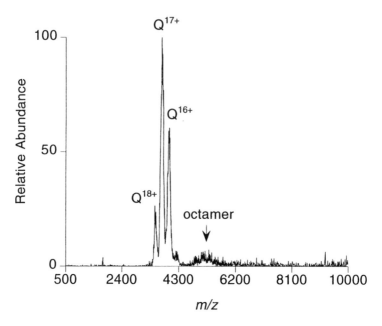

Fig. 2. Positive-ion ESI-MS of egg white avidin in 10 mM NH_4OAc (pH 6.7) acquired on the extended m/z range single-quadrupole mass spectrometer. ESI spectra were acquired with a capillary heating of 15 W (~174°C) and a $\Delta CS = 80$ V to observe the active form of the protein. Spectra were acquired from m/z 500 to 10,000, corresponding to approx 1.5 min/scan, with the accumulation of four averaged scans.

iting the correct stoichiometry, and do not arise from random aggregation from the electrospray process:

1. Changing the solution conditions to affect the associations present in solution should alter the appearance of the mass spectra. Taking advantage of the known solution pH dependence for the dimer–tetramer equilibrium of certain proteins, such as Concanavalin A (Con A) and adult human hemoglobin (Hb A_o), one can use ESI-MS to probe these interactions. For example, for the tetrameric protein, Con A, placing the protein in a pH 8.4 10-mM NH_4OAc solution produces mainly tetrameric ions in the mass spectra, which is expected from solution behavior. However, placing the protein in a pH 5.7 solution and utilizing the identical mild ESI interface conditions, as used for the pH 8.4 solution, produces mainly ions indicative of a dimer species, also consistent with known solution behavior *(4)*.

2. Chemical modification of a single component of the complex can also be used to give rise to an expected change in the mass spectra. Specifically, utilizing a derivatized form of Con A, succinyl-Con A, which should exist as a dimer in solution, produces predominately a dimer in the gas phase when analyzed under identical interface conditions *(4)*.

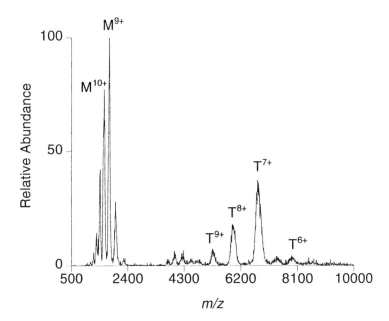

Fig. 3. Positive-ion ESI spectrum of avidin utilizing harsher interface conditions by increasing the capillary heating to 28 W (~276°C). Other experimental conditions were identical to those used to obtain Fig. 2.

3. Creating harsher conditions in the interface region by either increasing the heat applied to the metal inlet capillary or increasing the ΔCS voltage will generally dissociate the multimeric charge states into monomer species if the associations are noncovalent. By increasing the capillary heating to 28 W (~276°C), dissociation of the avidin tetramer into predominately monomer species (M^{13+} to M^{7+}) of relatively high charge state occurs (Fig. 3). There is also evidence of relatively low charge state ions at high *m/z* (5325–8000) indicative of a trimer species (T^{9+} to T^{6+}) (*see* Note 4).

Increasing the ΔCS voltage for a relatively labile tetrameric association, such as Hb A_0, can also result in dissociation of this heterotetramer, even under relatively gentle capillary heating conditions *(4)*. The more solution stable tetrameric associations of Con A and avidin are not as substantially affected by this variable, indicating the potential of ESI-MS for probing relative stabilities of noncovalent complexes.

4. Finally, MS/MS experiments using specific multimeric charge states should also readily dissociate the intact protein into its components.

3.2. Oligonucleotide Noncovalent Complexes: Low m/z Ions

3.2.1. Sequence-Specific Oligonucleotides

Negative-ion spectra of the annealed solution of the two sequence-specific 14-bp oligonucleotides (*see* Note 5) in a 10-mM ammonium acetate/10-mM

ammonium citrate buffer (see Section 2.1., step 4 and Note 6) were acquired on another single quadrupole mass spectrometer (m/z range 2000), as used previously for the analysis of oligonucleotide noncovalent complexes (6). Figure 4 illustrates the effects of conditions in the ESI interface region on the observation of the duplex in the mass spectra of these sequence-specific oligonucleotides. Spectra were acquired under different interface conditions by adjusting the temperature of the heated capillary and the capillary-skimmer offset voltage to generate extremely mild (Fig. 4A) or more severe (Fig. 4C) interface conditions. Under extremely gentle interface conditions (capillary heating of 10 W ~ 75°C, $\Delta CS = -50$ V), ions corresponding to the intact oligonucleotide duplex (Δ^{9-} to Δ^{5-}) are clearly observed, in addition to charge states at lower m/z indicative of single-stranded species (C^{n-} and G^{n-}). The Δ^{7-} ion of the duplex dominates at higher m/z for these interface conditions employed, but exhibits a significant amount of residual solvated and/or adducted species, as is evident from the wider peak (*see* Note 7). Harsher interface conditions can be used to effectively remove this solvation and/or adduction, but generally will also result in a decreased amount of the noncovalent complex being detected. This typical compromise between observation of the noncovalent complex and removal of solvated/adducted species to the charge states is well illustrated in Fig. 4. Increasing the ΔCS to -100 V promotes the removal of some of the solvated/adducted species (Fig. 4B). Utilizing even harsher interface conditions (capillary heating of 12 W ~ 105°C) results in little or no adduction to the duplex ions. However, it also induces a decrease in the relative abundance of the Δ^{7-} ion for the duplex, and a concomitant increase in the abundance for the single-stranded species.

3.2.2. Self-Complementary Oligonucleotides

At first glance, the mass spectra of duplexes formed by a self-complementary oligonucleotide rather than from two sequence-specific oligonucleotides (with different M_r values) can be a challenge to interpret. Specifically, assignment of ions originating from single-stranded or duplex species presents a difficulty, since the even charge state ions for single-stranded species appear at the same m/z as the duplex species bearing twice the monomer charge. Therefore, for self-complementary oligonucleotides, the appearance of "unique" (i.e., odd) charge states for the duplex must be used to confirm the presence of duplex species in mass spectra acquired on low-resolution instrumentation. The high-resolution mass measurement capability offered by FTICR does, however, allow differentiation between charge states at the same nominal m/z for such noncovalent complexes.

The negative-ion FTICR mass spectrum of the annealed 50 μM distamycin–100 μM oligonucleotide duplex in 30 mM ammonium acetate buffer solution

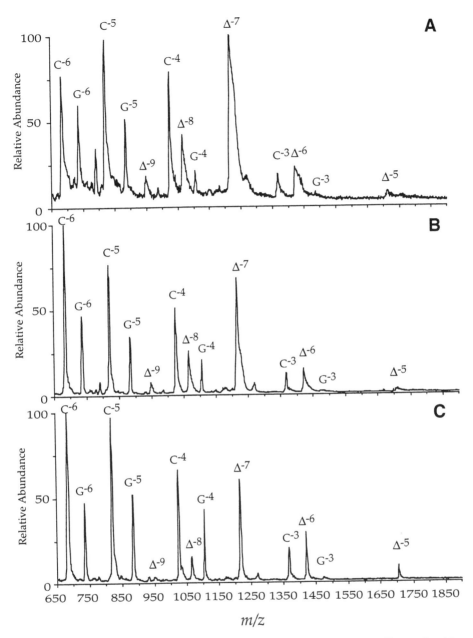

Fig. 4. Negative-ion ESI spectrum of two sequence-specific 14-bp oligonucleotides, 5'-dCCCCAAATTTCCCC-3' (strand C, M_r = 4104 Da) and 5'-dGGGGAAATTT GGGG-3' (strand G, M_r = 4424 Da), annealed in 10 mM ammonium acetate/10 mM ammonium citrate (pH 8.3). Spectra were acquired on a single-quadrupole mass spectrometer (m/z limit 2000) under various interface conditions ranging from extremely gentle to harsh: **(A)** capillary heating of 10 W (~75°C) and a ΔCS = –50 V, **(B)** capillary heating of 10 W (~75°C) and a ΔCS = –100 V, and **(C)** capillary heating of 12 W (~105°C) and a ΔCS = –100 V.

(Section 2.1., step 6) is shown in Fig. 5. A charge state distribution (6– to 4–) for the oligonucleotide duplex with and without distamycin (Dm) is observed (Fig. 5A). Isotopic resolution of the even charge states indicates the amount of contribution from either single-stranded or duplex species. For example, isotopic resolution of the peak at m/z 1214 indicates a contribution from both species, i.e., monomer^{3-} and Δ^{6-}, but primarily from single-stranded material (not shown). The intensities of the ions corresponding to the intact duplex, Δ^{5-}, and the 1:1 Dm:oligonucleotide duplex noncovalent complex $(\Delta{:}1Dm)^{5-}$ are nearly identical, consistent with the expected solution behavior for complete distamycin binding at these sample concentrations. The expanded regions for the 5– charge states of the intact oligonucleotide duplex and the 1:1 Dm:oligonucleotide duplex noncovalent complex illustrate the isotopic envelope and indicate no contribution from monomer, as would be expected for an odd charge state (Fig. 5B).

3.3. Summary

This chapter has attempted to demonstrate that observing intact noncovalent complexes by ESI-MS requires a balance between providing sufficient ion desolvation, while ensuring that the interface conditions are mild enough to preserve the complex. With some large noncovalent complex systems it may be inherently more difficult to provide ions by ESI at lower m/z (<3000) owing to the reduced extent of gas-phase charging arising from the more compact solution structures and Coulombic effects. Generally, some experimentation is required to determine the optimum solution and instrumental conditions (i.e., settings in the atmosphere–vacuum interface region, which also vary between instruments of different design) for the detection of intact noncovalent associations. This is because of differences in the noncovalent complex systems investigated (i.e., K_d values, M_r) and in the mass spectrometric instrumentation (i.e., type of electrospray interface used). Once determined, however, the conditions are generally suitable for complexes that are of similar composition. Thus, the potential exists to use ESI-MS for the initial screening of compounds for binding affinity to specific macromolecular targets.

4. Notes

1. Since the physiologically active forms of many large proteins are multimeric, conventional ESI-MS solutions (i.e., 5% acetic acid) cannot be employed in these instances. Placing these proteins in a 5% acetic acid solution results in the observation of ions corresponding to the individual subunits, not the active multimeric form of the protein, consistent with predicted solution behavior. Typically, solutions of 10 mM ammonium acetate at near neutral pH are used to preserve the predicted subunit associations, while simultaneously providing a

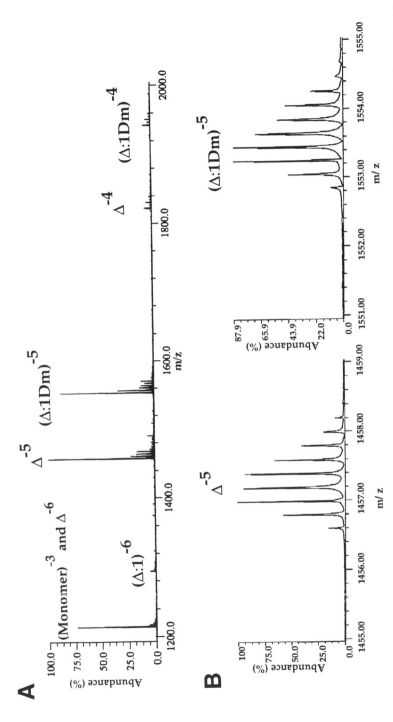

Fig. 5. (A) Negative-ion ESI-FTICR spectrum of an annealed solution containing 50 μM distamycin–100 μM oligonucleotide duplex formed by a 12-bp self-complementary oligonucleotide (5′-dCGCAAATTTGCG-3′, M_r = 3645.5 Da) in 30 mM ammonium acetate (pH 8.3); **(B)** expanded region of the Δ^{5-} charge states of the oligonucleotide duplex (Δ) and the 1:1 Dm:oligonucleotide duplex noncovalent complex.

volatile buffer effective for ESI. Specifically, a 10-mM NH$_4$OAc solution at pH ~ 8.0 can be used to preserve the tetrameric associations of proteins, such as Hb A$_o$ (M_r ~ 64.5 kDa), Con A (M_r ~ 102 kDa), and avidin (M_r ~ 64 kDa) (4). In addition, certain proteins, such as Con A and Hb A$_o$, exist in a dimer–tetramer equilibrium that is pH-dependent, with the dimer dominating in solution at lower pH (~5.0–6.0) (4). Typical protein concentrations range from 0.1–1 µg/µL, depending on the instrument sensitivity.

2. Capillary heating is reported as Watts applied across the capillary and the temperature is measured on the surface of the capillary.

3. In order to observe ions corresponding to the intact multimer of these proteins, relatively gentle conditions in the electrospray atmosphere–vacuum interface region must be utilized. The most important variables in affecting the observation of noncovalent species are the heat applied to the metal inlet capillary and the capillary-skimmer offset voltage (Fig. 1). Enough heating/activation must be utilized to provide sufficient desolvation for ion resolution without destroying the intact complex. Utilizing much lower capillary heating will not provide adequate desolvation and the charge state distribution for the tetramer will not be as well resolved.

4. The appearance of trimer charge states at high m/z (i.e., relatively low charge state) has been observed previously for dissociation of tetrameric proteins when substantial heating/activation has been applied in the interface (4). MS/MS studies have shown that the trimer species results from gas-phase dissociation of the tetramer, and once formed, proves to be very stable toward additional heating/activation in the interface.

5. Since the two oligonucleotides are of different M_r, all ions corresponding to formation of the sequence specific noncovalent duplex (M_r = 8528 Da) will appear at different m/z values than the ions for the single-stranded species, thereby facilitating assignment of charge states in the ESI spectra.

6. ESI-MS spectra of these oligonucleotides in aqueous unbuffered solution only exhibit ions corresponding to the single-stranded species even under gentle interface conditions (i.e., low ΔCS voltages and capillary temperatures), indicating that the presence of salts or buffers is required to stabilize duplex formation. Some oligonucleotide noncovalent complexes require higher salt concentrations than others to stabilize the complex, but one must experiment with various concentrations/buffers to find those that are also compatible with ESI-MS. The ones mentioned in this chapter have proven to be successful in analyzing several DNA noncovalent complexes (6).

7. Generally, a shift to higher m/z (i.e., lower charge state) is observed in the ESI mass spectra for oligonucleotides in buffered solutions compared to aqueous nonbuffered solution conditions, presumably owing to a more compact secondary structure and Coulombic constraints arising from duplex formation. Further, peaks corresponding to DNA noncovalent complexes typically exhibit broader peak widths owing to the presence of stabilizing counterions (i.e., Na) or water molecules. Therefore, a purification process for the oligonucleotide noncovalent

complex analysis is recommended such that the observed amount of adduction to peaks in the ESI spectra will be reduced and cleaner spectra can be obtained.

Acknowledgments

We thank the Laboratory Directed Research and Development of Pacific Northwest National Laboratory for support of this research through the US Department of Energy. Pacific Northwest National Laboratory is operated by Battelle Memorial Institute for the US Department of Energy, through contract DE-AC06-76RLO 1830.

References

1. Smith, R. D., Light-Wahl, K. J., Winger, B. E., and Loo, J. A. (1992) Preservation of non-covalent associations in electrospray ionization mass spectrometry: multiply charged polypeptide and protein dimers. *Org. Mass Spectrom.* **27**, 811–821.
2. Ganem, B. and Henion, J. D. (1993) Detecting noncovalent complexes of biological macromolecules: new applications of ion-spray mass spectrometry. *Chem. Tracts-Org. Chem.* **6**, 1–22.
3. Smith, R. D. and Light-Wahl, K. J. (1993) The observation of non-covalent interactions in solution by electrospray ionization mass spectrometry: promise, pitfalls and prognosis. *Biol. Mass Spectrom.* **22**, 493–501.
4. Light-Wahl, K. J., Schwartz, B. L., and Smith, R. D. (1994) Observation of the noncovalent quaternary associations of proteins by electrospray ionization mass spectrometry. *J. Am. Chem. Soc.* **116**, 5271–5278.
5. Winger, B. E., Light-Wahl, K. J., Ogorzalek Loo, R. R., Udseth, H. R., and Smith, R. D. (1993) Observation and implications of high mass-to-charge ratio ions from electrospray ionization mass spectrometry. *J. Am. Soc. Mass Spectrom.* **4**, 536–545.
6. Gale, D. C., Goodlett, D. R., Light-Wahl, K. J., and Smith, R. D. (1994) Observation of duplex DNA–drug noncovalent complexes by electrospray ionization mass spectrometry. *J. Am. Chem. Soc.* **116**, 6027,6028.
7. Bruce, J. E., Van Orden, S. L., Anderson, G. A., Hofstadler, S. A., Sherman, M. G., Rockwood, A. L., and Smith, R. D. (1995) Selected ion accumulation of noncovalent complexes in a Fourier transform ion cyclotron resonance mass spectrometer. *J. Mass Spectrom.* **30**, 124–133.
8. Gale, D. C. and Smith, R. D. (1993) Small volume and low flow-rate electrospray ionization mass spectrometry of aqueous samples. *Rapid Commun. Mass Spectrom.* **7**, 1017–1021.

10

Protein Secondary Structure Investigated by Electrospray Ionization

Carol V. Robinson

1. Introduction

The secondary structure of a protein is defined as the spatial arrangement of amino acid residues into elements of structure, such as α-helices and β-sheets. Such elements of secondary structure arise from hydrogen-bonding interactions, which impose conformations on the protein. The existence of distinct conformational states of proteins in solution has been demonstrated by a variety of biophysical techniques, including nuclear magnetic resonance (NMR) and circular dichroism (CD), and is known to be affected by conditions of pH and temperature. The application of mass spectrometry (MS) to the study of protein secondary structure requires a method in which differences in mass associated with conformations of the protein may be monitored. One such method, which has been successfully applied in NMR and more recently in MS, is that of hydrogen-deuterium exchange labeling. The practical considerations that underpin this method and the background to the technique are presented in this chapter.

Since the introduction of electrospray ionization (ESI)-MS *(1)* and its application to the study of proteins, it has become increasingly apparent that more information is available from the ESI spectrum than just the molecular weight of the protein. Thus, the changes in distribution of charge states in a spectrum are often interpreted as being the result of differences in folding of the protein *(2)*. For example, Fig. 1 shows the spectra obtained for lysozyme under identical tuning conditions on the mass spectrometer, but from different solvent conditions. Both confirm the molecular weight of the protein as 14,305 Da, but clearly demonstrate different charge-state distributions. The upper trace was obtained from 100% aqueous solution at pH 5.0 and shows a charge-state maxi-

From: *Methods in Molecular Biology, Vol. 61: Protein and Peptide Analysis by Mass Spectrometry*
Edited by: J. R. Chapman Humana Press Inc., Totowa, NJ

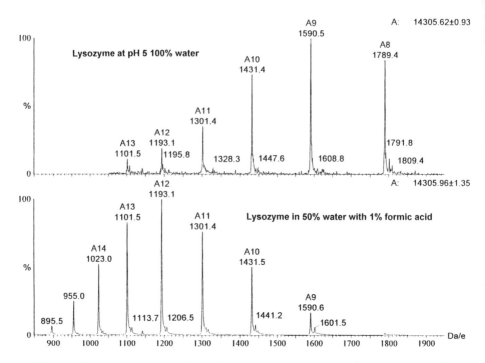

Fig. 1. ESI mass spectrum of hen lysozyme demonstrating the charge-state distributions obtained using different solvents, but identical source tuning conditions. The upper trace was obtained from 100% aqueous solution at pH 5.0, and the lower trace was obtained under standard electrospray solvent conditions of 50% acetonitrile and 1% formic acid.

mum at 9+, whereas the lower spectra, obtained from a solution containing 50% acetonitrile and 1% formic acid, exhibits a maximum at 12+. These charge-state differences can be interpreted in terms of the degree of folding of the protein molecule *(3)* in the solution before introduction into the ion source: the low-pH, high-organic-solvent solution unfolding the protein to a greater extent than the aqueous solution in which the protein is known to be in its native conformation. Thus, the unfolding of the protein disrupts salt bridges and exposes sites that are amenable to protonation, but that are then otherwise buried in the folded native-like structure.

Although these charge-state differences are certainly a reflection of the folding in the molecule, they are also known to be extremely sensitive to tuning conditions in the mass spectrometer, to slight variations in pH, and to counterion effects. It is therefore desirable to employ a more reliable method for observing conformational differences, such as hydrogen–deuterium exchange labeling. This method exploits the fact that elements of secondary structure that involve hydrogen bonding between amino acid side chains and protein-backbone amides in the core of the protein offer protection against hydrogen exchange, whereas regions not involved in secondary structure that are exposed to solvent on the surface of the protein will exchange more readily. Thus, measuring the change in mass of a protein with time can be used to obtain information about the secondary structure of a protein, since more tightly folded proteins would be expected to retain labeling in the core of the protein longer than less stable partially folded states .

The application of hydrogen-exchange labeling to the study of secondary structure by MS was first demonstrated by Katta and Chait *(4)* and has since been applied to the study of proteins in a variety of different environments. This combination of techniques has been used to study the native state of a protein *(5)*, the solution dynamics of proteins and peptides *(6)*, and to examine the protein folding pathways of lysozyme *(7)*, a three-disulfide derivative of lysozyme *(8)*, and a small four-helix bundle protein *(9)*. The rapid improvements in the ESI technique and the success of hydrogen-exchange labeling techniques for the study of native-state protein dynamics, conformational changes, protein folding pathways, and complex formation *(10)* suggest that the coupling of these two techniques is likely to enjoy a rich future in both basic and applied research.

2. Materials

2.1. Proteins

1. Hen egg white lysozyme and bovine α-lactalbumin (Sigma, St. Louis, MO) were purified by dialysis for 3 d against three changes of dilute HCl solution in H_2O at pH 3.8 to remove residual salts.
2. Human lysozyme was isolated and purified as described *(11)*.
3. Carboxymethylated lysozyme (CM^{6-127}) was prepared as described previously *(8)*.
4. The preparation of pulse-labeled samples has been detailed elsewhere *(7,12)*.

2.2. Solvents

1. Ultrapure water (Elga maxima system).
2. Formic acid (HPLC-grade).
3. Ammonia (Analar).
4. D_2O (Fluorochem).

2.3. Mass Spectrometry

Spectra were recorded on either a VG Platform quadrupole MS or a VG BioQ (VG Biotech, Altrincham, Manchester, UK) triple-quadrupole MS. Both instruments were equipped with ESI sources.

3. Methods and Discussion

3.1. ESI-MS

One of two sets of conditions was generally employed depending on the stability of the protein under study. In condition A, the pulse-labeled samples of hen and human lysozyme were analyzed on the VG BioQ triple-quadrupole mass spectrometer as follows:

1. Prepare a sample delivery solvent of 20% acetonitrile adjusted to pH 3.8 with formic acid.
2. Set the solvent delivery rate via a syringe pump to 10 μL/min.
3. Set the source temperature to 50°C.
4. Inject samples, at a concentration of 20 pmol/μL in a volume of 10 μL, using a Rheodyne loop injector.
5. Record spectra, calibrated against fully protonated hen lysozyme, over the mass range m/z 1100–1800.

Partially folded molten globule or A-states of α-lactalbumin, the less stable native state of carboxymethylated lysozyme, and denatured states, however, could not be examined under the conditions detailed above. A low-temperature mode of operation, condition B, is required to obtain spectra from these labile states. This was effected, on the VG Platform mass spectrometer as follows:

1. After preparation of the delivery solvent, typically immerse the solvent container in an ice/salt bath.
2. Set a solvent delivery rate of approx 10 μL/min, using a helium-pressurized flask (Michrom, BioResources, Auburn, CA).
3. Switch off the source heating and cool the nitrogen nebulizer gas by passing the gas through a copper coil immersed in an ice/salt bath.
4. Carry out steps 4 and 5 in condition A.

3.2. Preparation of Deuterated Proteins

The preparation of deuterated proteins is optimized for individual proteins, taking into account thermal stability and ability to withstand pH changes. The protocol described in Section 3.2.1. has been optimized for the deuteration of thermally stable proteins, whereas Section 3.2.2. describes the procedure for proteins that may be acid-denatured.

3.2.1. Preparation of Deuterated Proteins by Thermal Denaturation

Deuterated samples, prepared by thermal unfolding in D_2O, were obtained using the following protocol:

1. Dissolve the protein (1 mg) in D_2O (pH 3.8, 1 mL) and heat up to 75°C for 1 h.
2. Lyophilize the deuterated solution.

3. Repeat steps 1 and 2 at least three times to ensure complete deuteration. Both hen and human lysozyme are able to withstand high temperatures, and while they undergo thermal denaturation at this elevated temperature, both proteins spontaneously refold at reduced temperatures.

3.2.2. Preparation of Deuterated Proteins by Acid Denaturation

A number of proteins cannot withstand the extremes of temperature described in Section 3.2.1., and will aggregate or misfold after thermal denaturation. In these instances, exchange can be effected by lowering the pH. For example, bovine α-lactalbumin forms a molten globule state at pH 2.0, and hence, exchange is facilitated in this relatively open structure. The following protocol was employed for deuteration of acid-stable proteins.

1. Dissolve the protein (1 mg) in D_2O (1 mL) at pH 2.0, and incubate overnight at 20°C.
2. Lyophilize the deuterated solution and repeat the process two to three times to ensure a high level of deuteration.

3.3. Measurement of Hydrogen Exchange

3.3.1. Exchange from Deuterated Proteins (see Note 1)

Hydrogen exchange may be initiated either from lyophilized deuterated protein or by dilution from a concentrated solution of the deuterated protein in D_2O. Most consistent results, however, have been obtained from the latter process. The protocol for this procedure is as follows:

1. Dissolve the protein (1 mg) (*see* Note 2) in D_2O at pH 5.0 (*see* Note 3) (30 µL).
2. Initiate exchange by withdrawing 3 µL of this solution and carrying out a 100-fold dilution into 300 µL H_2O (*see* Note 4). Start a stopwatch at this point, since the dilution step represents initiation of exchange.
3. Withdraw samples (10 µL) for ESI analysis at several time points in the first hour (*see* Note 5), when exchange is rapid. Fewer time points are necessary to define the kinetic curve at later times.
4. Calculate results using the method given in Section 3.3.3.

3.3.2. Measurement of Exchange from Protonated Proteins

An alternative approach to the measurement of hydrogen-exchange kinetics involves dissolving the protein for study in H_2O at the appropriate pH and carrying out a dilution step into D_2O at the same pH. This process has obvious advantages for proteins that are difficult to deuterate, but suffers from the difficulty of ensuring that all traces of H_2O are removed from the electrospray interface, and that the gases used for drying droplets and nebulization are sufficiently dry to prevent back-exchange in the atmospheric pressure region.

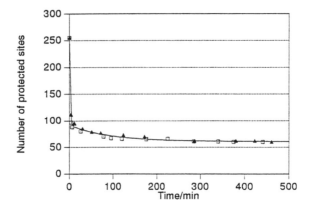

Fig. 2. Hydrogen-exchange kinetics of hen lysozyme obtained from deuterated protein into protonated exchange buffer (□) and from protonated protein into deuterated exchange buffer (▲). The number of protected sites decreases with time as exchange takes place and is calculated as described in the text (Section 3.3.3).

3.3.3. Calculation of Results

Hydrogen exchange from a deuterated protein is calculated by subtraction of the molecular weight of the undeuterated protein from the mass measured at time t, with a correction for the isotopic content of the exchange buffer. For example, a 100-fold dilution of a deuterated solution into protonated buffer gives rise to a solution that contains 1% D_2O. Thus, for lysozyme, where the total number of exchangeable sites is calculated as 255 by summing all the side-chain and backbone-exchangeable hydrogens, 2.55 (1% of the total number of exchangeable sites) will remain deuterated, after exchange is complete, owing to the isotopic content of the exchange buffer rather than to residual protection from exchange. A correction for this residual deuteration may be made at each time point by subtraction of 1% of the number of protected sites at time t.

The number of protected sites for hydrogen exchange from protonated to deuterated protein is calculated by subtraction of the measured mass at time t from the molecular weight of the fully deuterated protein (molecular weight + number of exchangeable sites). Figure 2 summarizes the data obtained for the hydrogen-exchange kinetics for lysozyme measured from H to D and D to H, and demonstrates the consistency of the two measurements. The results are plotted as the number of protected sites against time using Sigma Plot (Jandell Scientific GmbH, Erkarpth, Germany).

3.3.4. Measurement of Hydrogen-Exchange Pulse-Labeled Samples

The pulse-labeling technique has been adequately detailed elsewhere. It essentially involves unfolding the protein in a deuterated chemical denaturant,

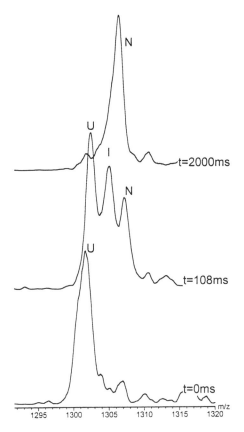

Fig. 3. Pulse-labeled samples of hen lysozyme after 0-, 108-, and 2000-ms refolding times. The peaks labeled U, I, and N represent unfolded, intermediate, and native folding populations, respectively. Each peak represents the 11+ charge state. The data shown are the raw data with minimal smoothing.

initiating folding of the protein by dilution into refolding buffer, labeling exposed regions of the protein by applying a short millisecond labeling pulse at high pH, quenching exchange at low pH, and allowing the protein to fold to its native state. Thus, by using a quench-flow apparatus, it is possible to vary the length of the folding time before applying the labeling pulse and thus to obtain a picture of the events that occur during protein-folding pathways.

For short refolding times, <10 ms, the hen lysozyme protein remains essentially unfolded before application of the labeling pulse. Longer folding times lead to essentially native-like protein. However, the observation of a clear intermediate state in the time course provided direct evidence for the folding of one of the two folding domains in lysozyme *(7)*. The results presented in Fig. 3

represent pulse-labeled samples of hen lysozyme, and show three distinct states: an unfolded state, U, without protection; an intermediate state, I; and the fully folded native state of the protein, N.

The observation of deuterium incorporation during pulse labeling relies not only on the success of the labeling process, but also on retaining the native conformation of the protein during work-up and analysis. Unfolding during these processes leads to loss of deuterium label and exchange with protonated solvent. The clean-up procedure for samples from pulse-labeling and their introduction into the mass spectrometer are critical to the success of the MS experiment and are detailed below:

1. Wash samples from the pulse-labeling process on a Centricon 1000 with ice-cold water at pH 3.0–7.0, depending on the stability of the protein.
2. Step 1 should be repeated 10 times to ensure complete removal of salts.
3. Additionally, analyze a fully deuterated sample of the protein under investigation to ensure that protection from hydrogen exchange can be observed under the mass spectrometric conditions being used. Although it is not possible to put an exact value on the levels of protection that would be expected, since this is highly variable and dependent on the structure of the protein, as a rough guide, for a 15-kDa protein, protection in the range of 35–70 sites would be reasonable.

3.3.5. Measurement of Hydrogen-Exchange Kinetics of a Protein Ligand Bound Within a Molecular Chaperone Complex

In a recent study *(10)*, the complex formed between the molecular chaperone GroEL and the bovine α-lactalbumin ligand was examined by hydrogen-exchange kinetics and ESI (*see* Fig. 4). Despite the large molecular mass of the oligomeric complex (800 kDa), dissociation in the gas phase was found to be a relatively facile process, so that the electrospray spectrum of the complex shows only the monomer of GroEL and the bovine α-lactalbumin ligand. Thus, the hydrogen-exchange kinetics of the ligand may be monitored, in real time, without dissociation of the complex prior to analysis. Given the hydrogen-exchange properties of the ligand, these could be compared with known conformational states: random coil, native state, and intermediate molten globule state, in free solution. Hence, the folding of the ligand within the chaperone complex could be identified as closely similar to that of a free molten globule state.

In order to ensure that the complex was retained until evaporation in the ion source, the electrospray interface was cooled as described under condition B in Section 3.1. and equilibrated with near neutral pH water.

3.4. Concluding Remarks

The attributes of the mass spectrometric method described in this chapter are the ability to study proteins in different conformational states, with a minimal sample size requirement in both real time and quenched exchange experiments.

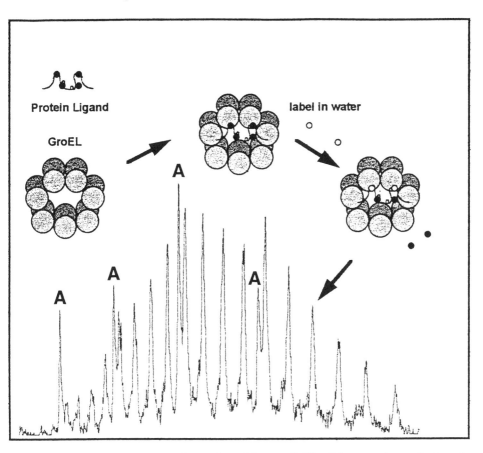

Fig. 4. Schematic representation (adapted from ref. *10*) of the experiment designed to monitor hydrogen exchange in a chaperone–protein substrate complex. The black and open circles represent deuterium and hydrogen, respectively. The charge states labeled **A** represent the series of multiply charged ions arising from the protein substrate; the remaining peaks are from the GroEL monomer.

Furthermore, the unique ability to dissociate complexes in the gas phase while retaining the hydrogen-exchange history is likely to be of importance in a number of areas involving protein–ligand complexes, not only in the molecular chaperones mentioned here, but in the whole field of molecular recognition in general.

4. Notes

1. One of the most common causes of failure for these experiments is a lack of hydrogen-exchange protection observed in the protein. This may be the result of a number of different factors, but the characteristics of the lack of protection may be indicative of the problem as detailed in Notes 6–8.

2. Sensitivity: The sensitivity of ESI-MS is currently at the micromolar level with approx 100 μL of a 20-μmol protein solution required for a complete time course of hydrogen exchange. Thus, a total of 2 nmol, or 20 μg for a 20-kDa protein, are generally required. However, the recent introduction of microspray promises to reduce this sample size requirement to the picomole level *(13)*.

3. Optimal pH for monitoring hydrogen exchange: The intrinsic rate for hydrogen exchange is at a minimum between pH 2.0 and 3.0, so that optimal conditions for observing hydrogen-exchange protection are generally below neutral. Because different conformational states are known to exist at different pHs, it is not generally possible to monitor exchange at the pH minimum, since a number of proteins are completely unfolded in this pH range. The choice of optimal pH is therefore a compromise between the stability of the protein at that pH and the minimum for exchange.

4. Choice of temperature for measuring exchange: Hydrogen exchange may be measured at elevated temperatures from very stable native states. For example, hydrogen exchange from the native state of hen lysozyme at room temperature is a slow process with some sites remaining protected for months. However, exchange can be measured at 69°C using a quench procedure in which samples are removed after an appropriate time interval and plunged into an ice-water bath. Under these conditions exchange is essentially complete within 4 min *(12)*. Alternatively, very labile molten globule states may be monitored on ice such that rapid exchange is slowed down to allow accurate monitoring of exchange *(10)*.

5. Time scale: The rapidity of the mass spectrometric approach, approx 2 min for the acquisition of a spectrum, enables monitoring of early events during exchange processes. In addition, the minimal data processing required for MS allows rapid and accurate measurements to be made. By comparison, similar experiments monitored by 1D- and 2D-NMR suffer from time-consuming data processing and a 4-h acquisition time for a 2D-NMR spectrum.

6. Lack of protection in initial mass determination: If a freshly dissolved, notionally fully deuterated protein is analyzed and found to have a fraction of sites protected from exchange, this may be indicative of an incomplete deuteration process. The appropriate deuteration procedure may be found by varying the pH and temperature of the deuteration step *(see* Section 3.2.).

7. Rapid loss of hydrogen-exchange protection: In some cases, the initial value for hydrogen-exchange protection is of the correct order, but subsequent time points show a rapid loss of exchange protection. This is, in general, indicative of an incorrectly folded state and would indicate that unfolding is occurring in the sample tube. In such cases, it is advisable to recheck the pH of the solution. Additional factors that could reduce protection include the temperature of the electrospray interface.

8. No loss of hydrogen-exchange protection: Some proteins appear to undergo very little hydrogen exchange, and a loss of protected sites is apparently not observed. This may be caused by aggregation of the protein at the pH selected for exchange or by the binding of a impurity to the protein. For example, it is well known that

metal attachment can occur in ESI-MS and is sometimes a problem in this type of experiment, since the increase in mass associated with the metal ion can be incorrectly interpreted in terms of protected sites. In such cases, the protein should be washed on a Centricon filter to remove salts or dialyzed extensively.

References

1. Fenn, J. B., Mann, M., Meng, C. K., Wong, S. F., and Whitehouse, C. M. (1989) Electrospray ionization for mass spectrometry of large biomolecules. *Science* **246**, 64–71.

2. Chowdhury, S. K., Katta, V., and Chait, B. T. (1990) Probing conformational changes in proteins by mass spectrometry. *J. Am. Chem. Soc.* **112**, 9012,9013.

3. Mirza, U. A. and Chait, B. T. (1994) Effects of anions on the positive ion electrospray ionization mass spectra of peptides and proteins. *Anal. Chem.* **66**, 2898–2904.

4. Katta, V. and Chait, B. T. (1991) Comformational changes in proteins probed by hydrogen exchange and electrospray mass spectrometry. *Rapid Commun. Mass Spectrom.* **5**, 214–217.

5. Katta, V. and Chait, B. T. (1993) Hydrogen/deuterium exchange electrospray ionization mass spectrometry: a method for probing protein conformational changes in solution. *J. Am . Chem. Soc.* **115**, 6317–6321.

6. Wagner, D. S., Melton, L. G., Yan, Y., Erickson, B. W., and Anderegg, R. J. (1994) Deuterium exchange of α-helices and β-sheets as monitored by electrospray ionization mass spectrometry. *Protein Sci.* **3**, 1305.

7. Miranker, A., Robinson, C. V., Radford, S. E., Aplin, R. T., and Dobson, C. M. (1993) Detection of transient folding populations by mass spectrometry. *Science* **262**, 896–899.

8. Eyles, S. J., Robinson, C. V., Radford, S. E., and Dobson, C. M. (1994) Kinetic consequences of the removal of a disulphide bond on the folding of hen lysozyme. *Biochemistry* **33**, 1534–1538.

9. Kragelund, B. B., Robinson, C. V., Knudsen, J., Dobson, C. M., and Poulsen, F. M. (1995) A simple protein structural motif folds efficiently and cooperatively. *Biochemistry* **34**, 7217–7224.

10. Robinson, C. V., Groß, M., Eyles, S. J., Ewbank, J. J., Mayhew, M., Hartl, F. U., Dobson, C. M., and Radford, S. E. (1994) Hydrogen exchange protection in GroEl-bound α-lactalbumin probed by mass spectrometry. *Nature* **372**, 645–651.

11. Hooke, S. D., Radford, S. E., and Dobson, C. M. (1994) The refolding of human lysozyme: a comparison with the structurally homologous hen protein. *Biochemistry* **33**, 5867–5876.

12. Radford, S. E., Dobson, C. M., and Evans, P. A. (1992) The folding of hen lysozyme involves partially structured intermediates and multiple pathways. *Nature* **358**, 302–307.

13. Wilm, M. and Mann, M. (1994) Electrospray and Taylor cone theory, "Dole's beam of macromolecules at last?" *Int. J. Mass Spectrom. Ion Processes* **136**, 167–180.

11

The Effect of Detergents on Proteins Analyzed by Electrospray Ionization

Rachele R. Ogorzalek Loo, Natalie Dales, and Philip C. Andrews

1. Introduction

Detergents or surfactants are widely used in protein chemistry *(1–3)*, principally to solubilize and stabilize proteins and to disaggregate protein complexes. Membrane-bound proteins, in particular, frequently require detergent treatment in order to dissolve. Some surfactants disrupt higher order structure, and others are employed as aids in protein refolding. They are also employed in both gel and capillary electrophoresis, and enhance peptide and protein recoveries from synthetic membranes in electroblotting and in electroelution of proteins from gels. Detergents have the reputation of being completely unsuitable for electrospray ionization-mass spectrometry (ESI-MS) of peptides and proteins; the reality is that some, but not all detergents, are unsuitable, and that in cases where one can be flexible in the choice of surfactant, ESI conditions can sometimes be found that will yield a useful mass spectrum *(4–9)*. It is still wise for mass spectrometrists to avoid detergents whenever possible; they add interfering ions and rarely improve signal-to-noise ratios. On the other hand, it is naive to believe that all biochemical problems can be solved without detergents. We present our studies on detergent effects hoping to provide a useful guide in detergent selection for ESI-MS applications.

2. Instrumentation and Materials

2.1. Mass Spectrometry

Positive-ion ESI mass spectra were acquired on a Vestec 201 single quadrupole mass spectrometer (Houston, TX) employing the Vestec atmospheric pressure/vacuum interface *(10,11)*. This interface employs nozzle-skimmer lens

From: *Methods in Molecular Biology, Vol. 61: Protein and Peptide Analysis by Mass Spectrometry*
Edited by: J. R. Chapman Humana Press Inc., Totowa, NJ

141

elements (in contrast to interfaces that use glass capillary *[12]* or metal capillary *[13]* ion-transport devices) and utilizes a heated chamber to assist in droplet desolvation, but no countercurrent or coaxial gas flows. For the studies reported here, the spray chamber was maintained at 55°C, and the electrospray needle was bent to enable off-axis sampling of the spray. The ESI lens block and heater block temperatures were maintained at 120 and 215°C. We have observed that this mass spectrometer generally displays lower charged (higher *m/z*) ions compared to protein mass spectra acquired with other instruments, and that it discriminates more against low-mol-wt components relative to higher mol-wt components.

Some studies were performed on a Fisons/VG Platform single quadrupole mass spectrometer (Danvers, MA). The Platform's pneumatically assisted electrospray interface *(14,15)* employs a nozzle-style inlet with nitrogen nebulizing gas (5–10 L/h) flowing coaxial to the electrospray needle which sprays into nitrogen bath gas (250–500 L/h) in a heated (60°C) source chamber.

Additional studies were carried out on a Finnigan MAT900Q magnetic sector mass spectrometer employing an Analytica of Branford electrospray source *(12)*. Samples were electrosprayed on-axis with this instrument.

Two solvent compositions, selected to reflect typical conditions for ESI-MS of proteins, were used: 4% acetic acid/50% CH_3CN/46% H_2O and 100% H_2O. The 100% H_2O solutions are particularly interesting because native protein conformations may be maintained and micellar properties are characterized under these conditions. Samples were introduced into the ESI source by continuous infusion (5–8 μL/min) or by flow injection, either manually (10 or 20-μL injection loop volume) or via autosampler (5-μL injection loop volume) at 5–20 μL/min. None of the ESI sources employed sheath flow *(16)* to spray 100% aqueous solutions, although the glass capillary interface on the MAT 900Q employed SF_6 *(17)*.

2.2. Other Instrumentation

Circular dichroism measurements on 0.1 mg/mL protein solutions in a 1-mm path length cell at 25°C were obtained by scanning a Jasco J-710 spectrometer from 300–180 nm.

2.3. Materials: Detergents

1. Triton-X 100 (reduced) (Aldrich, Milwaukee, WI).
2. Nonidet P40 (NP40), sodium taurocholate, and 3-(3-cholamidopropyl)dimethylammonio-1-propane sulfonate (CHAPS) (Sigma, St. Louis, MO).
3. Cetyl trimethylammonium bromide (CTAB), sodium cholate, lauryl dimethylamine oxide (LDAO), *n*-octyl sucrose, *n*-dodecyl sucrose, *n*-dodecyl-β-D-maltoside, *n*-dodecyl-β-D-glucopyranoside (*n*-dodecyl glucoside), *n*-decyl-β-D-glucopyranoside

(*n*-decyl glucoside), *n*-hexyl-β-D-glucopyranoside (*n*-hexyl glucoside), and Thesit (Calbiochem, La Jolla, CA).

4. Octyl-β-D-glucopyranoside (octyl glucoside) (Millipore, Bedford, MA).
5. 1-*S*-Octyl-β-D-thioglucopyranoside (octyl thioglucoside) (Pierce, Rockford, IL).
6. Tween-20 (Mallinckrodt, Paris, KY).
7. Sodium dodecyl sulfate (SDS) (Gibco BRL Life Technologies, Gaithersburg, MD).

Surfactant samples can be used without further purification. Stock surfactant solutions are prepared as 10% (w/v) and diluted further prior to analysis, except for *n*-dodecyl glucoside, *n*-decyl glucoside, and CTAB, which are prepared as 2% stock solutions. The 2% stock solutions are warmed slightly prior to removal of aliquots to dissolve the surfactants completely.

2.4. Materials: Proteins and Peptides

1. Horse heart myoglobin (M_r 16951), horse heart cytochrome-*c* (M_r 12,360), and bovine ubiquitin (M_r 8565) (Sigma).
2. The peptide KRTLRR (M_r 829) was synthesized in the University of Michigan Protein and Carbohydrate Structure Facility using standard 9-fluorenyl methoxy carbonyl (FMOC) protection strategies, and was purified by reversed-phase high-performance liquid chromatography (HPLC).

3. Results and Discussion

3.1. Behavior of Surfactants with Proteins in 4% Acetic Acid/50% CH₃CN/46% H₂O

Three effects must be considered in studying surfactant suitability:

1. Surfactant background ions, which can obscure the protein signal;
2. Suppression of the protein signal; and
3. Adduct formation.

A fourth effect, a shift in the charge envelope, is described in Sections 3.4.–3.6. Table 1 summarizes performance relating to effects 1 and 2 for the Vestec mass spectrometer from continuous infusion of a 6 pmol/μL horse heart myoglobin solution with 0.01–1.0% surfactant concentrations in 4% acetic acid/ 50% CH₃CN/46% H₂O. A wide range of anionic, cationic, zwitterionic, and nonionic surfactants has been explored. The nonionics include polyoxyethylenes, such as Tween-20, Thesit, Triton X, and NP40, and saccharides, such as *n*-hexyl and *n*-dodecyl glucoside. Table 1 indicates that protein signal suppression is a problem for ESI-MS of ionic surfactant solutions, as anticipated from the results of Ikonomou and colleagues *(18)*.

Performance was quite poor at 1.0% surfactant concentration. No protein signal was observed with the anionic, cationic, and zwitterionic surfactants, or with the nonionic polyoxyethylenes. Only a few percent of the protein signal observable without surfactant was recovered after the nonionic saccharides

Table 1
Effect of Surfactant on Myoglobin ESI-MS Sensitivity[a]
in 4% Acetic Acid/50% Acetonitrile/46% H_2O

	Detergent conc.[b]		
Detergent	1.0%	0.1%	0.01%
Anionic			
SDS	−	+	+
Taurocholate, sodium	−	+	+
Cholate, sodium	−	+++	+++++
Cationic			
CTAB	−	+	+++++
LDAO[c]	−	++++	+++++
Zwitterionic			
CHAPS	−	++++	+++++
Nonionic			
Tween-20	−	++	+++++
Thesit	−	+	+++++
Triton X-100	−	+	++++
NP40	−	+	++++
n-Octyl sucrose	+/5	++	+++++
n-Dodecyl sucrose	+/5	++	++++
n-Dodecyl maltoside	+/5	+++	+++++
Octyl glucoside	+/5	+++	+++++
Octyl thioglucoside	+/5	+++	+++++
n-Hexyl glucoside	+/1	++++	+++++
n-Dodecyl glucoside	+/1	+++++	+++++

[a]ESI-MS performance indicated by: −, 0% of the original intensity; +/5, 2–5%; +/1, 8–10%; +, <10%; ++, 10–20%; +++, 21–30%; ++++, 30–60%; +++++, >60%.
[b]Concentration in (w/v).
[c]LDAO is zwitterionic above pH 7.0.

were added. Under these conditions, *n*-hexyl glucoside and *n*-dodecyl glucoside were the best performers, yielding approx 10% of the protein signal strength relative to no surfactant; it should be noted that the signal-to-noise ratio for the glucosides at 1% was quite adequate for many protein studies.

Better results are obtained at lower surfactant concentrations. The best performer at 0.1% surfactant concentration, one of the most commonly used concentrations in protein chemistry, was *n*-dodecyl glucoside, yielding 70% of the signal obtained without surfactant. CHAPS and *n*-hexyl glucoside also provided good-quality spectra, yielding intensities in the 30–60% range. Background interference from the surfactant was not observed with 0.1% *n*-dodecyl glucoside.

At a 0.01% surfactant concentration, most surfactants were found suitable for ESI-MS on the Vestec instrument (Table 1), although SDS and sodium taurocholate, both strong anionic surfactants with sulfonic acid groups, were notable exceptions. They performed poorly under all continuous infusion conditions examined, displaying extensive surfactant multimers and strongly suppressing the protein signal. They should be avoided for MS. Excellent surfactants at this concentration included CHAPS, *n*-hexyl glucoside, *n*-octyl sucrose, and *n*-dodecyl maltoside, all of which yielded 100% of the signal obtained without surfactant. There is a choice of possible surfactants at concentrations in the 0.01% range, although the nonionic saccharides are the best choice owing to their lack of interfering ions and absence of adduct formation. The superior performance of the nonionic saccharides is consistent with the prediction that nonionic surfactants will not compete aggressively with the analyte for limited ion current. Chemical background is low with the nonionic saccharides, because the surfactants are not particularly efficient at ionizing and may contain less Na^+ than the polyoxyethylene mixtures.

Performance was also evaluated with a different ESI interface, the pneumatically assisted electrospray interface on the Fisons/VG mass spectrometer. Studies employing 1.0% concentrations of *n*-dodecyl glucoside and *n*-hexyl glucoside spraying 23 pmol/mL bovine ubiquitin in 4% acetic acid/50% acetonitrile yielded 13 and <2%, respectively, of the protein signals without surfactant. At the 0.1% level, *n*-dodecyl glucoside, octyl glucoside, *n*-hexyl glucoside, *n*-dodecyl sucrose, CHAPS, CTAB, LDAO, and sodium cholate were evaluated, yielding 40, 20, 20, 20, 10, 2, 0, and 0%, respectively, of the signal obtained without surfactant. With 0.01% surfactant concentrations of *n*-dodecyl glucoside, *n*-hexyl glucoside, and CTAB, signals equal to 50, 100, and 40% of those without surfactant were obtained, respectively. Although there were some differences in response between surfactants on the Vestec and VG mass spectrometers, the same general trends were observed in both. The pneumatically assisted electrospray instrument tended to yield more surfactant clusters than the heated-chamber electrospray instrument.

3.2. Behavior of Surfactants with Proteins in 100% H₂O

Electrospray in 100% aqueous solutions is of interest because the physical properties of surfactants are better characterized for aqueous conditions than for the acid/organic/water mixtures typically employed in ESI-MS. Moreover, aqueous solutions provide an opportunity to study proteins under native or near-native conditions, enabling the study of noncovalent interactions and protein conformation. Because critical micelle concentrations (CMCs) have been measured in aqueous solutions, we can determine if there is a relationship between surfactant performance in ESI and its CMC. The 100% aqueous stud-

Table 2
Effect of Surfactant on Ubiquitin ESI-MS Sensitivity[a] in H_2O

		Detergent conc.[b]	
Detergent	CMC, %[c]	0.1%	0.01%
Anionic			
SDS	0.2	+	+
Taurocholate, sodium	0.2	+	+
Cholate, sodium	0.4	+	++
Cationic			
CTAB	0.04	+	+++
Zwitterionic			
LDAO[d]	0.03	+	+++
CHAPS	0.3	+	++
Nonionic			
Tween-20	0.007	+	++++
Thesit	0.005	+	++++
Triton X-100	0.01	+	++++
NP40	0.02	++	++++
n-Octyl sucrose	1.0	+++	++++
n-Dodecyl sucrose	0.02	+++	++++
n-Dodecyl maltoside	0.01	++	++++
Octyl glucoside	0.4	+++	+++++
Octyl thioglucoside	0.3	++	+++++
n-Hexyl glucoside	7.0	++++	+++++
n-Dodecyl glucoside	0.005	++++	+++++

[a]ESI-MS performance indicated by: +, <10% of the original intensity; ++,
10–20%; +++, 21–30%; ++++, 30–60%; +++++, >60%.
[b]Concentration in (w/v).
[c]Data adapted from refs. *1* and *2*.
[d]LDAO is cationic below pH 3.0.

ies were carried out by continuous infusion of a 23 pmol/µL solution of bovine
ubiquitin in water. Ubiquitin was selected for these studies because of the ease
with which it can be sprayed from aqueous solutions.

At 1% concentration, only *n*-dodecyl glucoside consistently yielded significant ubiquitin ion signals on the Vestec and Fisons/VG mass spectrometers,
providing 20% of the signal obtained without surfactant. Performance was better at a 0.1% surfactant concentration, primarily with the saccharide-based
nonionics; many surfactants appeared to be suitable for ESI-MS analyses at
0.01% concentration. (*See* Table 2 for results on the Vestec mass spectrometer.) The saccharide-based nonionics are most suited to ESI-MS, but some

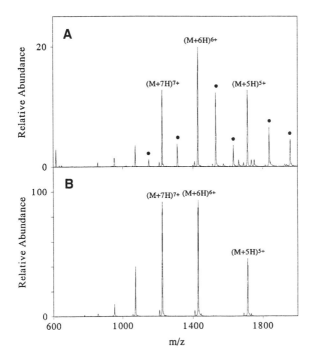

Fig. 1. ESI-MS of bovine ubiquitin in H_2O with **(A)** and without **(B)** 0.01% CHAPS. (•) Ubiquitin + 1 or 2 CHAPS adducts.

performance can be achieved with the polyoxyethylene nonionics at 0.01%. However, the polyoxyethylenes yield significant background ions and form adducts with ubiquitin, complicating the spectra acquired with these surfactants. As with the acid/organic/water mix, the SDS and sodium taurocholate performed poorly under all conditions. Figure 1 illustrates the ESI mass spectrum of ubiquitin in 0.01% CHAPS. CHAPS was a reasonable performer in the acid/organic/water mix, but only 20% of the signal obtained without surfactant was retrieved in 100% H_2O, and surfactant adducts were pronounced in water. Thesit also formed adducts with the protein ions, as did the other polyoxyethylene surfactants. In Fig. 2, both the singly charged polymer background and the adducts are observed. Because Thesit is polydisperse, the adducts give rise to many peaks in the spectrum. Clearly this background would be a problem with low-level protein components.

Based on the CMCs for *n*-hexyl and *n*-dodecyl glucoside (Table 2), it appears that CMC is not a good predictor of how well a surfactant will perform in ESI-MS experiments. The surfactant concentration of 0.1% exceeded the CMC of *n*-dodecyl glucoside, but was below that of *n*-hexyl glucoside, yet both

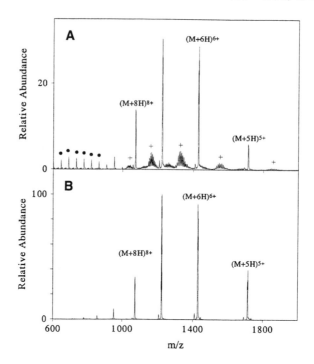

Fig. 2. ESI-MS of bovine ubiquitin in H_2O with **(A)** and without **(B)** 0.01% Thesit. (•) Thesit multimer ions; (+) ubiquitin + Thesit complexes.

surfactants performed similarly. On the Fisons/VG mass spectrometer, extensive surfactant clusters were observed for *n*-hexyl glucoside at concentrations of 0.01% and above, but not for *n*-dodecyl glucoside, also in contrast to predictions based on their CMC values.

3.3. General Considerations for Work with Surfactants

Analyte signal suppression was reduced when the surfactant-protein solutions were delivered by injection loop or autoinjector vs continuous infusion. The difference was attributed to dilution of the surfactant with the running solvent and possible interaction of the surfactant with the polyetherether ketone (PEEK) tubing. Evidence of surfactant interaction with the tubing was provided by the time dependence of mass spectra obtained during elution of the sample plug. Generally, the mass spectra obtained during the first 0.5–1.0 min of elution showed much less protein signal suppression, adduct formation, and surfactant background than spectra obtained later in the elution profile. Signal suppression increased with multiple injections of the surfactant/protein mix, e.g., a subsequent injection might not show high protein signals during the early portion of the elution profile. This behavior suggests that higher surfactant concentrations can be tolerated in flow injection, for a single injection, when using PEEK tubing. It may also explain the discrepancy between our studies, which found severe protein signal suppression with 0.01% SDS delivered by continuous infusion, and those of Kay and Mallet *(19)*, which reported 54% of cytochrome-*c* signal to be recovered with 0.01% SDS in an injection of 20 pmol of protein. Measurements for the figures and tables in this chapter were performed by continuous infusion, under conditions where we believe steady-state signal levels were measured. The tubing underwent a stripping procedure between measurements to elute any adsorbed surfactant or protein.

Several considerations are involved in surfactant selection for MS. Obviously the protein must be soluble in the surfactant and the surfactant must not seriously suppress the protein ion signal. Beyond these primary concerns is the need to minimize background ions arising from the surfactant. All of the surfactants produce some background ions, but in many cases, the background can be tolerated, particularly at 0.1% or lower surfactant concentrations. Background ions are relatively limited for the nonpolymeric surfactants. Dimers were observed with most of the surfactants listed in Table 3 at 1% concentration, but spectral congestion was generally not as much of a problem as signal suppression. Oligomer background ions were more apparent on the Fisons/VG mass spectrometer (pneumatically assisted ESI) than on the Vestec (no nebulizing gas), an observation that we attribute to differences in the interfaces of the instruments. Broad, unresolved background signals below *m/z* 500 were usually observed for 1% concentrations of the nonionic glucoside, maltoside, and sucrose detergents on the Vestec instrument. They may reflect solvent clusters indicating an unstable electrospray under these conditions.

3.4. Surfactants as Denaturants

Some surfactants are quite effective in denaturing proteins. Figure 3 illustrates mass spectra of myoglobin in 0.01% acetic acid/H_2O with (A) and without (B) 0.01% CTAB, a strongly denaturing surfactant. The dominant ion in the spectrum obtained without surfactant corresponds to the myoglobin–heme complex (M_r 17,568) *(20,21)*, whereas the surfactant spectrum shows primarily apo-myoglobin (M_r 16,951) in higher charge states; only a small amount of the binary myoglobin–heme complex is apparent, consistent with denaturation of myoglobin by the CTAB.

We have also used the less denaturing surfactant CHAPS to break up protein–protein interactions in noncovalent complexes of concanavalin A (Con A) dimer and tetramer (102 kDa) *(22,23)*. Con A is a dimer at pH 5.5 and a tetramer at pH > 7.0. Figure 4A (p. 152), obtained on the MAT900Q, shows that the noncovalent dimer is the primary species in the mass spectrum with monomer

Table 3
Background Ions and Other Observations from ESI-MS of Detergent Solutions

Detergent	M_r	Background ions and observations
SDS	288	1+, 2+, 3+-charged Na$^+$-attached SDS multimers extending to high m/z. Protein signal severely suppressed; SDS-protein adducts observed at low SDS concentrations
Taurocholate (sodium salt)	515 (free acid)	1+, 2+-charged Na$^+$-attached multimers; $(M + Na)^+$, $(M − H + 2Na)^+$ at m/z 538, 560, where M = free acid. Forms adducts with proteins.
Cholate (sodium salt)	409 (free acid)	Singly sodiated and protonated multimers of cholic acid extend to high m/z.
CTAB	284 (amine)	$(M + H)^+$, $(2M + Br + 2H)^+$ at m/z 285, 649 where M = cetyl trimethylamine. Denaturing detergent.
LDAO	229	$(M + H)^+$, $(2M + H)^+$, and ions at m/z 263, 487. Zwitterionic at pH > 7.0, cationic at pH < 3.0.
CHAPS	615	Singly charged protonated multimers extend to high m/z. Forms adducts with proteins, particularly in 100% aqueous solutions or at low temperatures (e.g., unheated spray chamber). Nondenaturing detergent effective at breaking protein–protein interactions.
Tween-20	1228 (average)	Polymer background m/z 500–1200 . Severe signal suppression. Forms adducts with proteins.
Thesit	583 (average)	Polymer background m/z 400–900.
Triton X-100 (reduced)	631 (average)	Polymer background m/z 500–1000. Severe signal suppression. Forms adducts with proteins.
Nonidet P40	603 (average)	Polymer background m/z 400–1000. Severe signal suppression. Forms adducts with proteins.
n-Octyl sucrose	468	$(M + Na)^+$, $(M + K)^+$ at m/z 491, 507.
n-Dodecyl sucrose	524	$(M + Na)^+$, $(M + K)^+$ at m/z 547, 563.
n-Dodecyl maltoside	511	$(M + Na)^+$, $(2M + Na)^+$ at m/z 534, 1045.
Octyl glucoside	292	$(M + Na)^+$, $(2M + Na)^+$ at m/z 315, 608. 1% solution gives intense background for m/z < 500.
Octyl thioglucoside	308	$(M + Na)^+$, $(2M + H)^+$, $(2M + Na)^+$ at m/z 331, 618, 640. 1% solution gives intense background for m/z < 500.
n-Hexyl glucoside	264	$(M + Na)^+$, $(2M + Na)^+$ at m/z 287, 551.
n-Dodecyl glucoside	348	$(M + Na)^+$

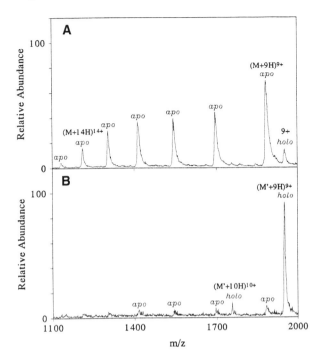

Fig. 3. Horse-heart myoglobin in 0.01% acetic acid/H$_2$O with **(A)** and without **(B)** 0.01% CTAB.

and tetramer also being observed. In the presence of 0.01% CHAPS (Fig. 4B), the dimer and tetramer are greatly reduced. This behavior is consistent with the detergent properties of CHAPS; it is considered to be an effective, but mild detergent often used to disaggregate proteins.

Charge state distributions can be affected by many things, including interface voltages and the presence of counterions *(24)*. Thus, one tries to hold most parameters constant during an experiment, varying only the parameter being explored. Even so, there can be some question regarding whether an ESI-MS charge shift observed is properly interpreted as a denaturation or as something other than a higher order structural change. When the denaturation is accompanied by a change in mass, as in the myoglobin/apomyoglobin data of Fig. 3, or the ConA data of Fig. 4, the interpretation is clear. When no mass change occurs, further care must be exercised. One guide in interpretation is to look for the appearance of a bimodal charge distribution, or a change in the relative contribution of charge states in a bimodal distribution.

Using these guidelines, the cytochrome-*c* data of Fig. 5 illustrate clear changes in a bimodal charge distribution. After heating without surfactant, a

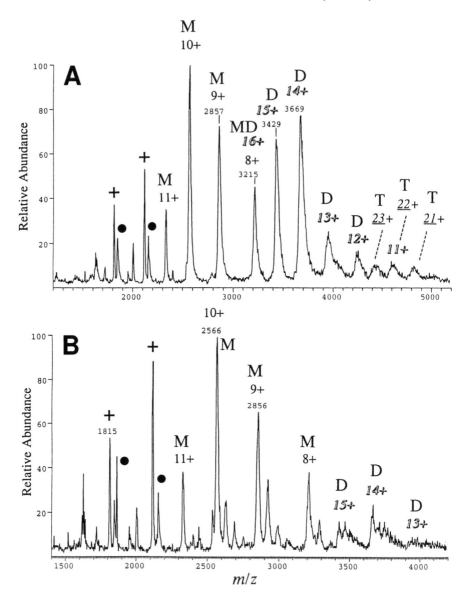

Fig. 4. ESI-MS of Con A (M_r 25,601) in 5 m*M* ammonium acetate, pH 6.0, without **(A)** and with **(B)** 0.1% CHAPS. (M) monomer; (D) dimer; (T) tetramer; (•) fragment 1–118, (+) fragment 119–237.

bimodal charge distribution is observed, showing intense 7+ and 8+, and weak 10+ to 14+ charge states, indicating that only some of the cytochrome-*c* unfolded fully on heating or that the cytochrome-*c* unfolded fully on heating,

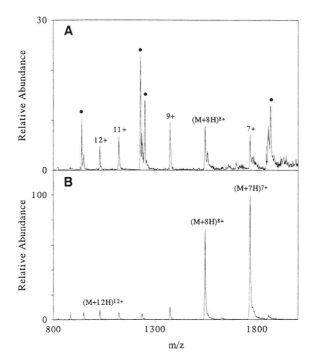

Fig. 5. ESI-MS of horse heart cytochrome-*c* heated off-line in H_2O with **(A)** and without **(B)** 0.1% CHAPS. (•) Surfactant-related ions.

but is quickly refolding. In the presence of surfactant, the bimodal distribution is mostly gone and the 7+ and 8+ charge states are not significantly more intense than 9+ to 12+, indicating that the protein is either folding more slowly in the surfactant solution or denatures more readily in surfactant. This observation is consistent with the use of surfactants to alter protein folding.

3.5. Nonionic Saccharides Alter Charging in 100% Aqueous Solutions

Bovine ubiquitin was sprayed from an aqueous solution without acid, but with 0, 0.01, 0.1, and 1% *n*-hexyl glucoside (Fig. 6). The charge state of maximum intensity shifted from 5+ in surfactant-free solution to 9+ in 1% surfactant. The highest charge state increased from 8+ to 11+, and the mean charge state increased from 5.5–9.2 over that same concentration range. (The mean charge state is obtained by weighting all charge states observed by their relative intensities. Any variation in ion transmission or detection with *m/z* is neglected.) Aqueous solutions of horse heart cytochrome-*c* show behavior similar to ubiquitin, with the mean charge state shifting from 7.4 to 7.7 to 8.9 at 0,

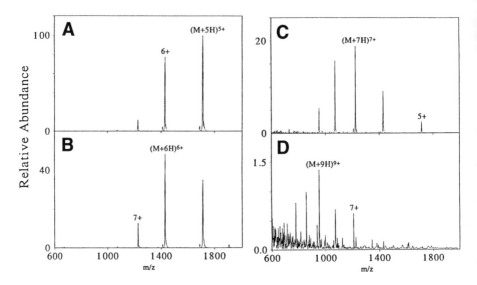

Fig. 6. ESI-MS of aqueous bovine ubiquitin (20 pmol/μL) in the absence of acid
with **(A)** 0%, **(B)** 0.01%, **(C)** 0.1%, and **(D)** 1.0% *n*-hexyl glucoside.

0.01, and 0.1% *n*-hexyl glucoside. The shift in charge state of maximum intensity on addition of *n*-hexyl glucoside holds for other nonionic saccharide surfactants also, as was explored by varying the alkyl constituent of glucoside
surfactants. At a 0.1% surfactant concentration, *n*-hexyl and *n*-decyl glucoside yielded similar charge distributions for horse heart cytochrome-*c*, but
the spectrum for the protein in *n*-dodecyl glucoside clearly showed less charge
shifting than for the other alkyl chain surfactants. ESI-MS charge distributions
were repeatedly examined for *n*-hexyl glucoside and *n*-dodecyl glucoside
addition to aqueous solutions of cytochrome-*c* and ubiquitin (Table 2). In comparing *n*-hexyl and *n*-dodecyl glucoside, smaller increases were consistently
observed for *n*-dodecyl glucoside containing solutions at surfactant concentrations ranging from 0.01 to 1.0%. The shifts in charging that we observe are not
simply the result of suppressing low-charge-state ions to a larger extent. We
observe that high-charge-state intensities actually increase when 0.01% surfactant is added.

Questions about the source of this charge shift arise immediately. Could a
change in protein conformation be responsible? We have demonstrated that
myoglobin denaturation by CTAB can be observed by ESI-MS. Because the
data presented so far were obtained in aqueous solutions without addition of
acid, the possibility for a shift from native or near-native conformation to a
denatured one cannot be discounted. However, although surfactants, such as

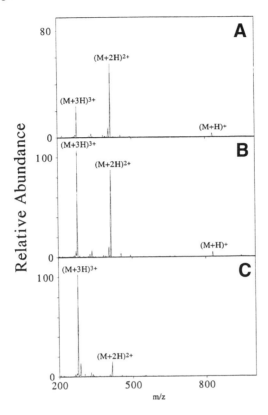

Fig. 7. Charge-state distributions for aqueous solutions of KRTLRR with **(A)** 0%, **(B)** 0.01%, and **(C)** 0.1% *n*-hexyl glucoside.

SDS and CTAB, are well-known for their ability to denature proteins, other surfactants, such as the nonionic saccharides, are generally considered nondenaturing *(1–3)*.

The possibility that higher order structural changes might be responsible for the increased charging was explored further by examining ESI charge distributions from a peptide without higher order structure. Because very short peptides are thought to have no higher order structure, we examined surfactant effects with the six-residue peptide KRTLRR. Note that this peptide is extremely hydrophilic and is unlikely to interact wth the hydrophobic "tail" of the detergents. Figure 7 illustrates the dramatic increase in charging observed for ESI of an aqueous solution of 150 pmol/μL of the peptide with increasing concentrations of *n*-hexyl glucoside. Clearly, increased charging in the presence of nonionic saccharides occurs in polypeptides unlikely to have secondary structure.

Table 4
Effect of Surfactants on Ubiquitin Charge State Distributions
for Solutions in 4% Acetic Acid/50% Acetonitrile

Conditions[a]	Most intense charge state	Mean charge state	Range of charge states
No surfactant	11+/12+	10.9	6+ to 13+
0.01% ndodg	11+/12+	11.3	7+ to 13+
0.01% nhg	11+/12+	11.3	7+ to 13+
0.01% CTAB	11+	10.9	7+ to 13+
0.1% ndodg	12+	12.0	9+ to 14+
0.1% nog[b]	12+/13+	12.3	10+ to 14+
0.1% LDAO[c]	—	—	—
0.1% sodium cholate[b]	11+/12+	11.1	9+ to 13+
0.1% nhg[b]	12+	12.1	10+ to 14+
0.1% CHAPS[b]	11+	10.8	8+ to 13+
0.1% nds	12+	11.8	10+ to 13+
0.1% CTAB	11+	10.3	8+ to 13+
1.0% ndodg[b]	12+/13+	12.3	10+ to 14+
1.0% nhg[b]	12+	12.3	10+ to 14+[d]

[a]ndodg, n-dodecyl glucoside; nds, n-dodecyl sucrose; nhg, n-hexyl glucoside; nog, n-octyl glucoside.
[b]See surfactant oligomers observed.
[c]Complete suppression of analyte signal.
[d]Limited by low signal-to-noise ratio.

Circular dichroism (CD) spectroscopy provides another way to address the question of whether conformational changes contribute to the differences observed in charging. Aqueous solutions of cytochrome-c and ubiquitin were examined with and without 0.1% n-hexyl glucoside and n-dodecyl glucoside. For both proteins, there were no differences in the CD spectra, indicating that these detergents caused no significant changes in their structures.

3.6. Nonionic Saccharides Alter Charging Under Denaturing Conditions

Bovine ubiquitin solutions (20 pmol/μL) in the 4% acetic acid/50% acetonitrile solvent with and without surfactant were sprayed with the pneumatically assisted ESI source (Table 4). The most intense charge state, highest charge state, and mean charge state shifted to higher charge for the nonionic saccharide surfactants. Other surfactants, including cationic CTAB and zwitterionic CHAPS, did not show increased charging. (For 0.1% CTAB, the mean charge state actually decreased relative to that observed in H_2O, possibly because of proton transfer to tertiary amine impurities.) At the concentrations employed,

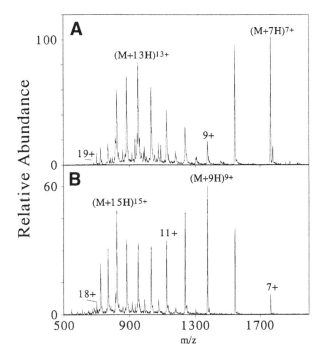

Fig. 8. Electrospray mass spectrum of cytochrome-*c* in 0.1% acetic acid/3% CH_3CN/97% H_2O with **(A)** 0%, and **(B)** 0.1% *n*-hexyl glucoside.

LDAO suppressed signals too strongly to allow the evaluation of charge-state shifts. Clearly, charge distributions shift in the presence of nonionic saccharides for denatured protein solutions, as well as for native or near-native solutions. Additional experiments were performed with KRTLRR in the following solution compositions: 90% CH_3CN/10% H_2O and 50% CH_3CN/50% H_2O. The charge distributions obtained with organic solution compositions did not differ from those obtained earlier for 100% aqueous solutions (Fig. 7), demonstrating again that this increase in charging is not limited to aqueous solution compositions.

Another way to address the question of whether conformational changes contribute to the observed charge shifts is to start with a solution containing more than one conformer, as indicated by a bimodal charge state distribution. Figure 8 illustrates spectra obtained for cytochrome-*c* in 0.1% acetic acid/3% acetonitrile without and with 0.1% *n*-hexyl glucoside. The maxima of both envelopes shift to higher charge in the presence of surfactant. Because both spectra have bimodal charge envelopes, we conclude that the shift in charge states is not a consequence of one conformer converting to the other.

Similar behavior was observed when bovine ubiquitin in 3% acetic acid/9% acetonitrile was sprayed with and without 0.1% n-hexyl glucoside. Without surfactant, the two charge envelopes peaked at 6+ and 8+/9+; $(M + 11H)^{11+}$ was the highest charge state observed. The equivalent solution, but containing 0.1% n-hexyl glucoside, yielded a charge distribution that peaked at $(M + 8H)^{8+}$, but with $(M + 9H)^{9+}$ and $(M + 10H)^{10+}$ only slightly weaker in intensity; $(M + 12H)^{12+}$ was the highest charge state observed. Again it appeared that the shift in charge states was not a consequence of one conformer converting to another, as both charge envelopes shifted to higher charge.

4. Conclusions

Several of the surfactants examined in this study should be useful in the mass spectrometric analysis of peptides and proteins, including membrane proteins (5), that are difficult to solubilize, as well as in the coupling of separation techniques, such as micellar electrophoresis, to ESI-MS (7,25), and in improving protein recoveries from analytical or purification steps preceding MS. This study, in applying surfactants to protein denaturation and protein–protein interactions, demonstrates that surfactants can be used in conjunction with ESI-MS to study higher order protein structure. Examples of detergent-induced conformational effects uncovered include the effect of CTAB on myoglobin and CHAPS on Con A dimers and tetramers. Surfactants and conditions suitable for ESI-MS of proteins are identified and the unusual effect of surfactants on charge populations is described. In the light of the ability of nonionic surfactants to increase charging independent of higher order structure, care must be exercised in drawing conclusions regarding protein structure and denaturation from ESI-MS studies in the presence of surfactant. Identification of conformational charges requires the presence of a bimodal charge-state distribution, a change in quaternary structure, or confirmation by other analytical methods, such as circular dichroism, analytical ultracentrifugation, light scattering, or nuclear magnetic resonance studies.

Under the range of conditions explored, nonionic saccharide surfactants, especially n-dodecyl glucoside, yielded the strongest ESI signals without serious chemical background, whereas SDS and sodium taurocholate performed poorly. The suppression of analyte signal observed for ionic surfactants, as compared to nonionic, is consistent with a competition between surfactant and analyte ions for ESI current. Protein signal levels and oligomer background with various surfactants were ESI-interface-dependent, although general trends were similar between instruments. Laboratories that have more than one ESI mass spectrometer should assess whether the performance of one instrument is superior for surfactant-containing solutions. Off-axis vs on-axis spraying should also be evaluated.

Acknowledgments

We acknowledge Joseph A. Loo for helpful discussions, skillful assistance with the manuscript, and our collaboration on the CHAPS/Con A experiments. J. E. Hoover was a catalyst for ideas. This work was supported in part by a grant from OVPR, University of Michigan, Ann Arbor.

References

1. Neugebauer, J. (1990) Detergents: an overview. *Methods Enzymol.* **182,** 239–282.
2. Neugebauer, J. (1992) A guide to the properties and uses of detergents in biology and biochemistry. Doc. No. CB0068-0892, Calbiochem-Novabiochem Corp, LaJolla, CA.
3. Bollag, D. M. and Edelstein, S. J. (1991) *Protein Methods.* Wiley-Liss, New York.
4. Ogorzalek Loo, R. R., Dales, N., and Andrews, P. C. (1994) Surfactant effects on protein structure examined by electrospray ionization mass spectrometry. *Protein Sci.* **3,** 1975–1983.
5. Barnidge, D. R., Dratz, E., Sunner, J., and Jesaitis, A. J. (1994) Analysis of Hydrophobic Protein Samples Containing Detergents by Electrospray Ionization Mass Spectrometry. *Proceedings of the 42nd ASMS Conference on Mass Spectrometry and Allied Topics,* Chicago, IL, p. 671.
6. Kirby, D., Greeve, K. F., Foret, F., Vouros, P., and Karger, B. L. (1994) Capillary Electrophoresis-Electrospray Ionization Mass Spectrometry Utilizing Electrolytes Containing Surfactants. *Proceedings of the 42nd ASMS Conference on Mass Spectrometry and Allied Topics,* Chicago, IL, p. 1014.
7. Varghese, J. and Cole, R. B. (1993) Cetyltrimethylammonium chloride as a surfactant buffer additive for reversed-polarity capillary electrophoresis-electrospray mass spectrometry. *J. Chromatogr. A* **652,** 369–376.
8. Loo, J. A. and Pesch, R. (1994) Sensitive and selective determination of proteins with electrospray ionization magnetic sector mass spectrometry and array detection. *Anal. Chem.* **66,** 3659–3663.
9. Loo, J. A. and Ogorzalek Loo, R. R. (1995) Applying charge discrimination with electrospray ionization-mass spectrometry to protein analysis. *J. Am. Soc. Mass Spectrom.* **6,** 1098–1104.
10. Allen, M. H. and Vestal, M. L. (1992) Design and performance of a novel electrospray interface. *J. Am. Soc. Mass Spectrom.* **3,** 18-26.
11. Andrews, P. C., Allen, M. H., Vestal, M. H., and Nelson, R. W. (1992), Large-scale protein mapping using infrequent cleavage reagents, LD TOF MS, and ESMS, in *Techniques in Protein Chemistry III* (Angeletti, R. H., ed.), Academic, San Diego, CA, pp. 515–523.
12. Whitehouse, C. M., Dreyer, R. N., Yamashita, M., and Fenn, J. B. (1985) Electrospray interface for liquid chromatographs and mass spectrometers. *Anal. Chem.* **57,** 675–679.
13. Chowdhury, S. K., Katta, V., and Chait, B. T. (1990) An electrospray-ionization mass spectrometer with new features. *Rapid Commun. Mass Spectrom.* **4,** 81–87.

14. Bruins, A. P., Covey, T. R., and Henion, J. D. (1987) Ion spray interface for combined liquid chromatography/atmospheric pressure ionization mass spectrometry. *Anal. Chem.* **59,** 2642–2646.
15. Covey, T. R., Bonner, R. F., Shushan, B. I., and Henion, J. D. (1988) The determination of protein, oligonucleotide and peptide molecular weights by ion-spray mass spectrometry. *Rapid Commun. Mass Spectrom.* **2,** 249–256.
16. Smith, R. D., Loo, J. A., Edmonds, C. G., Barinaga, C. J., and Udseth, H. R. (1990) New developments in biochemical mass spectrometry: electrospray ionization. *Anal. Chem.* **62,** 882–899.
17. Ikonomou, M. G., Blades, A. T., and Kebarle, P. J. (1991) Electrospray mass spectrometry of methanol and water solutions suppression of electric discharge with SF_6 gas. *J. Am. Soc. Mass Spectrom.* **2,** 497–503.
18. Ikonomou, M. G., Blades, A. T., and Kebarle, P. (1990) Investigations of the electrospray interface for liquid chromatography/mass spectrometry. *Anal. Chem.* **62,** 957–967.
19. Kay, I. and Mallet, A. I. (1993) Use of an on-line liquid chromatography trapping column for the purification of protein samples prior to electrospray mass spectrometric analysis. *Rapid Commun. Mass Spectrom.* **7,** 744–746.
20. Katta, V. and Chait, B. T. (1991) Observation of the heme-globin complex in native myoglobin by electrospray-ionization mass spectrometry. *J. Am. Chem. Soc.* **113,** 8534,8535.
21. Loo, J. A., Giordani, A. B., and Muenster, H. (1993) Observation of intact (heme-bound) myoglobin by electrospray ionization on a double-focusing mass spectrometer. *Rapid Commun. Mass Spectrom.* **7,** 186–189.
22. Light-Wahl, K. J., Winger, B. E., and Smith, R. D. (1993) Observation of the multimeric forms of concanavalin-A by electrospray ionization mass spectrometry. *J. Am. Chem. Soc.* **115,** 5869,5870.
23. Loo, J. A., Ogorzalek Loo, R. R., and Andrews, P. C. (1993) Primary to quaternary protein structure determination with electrospray ionization and magnetic sector mass spectrometry. *Org. Mass Spectrom.* **28,** 1640–1649.
24. Mirza, U. A. and Chait, B. T. (1994) Effects of anions on the positive ion electrospray ionization mass spectra of peptides and proteins. *Anal. Chem.* **66,** 2898–2904.
25. Matsubara, N. and Terabe, S. (1992) Separation of closely related peptides by capillary electrophoresis with a nonionic surfactant. *Chromatographia* **34,** 493–496.

12

Posttranslational Modifications
Analyzed by Automated Protein Ladder Sequencing

Rong Wang and Brian T. Chait

1. Introduction

Protein ladder sequencing *(1)* is a method for obtaining information concerning the primary structures of proteins. The method involves two steps. In the first step, a nested set of fragments of the polypeptide chain is generated by controlled stepwise chemical degradation at the amino terminus (Fig. 1). In the second step, the molecular masses of the set of peptides are measured by matrix-assisted laser desorption/ionization time-of-flight mass spectrometry (MALDI-TOFMS) *(2)*. This measurement results in a mass spectrum that provides information on the sequence and identities of amino acid residues, both unmodified and modified.

The identities of amino acid residues are determined by the mass differences between consecutive peaks in the mass spectrum, and the sequence is defined by their order of occurrence. Posttranslational modifications introduce changes in the masses of the amino acid residues that are modified. Measurement of the molecular masses of such modified residues provides information on the site and nature of the modifications. For example, phosphorylation of serine, threonine, or tyrosine will increase the molecular mass of the modified amino acid residue by 80 Da, whereas deamidation of asparagine or glutamine will decrease the molecular mass by 1 Da.

There are several ways to generate sets of peptide fragments suitable for protein ladder sequencing *(1,3–5)*. This chapter describes a method which uses an automatic gas-phase protein sequencer to produce protein sequencing ladders by controlled partial Edman degradation (*see* Note 1). The resulting peptide ladders are used for identifying posttranslational modifications.

From: *Methods in Molecular Biology, Vol. 61: Protein and Peptide Analysis by Mass Spectrometry*
Edited by: J. R. Chapman Humana Press Inc., Totowa, NJ

```
Starting peptide:    SRASS----------
          Step 1     RASS----------
          Step 2     ASS----------
          Step 3     SS----------
          Step 4     S----------
```

Fig. 1. Nested set comprising starting peptide and four fragments generated by stepwise partial degradation.

2. Materials

2.1. Instruments

1. Sequencer: An ABI model 471A protein sequencer (Applied Biosystems, a Division of Perkin-Elmer Corporation, Foster City, CA) is used in our laboratory to perform the ladder-generating degradation reactions. Other models of automated protein sequencer should also be suitable for this purpose.
2. Mass spectrometer: MALDI-TOF mass spectrometer *(6)* (*see* Note 2).

2.2. Chemical Reagents

2.2.1. Peptide-Supporting Membrane

Immobilon-CD transfer membrane (Millipore, Bedford, MA).

2.2.2. Sequencing Reagents

1. 5% Phenyl isothiocyanate (PITC) in *n*-heptane (R1).
2. 12.5% Trimethylamine (R2).
3. Trifluoroacetic acid (TFA) (R3).
4. Ethyl acetate (S2).

Items 1–4 are obtained from Applied Biosystems and used without further purification.

2.2.3. Mass Spectrometric Reagents and Solvents (see Note 3)

1. Matrix: α-cyano-4-hydroxycinnamic acid *(7)* (Aldrich, Milwaukee, WI).
2. Methanol (Fisher Scientific, Pittsburgh, PA).
3. TFA (Applied Biosystems).
4. Acetonitrile (Applied Biosystems).
5. Water.

3. Methods

3.1. Measuring the Molecular Mass of the Peptide to Be Sequenced

The molecular mass of the peptide (*see* Note 4) is measured by MALDI-TOFMS (*see* Note 5).

1. Prepare a saturated matrix solution of α-cyano-4-hydroxycinnamic acid in water/
   acetonitrile (2:1 [v/v]), containing 0.1% TFA.
2. Mix the peptide solution with the matrix solution to give a final peptide concentration of 1 μM (*see* Note 6).
3. Apply a small volume (0.5 μL) of the resulting solution onto the sample probe,
   and dry at room temperature.
4. Insert the sample probe into the mass spectrometer and acquire a mass spectrum
   in positive- (or negative-) ion mode.

3.2. Peptide Immobilization

The degradation reaction is carried out on peptides that have been immobilized on an Immobilion-CD membrane.

1. Cut the membrane into small pieces: 2 × 2 mm² for peptide amounts in the range
   1–10 pmol and 5 × 5 mm² for peptide amounts >10 pmol.
2. Prior to use, precondition the membrane with methanol/12.5% trimethylamine
   (1:1 [v/v]) (2 μL for the 2 × 2 mm² membrane or 5 μL for the 5 × 5 mm² membrane).
3. While the membrane is still moist, apply the peptide sample (1–50 pmol in a
   volume of 2–5 μL) onto the membrane and dry under a nitrogen flow (*see* Note 7).
4. Load the membrane containing the peptide into the reaction cartridge of the protein sequencer in readiness for the degradation reaction.

3.3. Peptide Ladder Generation

The peptide degradation reaction is carried out in an ABI 471A protein
sequencer. A modified sequencing cycle (Table 1) is used to produce the peptide sequencing ladder. Partial degradation is achieved by using a short cou-
pling-reaction time (3 min).

The sequencing program ("run file") consists of two cycle subprograms ("cycle
files"), viz. the preconditioning cycle (Table 2) and the partial degradation cycle
(Table 1). The preconditioning cycle is used only once to remove residual air
from the membrane. Following the preconditioning cycle, the partial degradation cycle is repeated the desired number of times to produce the ladder peptides.

In the partial degradation cycle program (Table 1), the coupling reaction is
achieved in steps 1–10 at 55°C. The coupling reaction is followed by washing
three times with ethyl acetate (steps 12–21) to remove nonreacted PITC and
side products. The cleavage reaction is carried out in steps 23–27 at 46°C (*see*
Notes 8 and 9). After cleavage, several steps are used to clean TFA from the system. Steps 11 and 36 set the temperature of the reaction cartridge (*see* Note 10).

3.4. Peptide Mixture Extraction

1. After completing the desired number of cycles of the degradation reaction, withdraw the sample-containing membrane from the sequencer, cut it into small pieces
   (about 1 × 1 mm²), and transfer these into a 0.6-mL microcentrifuge tube.

Table 1
Partial Degradation Cycle
Used for Generating Peptide Sequencing Ladder

Step	Cartridge function	Flask function	Parameter	Elapsed time
1	4 PREP R2	39 WAIT	5	00:00:05
2	5 DELIV R2	39 WAIT	10	00:00:15
3	1 PREP R1	39 WAIT	5	00:00:20
4	2 DELIV R1	39 WAIT	2	00:00:22
5	29 DRY CART	39 WAIT	5	00:00:27
6	5 DELIV R2	39 WAIT	90	00:01:57
7	2 DELIV R1	39 WAIT	2	00:01:59
8	29 DRY CART	39 WAIT	5	00:02:04
9	5 DELIV R2	39 WAIT	83	00:03:27
10	29 DRY CART	39 WAIT	10	00:03:37
11	31 RXN HEATER	39 WAIT	46	00:04:23
12	13 PREP S2	39 WAIT	5	00:04:28
13	14 DELIV S2	39 WAIT	9	00:04:37
14	39 WAIT	39 WAIT	5	00:04:42
15	29 DRY CART	39 WAIT	5	00:04:47
16	14 DELIV S2	39 WAIT	9	00:04:56
17	39 WAIT	39 WAIT	5	00:05:01
18	29 DRY CART	39 WAIT	5	00:05:06
19	14 DELIV S2	39 WAIT	9	00:05:15
20	39 WAIT	39 WAIT	5	00:05:20
21	29 DRY CART	39 WAIT	120	00:07:20
22	39 WAIT	39 WAIT	95	00:08:55
23	7 PREP R3	39 WAIT	5	00:09:00
24	9 LOAD R3	39 WAIT	8	00:09:08
25	29 DRY CART	39 WAIT	5	00:09:13
26	39 WAIT	39 WAIT	60	00:10:13
27	29 DRY CART	39 WAIT	5	00:10:18
28	15 LOAD S2	39 WAIT	5	00:10:23
29	30 BLK FLUSH	39 WAIT	5	00:10:28
30	14 DELIV S2	39 WAIT	9	00:10:37
31	39 WAIT	39 WAIT	5	00:10:42
32	29 DRY CART	39 WAIT	5	00:10:47
33	14 DELIV S2	39 WAIT	9	00:10:56
34	39 WAIT	39 WAIT	5	00:11:01
35	29 DRY CART	39 WAIT	5	00:11:06
36	31 RXN HEATER	39 WAIT	55	00:12:01
37	29 DRY CART	39 WAIT	140	00:14:21

Table 2
Preconditioning Cycle
Used to Generate Peptide Sequencing Ladder

Step	Cartridge function	Flask function	Parameter	Elapsed time
1	29 DRY CART	39 WAIT	30	00:00:30
1	4 PREP R2	39 WAIT	5	00:00:35
2	5 DELIV R2	39 WAIT	10	00:00:45

2. Extract the peptide mixture (sequencing ladder) with 10 µL of 2.5% TFA in acetonitrile/water (6:4 [v/v]) for 3 min (with sonication; *see* Note 11).
3. Remove the membrane and add the peptide solution to the matrix solution (*see* Section 3.5. and Note 12).

3.5. Measuring the Molecular Masses of the Ladder Peptides

1. Mix 1–2 µL of the extracted peptide solution with 2 µL of the matrix solution (*see* Sections 2.2.3. and 3.1.).
2. Apply 1 µL of this peptide/matrix solution on the sample probe tip, and dry at room temperature.
3. Acquire mass spectra in the positive-ion mode using a MALDI-TOF mass spectrometer. It is important to note that the mass spectrometric response is dependent on the amino acid composition of the ladder peptides (*see* Note 13).

3.6. Read-Out of the Amino Acid Sequence and the Amino Acid Modifications

The amino acid sequence is determined by calculating the mass differences between consecutive peaks. In the example shown in Fig. 2A, peak (a) corresponds to a mass of 977.2 Da (*see* Note 14) and peak (b) corresponds to a mass of 1064.3 Da. The mass difference is calculated as $1064.3 - 977.2 = 87.1$ (Da). The identity of this amino acid residue (Ser) is determined by comparing the calculated mass difference with the amino acid residue masses listed in Table 3 (*see* Note 15). In a similar manner, the mass differences between the other pairs of adjacent peaks are 87.1 (Ser), 71.2 (Ala), 156.3 (Arg), 87.0 (Ser), and so on. The amino acid sequence (N-terminal to C-terminal) is determined by the order of the residues identified from the sequencing ladder—in the direction of high to low mass (i.e., … SRASS …).

Modified amino acid residues can be identified using the same procedure as that just described (*see* Note 16). In the example shown in Fig. 2B, the mass difference between (b) and (c), $1231.5 - 1064.3 = 167.2$, does not correspond to any of the amino acid residue masses given in Table 3, indicating that the amino acid is modified. The identification of the modification can be made by reference to Table 4 (p. 168). In the present example, the modification is unambiguously identified as phosphorylation of a serine residue (pS). The mass dif-

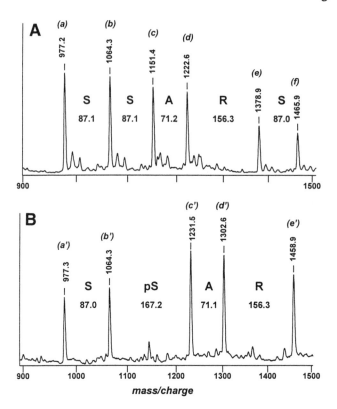

Fig. 2. Partial protein ladder-sequencing data from a peptide that is a substrate for protein kinase C. **(A)** Peptide without phosphorylation and **(B)** peptide with phosphorylation. The amino acid identities and modifications are determined by measuring the mass differences between consecutive peaks (e.g., [a] and [b], [b] and [c]) and comparing these mass differences with the amino acid residue masses (Table 3) and modified amino acid residue masses (Table 4). The amino acid sequence is determined by the order of residues from the high-mass to low-mass end of the spectrum (e.g., … SRASS … for spectrum A and … RApSS … for spectrum B). The mass shift caused by the modification is seen by comparing corresponding peaks in (A) and (B) (b with b', c with c', …).

ferences between the other peaks in the spectrum demonstrate that no other modifications are present in this portion of the sequence.

The modified amino acid residues can be identified with special facility if both modified and unmodified peptides are available and can be subjected to ladder sequencing. As a result of the modification, the molecular masses of all of the ladder peptides that contain the modified amino acid residue will shift by an amount corresponding to the modification. This shift can readily be detected

Table 3
Masses and Compositions
of the 20 Commonly Occurring Amino Acid Residues

Average mass	Name	Three-letter code	One-letter code	Composition
57.05	Glycine	Gly	G	C_2H_3NO
71.08	Alanine	Ala	A	C_3H_5NO
87.08	Serine	Ser	S	$C_3H_5NO_2$
97.12	Proline	Pro	P	C_5H_7NO
99.13	Valine	Val	V	C_5H_9NO
101.10	Threonine	Thr	T	$C_4H_7NO_2$
103.14	Cysteine	Cys	C	C_3H_5NOS
113.16	Isoleucine	Ile	I	$C_6H_{11}NO$
113.16	Leucine	Leu	L	$C_6H_{11}NO$
114.10	Asparagine	Asn	N	$C_4H_6N_2O_2$
115.09	Aspartic acid	Asp	D	$C_4H_5NO_3$
128.13	Glutamine	Gln	Q	$C_5H_8N_2O_2$
128.17	Lysine	Lys	K	$C_6H_{12}N_2O$
129.12	Glutamic acid	Glu	E	$C_5H_7NO_3$
131.19	Methionine	Met	M	C_5H_9NOS
137.14	Histidine	His	H	$C_6H_7N_3O$
147.18	Phenylalanine	Phe	F	C_9H_9NO
156.19	Arginine	Arg	R	$C_6H_{12}N_4O$
163.18	Tyrosine	Tyr	Y	$C_9H_9NO_2$
186.21	Tryptophan	Trp	W	$C_{11}H_{10}N_2O$

by comparing the spectra of the modified and unmodified peptides (compare Fig. 2A,B).

4. Notes

1. A manual method for carrying out the ladder-generating chemistry is described in detail elsewhere *(4)*.
2. The TOF mass spectrometer is a custom linear instrument with a 2-m flight path. Peptide samples are loaded onto a multiple-sample probe containing 10 sample positions and then irradiated with pulses of laser light (wavelength 355 nm, pulse duration 10 ns) from an Nd-YAG laser source (HY 400, Lumonics, Inc., Kanata, Ontario, Canada). Ions desorbed from the probe are accelerated in two stages to an energy of 30 keV. An electrostatic particle guide is used in the flight tube to improve the ion transmission. The ions are detected by a hybrid microchannel plate/gridded discrete-dynode electron multiplier detector. The output is amplified by a fast preamplifier (Model 9305, EG&G Ortec, Oak Ridge, TN) and fed into a LeCroy Model 7200A digital oscilloscope (LeCroy, Chestnut Ridge, NY).

Table 4
Residue Masses of Some Chemically
and Posttranslationally Modified Amino Acids (see Note 16)

Modification	Amino acid	Modified amino acid	Mass shift[a]	Modified amino acid residue mass[b]
O-linked glycosylation[c]	Ser	Dehydroalanine (Dha)	-18	69.06
O-linked glycosylation[c]	Thr	Dehydroamino-2-butyric acid (Dhb)	-18	83.09
Hydroxylation	Pro	γ-Hydroxy-Pro (Hyp)	16	113.12
Hydroxylation		β-Hydroxy-Pro	16	113.12
Hydroxylation	Asn	β-Hydroxy-Asn (βHyn)	16	130.10
Hydroxylation	Asp	β-Hydroxy-Asp (βHya)	16	131.09
Hydroxylation	Lys	δ-Hydroxy-Lys (Hyl)	16	144.17
Oxidation	Met	Met-Sulfoxide	16	147.19
Oxidation	Cys	Cysteic acid	16	151.14
Carboxylation	Asp	β-Carboxy-Asp	44	159.09
Phosphorylation	Ser	Phosphoryl-Ser	80	167.06
Carboxylation	Glu	γ-Carboxy-Glu (Gla)	44	173.12
Phosphorylation	Thr	Phosphoryl-Thr	80	181.08
Phosphorylation	Asp	Phosphoryl-Asp	80	195.07
Phosphorylation	Lys	Phosphoryl-Lys	96	208.15
Phosphorylation	His	Phosphoryl-His	96	217.12
Phosphorylation	Tyr	Phosphoryl-Tyr	80	243.16
Sulfation	Tyr	Tyrosine sulfate	80	243.24
O-linked N-acetylglucosamine	Ser	GlcNAc-Ser	203	290.27
O-linked N-acetylglucosamine	Thr	GlcNAc-Thr	203	304.30
Prenylation	Cys	Farnesyl-Cys	204	307.50
N-linked glycosylation[d]	Asn	N-GlcNAc-Asn	203	317.29
Palmitoylation	Cys	Palmitoyl-Cys	238	341.56
Prenylation		Geranylgeranyl-Cys	272	375.62
N-linked glycosylation[d]	Asn	N-Fuc-GlcNAc-Asn	349	463.44

[a]Nominal mass.

[b]The mass shifts and average masses are calculated using the atomic weights of the elements: C = 12.011, H = 1.00794, N = 14.00674, O = 15.9994, P = 30.97376, S = 32.066.

[c]β-Elimination caused by deglycosylation by dilute alkali solution (nonreductive medium) in mild conditions.

[d]Released by endo-H digestion (from high-mannose type or hybrid type N-linked glycosylations) or endo-F digestion of N-linked glycosylation.

Commercial MALDI-TOF mass spectrometers should also be suitable for reading out the masses of the ladder peptides.

3. The reagents and solvents used for mass spectrometric sample preparation should be of the highest quality available. Water is obtained by purifying distilled water through a Milli-Q UV Plus water system.

4. The molecular mass measurement helps to determine the appropriate number of sequencing cycles for the analysis. Ions originating from the matrix produce an intense background in the mass spectrum below m/z 500. Thus, for example, if the molecular mass of the peptide to be sequenced is ~1000 Da, the maximum number of residues that can readily be sequenced is ~5. The molecular mass is also helpful in the interpretation of the sequence data. For example, the number of lysine residues can be determined by the mass shift, since each lysine residue will react with one molecule of phenyl isothiocyanate and cause a mass shift of 135 Da. In addition to the determination of molecular mass, the measurement provides useful information regarding the purity of the peptide sample.

5. The molecular mass of the peptide of interest can be measured with other types of mass spectrometer (e.g., electrospray ionization quadrupole mass spectrometer).

6. When sufficient peptide sample is available, a 10-μM solution is prepared with 0.1% TFA in water/acetonitrile (2:1 [v/v]). One microliter of peptide solution is mixed with 9 μL of matrix solution to give a final peptide concentration of 1 μM. When very limited amounts of peptide are available, the peptide solution can be mixed with the matrix solution directly on the probe. In this case, 1 μL of peptide solution is rapidly added to 1 μL of matrix solution on the probe. Care must be taken to avoid precipitation of the matrix prior to addition of the peptide solution.

7. For dilute samples, the peptide solution can be applied to the membrane in multiple loadings.

8. The amount of TFA delivered should be tuned very carefully. Excessive amounts of TFA will cause peptide sample loss from the Immobilon-CD membrane.

9. High temperature during the cleavage reaction will cause dehydration side reactions as well as the oxidation of methionine residues.

10. With a cooling device, the total cycle time can be further reduced to <10 min. Steps 21, 22, and 37 (Table 1) are used to achieve the temperature changes.

11. Sonication for an extended time period can produce unwanted side reactions.

12. If necessary, the peptide solution can be concentrated using a SpeedVac (Savant Instrument Inc., Farmingdale, NY). To avoid sample loss, the solution should not allowed to dry completely.

13. The response of MALDI-MS depends on the amino acid composition of the ladder peptides. In particular, the sensitivity is enhanced when the peptide contains basic amino acid residues. Conversely, the sensitivity is reduced when no basic amino acid residues are present. In general, the order of sensitivity is Arg-containing peptides > Lys-containing peptides > peptides that do not contain basic groups, except the free N-terminal amino group > peptides containing no basic functionality. Thus, the modification on ε-amino groups of lysine residues by phenyl isothiocyanate may decrease the mass spectrometric response. This may become a serious problem for peptides that do not contain arginine residues.

14. Formally, the measured value is a mass-to-charge ratio. However, under the conditions used here, the charge is unity.
15. Leucine and isoleucine have identical residue masses (113 Da). These isobaric amino acid residues cannot be distinguished by the present ladder sequencing method. Glutamine and lysine have the same nominal residue masses (128 Da). However, they can be distinguished since the ε-amino group of the lysyl residue is modified by PITC, resulting in a mass shift of 135 (residue mass of PTC-lysine = 263.35 Da).
16. Detection of a posttranslational modification requires that the modification is not labile to the ladder-sequencing chemistry. Ambiguous results may result from posttranslationally modified amino acid residues that have similar masses to natural amino acid residues.

References

1. Chait, B. T., Wang, R., Beavis, R. C., and Kent, S. B. H. (1993) Protein ladder sequencing. *Science* **262,** 89–92.
2. Hillenkamp, F., Karas, M., Beavis, R. C., and Chait, B. T. (1991) Matrix-assisted laser desorption/ionization mass spectrometry of biopolymers. *Anal. Chem.* **63,** 1193A–1203A.
3. Wang, R., Chait, B. T., and Kent, S. B. H. (1994) Protein ladder sequencing: towards automation, in *Techniques in Protein Chemistry V* (Crabb, J. W., ed.), Academic, San Diego, CA, pp. 19–26.
4. Wang, R. and Chait, B. T. (1996) N-terminal protein ladder sequencing, in *Methods in Molecular Biology, Vol. 64: Protein Sequencing Protocols* (Smith, B. J., ed.), Humana Press, Totowa, NJ, in press.
5. Bartlet-Jones, M., Jeffery, W. A., Hansen, H. F., and Pappin, D. J. C. (1994) Peptide ladder sequencing by mass spectrometry using a novel, volatile degradation reagent. *Rapid Commun. Mass Spectrom.* **8,** 737–742.
6. Beavis, R. C. and Chait, B. T. (1989) Factors affecting the ultraviolet laser desorption of proteins. *Rapid Commun. Mass Spectrom.* **3,** 233–237.
7. Beavis, R. C., Chaudhary, T., and Chait, B. T. (1992) α-Cyano-4-hydroxycinnamic acid as a matrix for matrix-assisted laser desorption mass spectrometry. *Organic Mass Spectrom.* **27,** 156–158.

13

Rapid Analysis of Single-Cysteine Variants of Recombinant Proteins

Thomas W. Keough, Yiping Sun, Bobby L. Barnett, Martin P. Lacey, Mark D. Bauer, Ellen S. Wang, and Christopher R. Erwin

1. Introduction

Protein engineering methods have been widely used to study individual structural factors that contribute to protein stability *(1,2)*. An important goal of that research is to enhance the commercial or medicinal utility of wild-type (WT) proteins by increasing their stability in a rational, step-by-step fashion *(3)*. We recently characterized a series of single-cysteine variants of subtilisin BPN', a proteolytic enzyme used in commercial laundry formulations. They had been prepared *(4)* in part because random mutagenesis experiments indicated that some of these variants were more stable than the WT enzyme *(3)*. We wanted to compare the stabilities of the mutants with those observed after engineering single disulfide bonds into the subtilisin BPN' backbone *(5,6)*.

In this chapter, the use of mass spectrometry (MS) for the rapid characterization of single-cysteine variants of subtilisin BPN' is illustrated. This same general approach is also applicable to the characterization of other single-cysteine containing proteins, such as those that have been engineered to provide specific sites for spectroscopic labeling for detailed structure or topography studies *(7–10)*.

Matrix-assisted laser desorption/ionization (MALDI) *(11,12)* and ionspray MS *(13)* were used to characterize the subtilisin BPN' variants, which had the cysteine group located at various positions along the enzyme backbone. The MS methods allow verification that the desired mutations were properly made, and allow one to determine quickly whether the lone sulfhydryl group is free or has chemically reacted in some way. This latter information is important

From: *Methods in Molecular Biology, Vol. 61: Protein and Peptide Analysis by Mass Spectrometry*
Edited by: J. R. Chapman Humana Press Inc., Totowa, NJ

because, for some proteins, the sulfhydryl groups react quickly with endogenous sulfur-containing compounds, either during fermentation *(3)* or purification *(14)*. The oxidation state of the thiol group affects its subsequent reactivity, and may also affect the overall stability of the recombinant protein *(3)*. In this study, a strong correlation was found to exist between the calculated surface accessibilties of the cysteines in the various mutant enzymes and the extent of unwanted cysteine-bound products formed during enzyme preparation.

2. Materials

2.1. MALDI Matrix and Mass Calibration Standards

1. Sinapinic acid (Aldrich, Milwaukee, WI, cat. no. D13,460-0).
2. Bovine trypsinogen (Sigma, St. Louis, MO, cat. no. T1143).
3. Horse-heart cytochrome-*c* (Sigma, cat. no. C-2506).
4. Horse-heart myoglobin (Sigma, cat. no. M-1882).
5. Polypropyleneglycol (PPG) calibration solution: Dissolve 0.4 g PPG of average mol wt 200 (Aldrich), 0.1 g PPG of average mol wt 1000 (Aldrich), and 0.014 g PPG of average mol wt 425 in 1 L of 50:50 water/methanol containing 0.1% formic acid.

Sinapinic acid was stored at 4°C and the enzymes were stored at −15°C when not in use.

2.2. Enzyme Digestion Reagents

1. 97% pure cyanogen bromide (CNBr) (Aldrich, cat. no. C9,149-2).
2. *N*-Tosyl-L-phenylalanine chloromethyl ketone (TPCK)-treated trypsin (Sigma, cat. no. T-8642).
3. 99% Pure dithiothreitol (DTT) (Aldrich, cat. no. 15,046-0).

The TPCK-treated trypsin was stored at −15°C, and the other reagents were stored at 4°C, when not in use.

2.3. Reagents and Solvents

1. Deionized water (18 MΩ) was prepared on a Millipore water purification system.
2. Methanol (Baker, Phillipsburg, NJ), high-performance liquid chromatography (HPLC) grade.
3. Acetonitrile (Baker), HPLC grade.
4. 0.1*N* HCl (lot no. 925734-24) (Fisher, Fairlawn, NJ) certified.
5. 99% Pure ammonium acetate (Baker), ACS reagent grade.
6. Trifluoroacetic acid (TFA) (Aldrich, cat. no. 29,953-7).
7. 30% Hydrogen peroxide (Baker).
8. 88% Formic acid (Baker).
9. 0.1% Formic acid prepared by suitable dilution of item 8.

2.4. MALDI Instrumentation

MALDI experiments were carried out on a Vestec VT 2000 linear time-of-flight mass spectrometer (PerSeptive Biosystems, Vestec Division, Houston, TX). Protein samples were irradiated with the frequency-tripled output of a Q-switched Nd:YAG laser (Lumonics HY-400, 355 nm, 10-ns pulse width) operated at a repetition rate of 10 Hz. Desorbed secondary ions were accelerated to 35 kV in two stages, and then drifted down a 2-m flight tube before detection with a 20-stage electron multiplier. The mass spectra were recorded with a transient recorder (TR8828D) operated at 5-ns time resolution and a CAMAC crate controller (model 6010), both from LeCroy (Chestnut Ridge, NY). Typically, spectra produced from 50 individual laser pulses were added together in the transient recorder using LeCroy-supplied software. These spectra were then imported into a Compaq 386/33 (Houston, TX), using software written in-house, and were processed using Lab Calc® (Galactic Industries Corp., Salem, NH), a commercially available software package.

2.5. Ionspray Instrumentation

All ionspray experiments were carried out on a PE Sciex API III triple-quadrupole mass spectrometer (PE Sciex, Thornhill, Canada) using a delivery solvent of methanol:water (1:1) with 0.1% formic acid at a flow rate of about 5 μL/min. The instrument was calibrated over the m/z range of 50–2400 using polypropylene glycols. Samples were introduced with a Rheodyne 7125 loop injector having a 50 μL loop. For the tandem mass spectrometry (MS/MS) experiments, pure Ar was used as the collision gas at a thickness of 300×10^{-14} atoms/cm^2. This type of instrument is described more fully in Chapter 2.

2.6. HPLC Instrumentation

Peptide purifications were carried out with a Waters 600MS HPLC equipped with a 490E programmable multiple wavelength detector operated at 216 and 254 nm (Waters Corporation, Milford, MA).

3. Methods

3.1. MALDI Determination of the Molecular Mass of Intact Recombinant Proteins

1. Dissolve the purified recombinant proteins in 0.1% TFA to a concentration of 1–5 μM. Prepare a separate solution of a known protein, bovine trypsinogen (M_r = 23,981), in the same fashion for use as an internal mass calibration standard (*see* Note 1).
2. Freshly prepare a solution of sinapinic acid at a concentration of 10 g/L in 0.1% TFA/acetonitrile (70/30 [v/v]) according to the procedure of Beavis and Chait *(11)* (*see* Note 2).

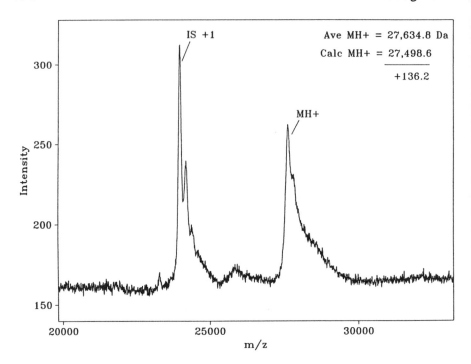

Fig. 1. The high-mass portion of the MALDI mass spectrum from a variant of sub-tilisin BPN' containing a single-cysteine residue.

3. Mix equal volumes (2–10 µL) of the two solutions from step 1 in a separate microcentrifuge tube and then mix 2 µL of the resulting solution with an equal volume of the sinapinic acid solution from step 2.
4. Pipet 1 µL of the final solution onto the end of the stainless-steel solids probe used for MALDI sample introduction and allow it to dry at ambient temperature.
5. Prior to analysis, dip the sample-loaded probe tip into cold deionized water to wash away any excess alkali salts.

Previous studies *(15)* have shown that the accuracy of this MALDI mass measurement procedure is about ± 0.01% for proteins having molecular masses < 30,000 Da. The high-mass portion of the MALDI mass spectrum obtained from a recombinant subtilisin BPN' containing 10 modifications, including S204C, illustrates the quality of data that is readily achieved with this method (Fig. 1). The singly (IS +1) and doubly charged (not shown) molecular ions of bovine trypsinogen were used to calibrate the mass scale, allowing accurate determination of the molecular mass (MH^+ = 27,635 Da) of the recombinant protein. The experimentally measured molecular mass was compared to the value calculated, assuming that all of the modifications had been correctly made. In this particular example, the measured molecular mass was about 136 Da greater than the expected value. This mass difference is much greater than experimental error (3–5 Da), indicating that the primary structure of the protein was not entirely correct. The digestion experiments discussed in the next section were designed to identify the specific structural modification(s) producing this increase in molecular mass.

3.2. Protein Mapping Procedures
to Identify Structural Modifications in Recombinant Proteins

The exact chemical and enzymatic digestion procedures used to map a particular recombinant protein will vary, depending on the primary structure of the protein and the information required. The overall procedure used for the recombinant protein, whose preliminary analysis was discussed in Section 3.1., involves several steps:

1. CNBr digestion.
2. HPLC isolation of the CNBr digestion products.
3. Identification of the modified CNBr fragment.
4. Tryptic subdigestion of the modified CNBr fragment.
5. Identification of the modified tryptic peptide.
6. MS sequencing of the modified tryptic peptide to locate precisely the site of modification.

Both the rationale and the detailed experimental conditions used for each step of this procedure are discussed in turn.

3.2.1. CNBr Digestion

The recombinant protein of interest contains 275 amino acid residues *(16,17)*, and has two methionines, located at residues 119 and 199, that are susceptible to CNBr digestion. CNBr digestion would be expected to produce three fragments, C1 = 1–119, C2 = 120–199, and C3 = 200–275. Moreover, fragment C1 should contain two of the desired amino acid modifications, fragment C2 should contain four of the modifications, and fragment C3 should contain the remaining four. Mass measurement of the three CNBr digestion products would then allow localization of the mass increment, ideally added to only one of the three products, and verification that the modifications in the other CNBr fragments had been correctly made.

1. Dissolve the protein (0.5–1 nmol) in 200 μL of 0.1*N* HCl in a microcentrifuge tube (*see* Note 3).
2. Add a small (unweighed) crystal of CNBr to the tube, which is then wrapped in aluminum foil.
3. Vortex the resulting solution thoroughly and heat to 37°C for the 2-h reaction period.
4. Quench the reaction by adding 500 μL of deionized water, and dry the solution with a SpeedVac.

3.2.2. HPLC Isolation and Identification of the Modified CNBr Fragment

1. Dissolve the dried CNBr digestion products in 200 μL of 0.05% TFA.
2. Separate using reversed-phase HPLC with a Waters Delta-Pak C4 column (3.9 mm × 15 cm, 300-Å pore size, and 5-μm particle size) and using a linear acetonitrile/water gradient of 0–90% B for 60 min. The composition of A was 0.05% TFA in H_2O, whereas the composition of B was 0.05% TFA in acetonitrile.
3. Collect the HPLC fractions manually into separate microcentrifuge tubes, and dry in a SpeedVac.
4. Redissolve the isolated fractions in 50 μL of deionized water and take an aliquot for MALDI molecular mass measurement using the procedure discussed in Section 3.1., with the exception that myoglobin (M_r = 16,951.7) or cytochrome-c (M_r = 12,360.7) is used as the internal mass standard for these latter measurements (*see* Note 1).

The molecular mass measurements indicated that fragments C1 (MH^+ calculated = 12,001.3 Da, observed = 12,003.0) and C2 (MH^+ calculated = 7609.4 Da, observed = 7612.0) had the expected masses and probably had the correct structures. The MALDI spectrum of fragment C3 indicated it contained a small amount (10%) of the desired product, but most of the fragment weighed about 129 Da more than expected (MH^+ calculated = 7829.8 Da, observed = 7958.6). This measurement is in reasonable agreement with the mass shift obtained from MALDI analysis of the intact enzyme. Thus, the CNBr mapping procedure localized the structural modification to fragment C3, i.e., residues 200–275.

3.2.3. Tryptic Subdigestion of the Modified CNBr Fragment (see Note 4)

The CNBr fragment C3 contains four lysines and one arginine that are susceptible to specific cleavage with trypsin. Complete tryptic digestion will produce six fragments whose structures and molecular masses are summarized in Table 1. Two of the expected mutations, including S204C, occur in tryptic fragment T1 (200–213). Again, comparison of the calculated and observed molecular masses should allow localization of the protein modification(s) to a relatively small peptide. Localization of the modification(s) to the precise amino acid residue(s) should then be possible by sequencing the modified peptide.

1. Mix the isolated CNBr fragment C3 with 200 mL of 0.1M NH₄Ac buffer (pH = 8.0). Add TPCK-treated trypsin (2 mL at a concentration of 1 mg/mL in NH₄Ac buffer), and react at 37°C for 30 min.

3.2.4. Ionspray MS and Sequencing of the Modified Tryptic Peptide

1. Redissolve the mixture of tryptic fragments in 30 μL of methanol/water (1:1) with 0.1% formic acid for direct analysis with ionspray MS.

Table 1
Summary of Mass Measurements from a Series of Tryptic
Peptides Formed by Digestion of the CNBr Fragment C3

| Structure | MH$^+$ | | Error, Da |
	Calculated	Measured	
T1 200–213	1413.6	1412.2	−1.4
		1533.0	+119.4
		1546.8	+133.2
T2 214–237	2241.6	2241.6	0.0
T3 238–247	1253.4	1253.3	−0.1
T4 248–256	981.0	981.2	+0.2
T5 257–265	1050.2	1049.0	−1.2
T6 266–275	985.1	984.4	−0.7

In this experiment, all six of the expected tryptic peptides were observed. Comparison of the calculated and measured molecular masses (Table 1) indicates that fragments T2–T6 (residues 214–275) were not modified, but fragment T1 corresponds to at least three components: a minor unmodified component, and the major species weighing 119 and 133 Da more than the calculated value. Again, these ionspray measurements are consistent with the MALDI results obtained from the CNBr fragment C3 and the intact molecule. The apparent shift in the mass of the unwanted modification(s) from 136 (intact molecule) to 129 (C3) to 119 and 133 Da (T1) is not surprising. This simply reflects a refinement of the measurements as we proceed to lower molecular mass fragments and employ a higher resolution measurement technique.

The tryptic fragment T1 containing the 119-Da mass increment was sequenced using ionspray MS/MS, and the resulting spectrum is given in Fig. 2. The expected sequence of the corresponding unmodified peptide (200–213) is also shown. Nearly complete sequence information from the N-terminus was obtained, and the spectrum localizes the modification to C204. No modifications were observed for any of the other amino acids. MS sequencing of the tryptic fragment T1 containing the 133-Da mass shift (not shown) also localizes the modification to C204.

3.2.4.1. Performic Acid Oxidation (*see* Note 5)

Proof that the modifications to the cysteine residue involved bond formation at the sulfur atom was obtained by reanalysis of the modified peptide after performic acid oxidation. This reaction was carried out as follows:

1. Dissolve a portion of the dried peptide in 5–10 μL of a mixture containing 30% H$_2$O$_2$ and 88% formic acid in the ratio of 1:19.
2. React for 30 min at room temperature.

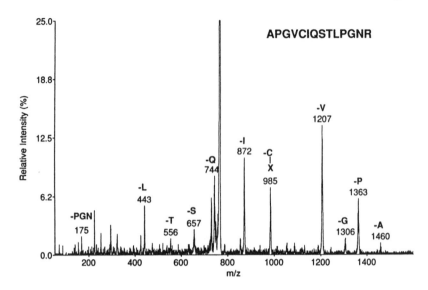

Fig. 2. Ionspray MS/MS spectrum of the tryptic peptide containing the unwanted modification.

In this case, the masses of the modified peptides (1533.0 and 1546.8 Da) shifted to 1461.6 Da, reflecting the loss of the sulfur-bound substituents and complete oxidation of the resulting sulfhydryl group. Formation of a Cys-Cys disulfide bond is one possible interpretation for the 119-Da mass increment. However, no attempt was made to identify the Cys-bound components.

3.3. DTT Reduction for Early Verification of Cys Modification (see Note 5)

Verification that any observed unwanted modifications involve reaction at the sulfhydryl group of the lone Cys can be accomplished much earlier in the analysis by DTT reduction of the intact enzyme. Typical results are shown in Fig. 3. The broad high-mass molecular-ion envelope of the modified protein, observed before DTT reduction, shifts to the expected lower molecular mass once the modifications have been released. Notice that the width of the molecular-ion envelope also decreases. This occurs because the unresolved overlapping components comprising the "molecular ion" of the modified protein collapse to a single component after release of the modifications. DTT reduction of the recombinant protein was accomplished as follows:

1. Dissolve 0.5–1 nmol of the protein in approx 10 μL of a freshly prepared solution of DTT (10 mM DTT in a 0.06M NH$_4$Ac buffer, pH = 8.0).
2. React at 37°C for 30 min.

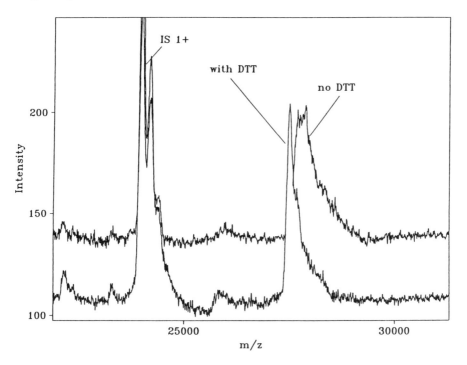

Fig. 3. Comparison of portions of the MALDI mass spectra of a single-cysteine variant of subtilisin BPN' before (upper) and after (lower) reduction with DTT.

3.4. Correlation of the Extent of Chemical Reaction and the Surface Accessibilities of the Cysteine Residues

A series of nine single-cysteine variants of subtilisin BPN' were characterized, and typical MALDI mass spectra, spanning the molecular-ion region, are given in Fig. 4. Six of the recombinant proteins showed narrow molecular-ion envelopes (lower), whereas the remaining three had broad molecular-ion envelopes shifted to higher mass, as is characteristic of the unwanted products formed by chemical reaction at the cysteine residues (upper). All of the molecular-mass calculations and measurements are summarized in Table 2. The molecular-mass measurements for six of the variant enzymes agreed within ±0.02% of the calculated values, indicating that the appropriate amino acid residues had been replaced with cysteines having free sulfhydryl groups. The remaining three enzymes (1, 4, and 7) are modified, as indicated by increases in their molecular masses and the widths of their molecular-ion envelopes.

The surface accessibilities of the cysteine residues in the various subtilisin BPN' mutants were estimated by use of the computer program ACCESS *(18)*

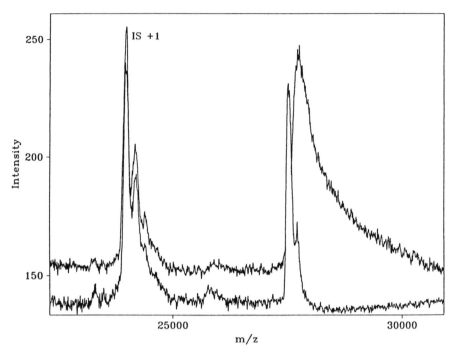

Fig. 4. Comparison of portions of the MALDI mass spectra of two single-cysteine variants of subtilisin BPN'. One variant (upper) shows the unwanted sulfhydryl modifications, whereas the other (lower) does not.

Table 2
Summary of MALDI Mass Measurements, Chemical Reactivities, and Exposed Cysteine Surface Areas for a Series of Single-Cysteine Variants of Subtilisin BPN'

Mutant	MH^+ Calculated	Measured	Error[a]	Substitution	Exposed area (Å^2)
1	27,550.7	27,743.6	[b]	Yes	41.1
2	27,538.4	27,534.0	−0.016	No	0.0
3	27,566.7	27,567.6	+0.003	No	0.0
4	27,550.7	27,775.0	[b]	Yes	28.0
5	27,506.7	27,512.8	+0.022	No	4.9
6	27,540.7	27,538.6	−0.008	No	7.6
7	27,509.7	27,699.2	[b]	Yes	104.5
8	27,566.7	27,563.9	−0.010	No	0.0
9	27,523.7	27,523.0	−0.000	No	5.3

[a]Expressed as percentage of the molecular mass of the enzyme.
[b]Could not be accurately measured because the peak was too broad.

and the tertiary structure of subtilisin BPN', which is known to a resolution of 2 Å *(19)*. Basically, the program mathematically rolls a sphere across the surface of the protein. For these calculations, the diameter of the sphere was set to 1.5 Å, the diameter of a water molecule. Any area of the protein that the sphere contacted was considered accessible and was integrated. For the sake of these estimations, the surface area (main + side chain) of the cysteine residues in each of the variant enzymes was assumed to be the same as that of the amino acid residue it replaced in the WT enzyme. In other words, it was assumed that each mutation produced a negligible change in the tertiary structure of subtilisin BPN'. The cysteine residues in the three variants that showed the unwanted sulfur-bound products have large exposed surface areas ranging between 28 and 105 Å^2. The cysteine residues in the remaining six variants, which showed no evidence for reaction at the sulfhydryl group, have exposed surface areas <8 Å^2. The correlation between the surface accessibility calculations and the MALDI mass spectra is striking.

4. Notes

1. Bovine trypsinogen was used as the internal mass standard for molecular-mass measurement of the intact recombinant proteins. Lower mass internal standards (myoglobin and cytochrome-*c*) were used for analysis of the CNBr fragments. We use internal standards having masses close to those of the analytes to improve mass-measurement accuracy.

2. The solutions of the UV-absorbing matrices were prepared fresh daily. They were stored in the dark at ambient temperature when not in use.

3. The CNBr digestions were carried out in 0.1*N* HCl, not formic acid or TFA, to avoid formylation or trifluoroacetylation of free amines. MS is sensitive to subtle changes in protein molecular mass, and the mapping studies would be complicated by introducing these additional structural modifications.

4. We characterized the CNBr digestion products after off-line HPLC purification. This approach was taken because we knew that subdigestion would be necessary to localize the unwanted modifications. Direct infusion ionspray MS was used to characterize the complete mixture produced by tryptic subdigestion of the modified CNBr fragment C3. This was done because all of the tryptic fragments were detected in the ionspray mass spectrum and, unlike Edman degradation, the subsequent MS/MS sequencing experiment does not require purified materials for sequencing.

5. Performic acid oxidation (not DTT reduction) of the modified tryptic peptide was used to verify the presence of sulfur-bound modifications because the resulting mixture of products was very clean (no added buffer), facilitating direct analysis by ionspray MS. DTT reduction was used prior to MALDI analysis of recombinant enzymes because intact enzymes contain too many additional sites that are susceptible to oxidation. These additional side products would complicate any subsequent MS analyses.

References

1. Alber, T. (1989) Mutational effects on protein stability. *Annu. Rev. Biochem.* **58,** 765–798.
2. Pantoliano, M. W., Ladner, R. C., Bryan, P. N., Rollence, M. L., Wood, J. F., and Poulos, T. L. (1987) Protein engineering of subtilisin BPN': enhanced stabilization through the introduction of two cysteines to form a disulfide bond. *Biochemistry* **26,** 2077–2082.
3. Pantoliano, M. W., Whitlow, M., Wood, J. F., Dodd, S. W., Hardman, K. D., Rollence, M. L., and Bryan, P. N. (1989) Large increases in general stability of subtilisin BPN' through incremental changes in the free energy of unfolding. *Biochemistry* **28,** 7205–7212.
4. Brode, P. F., Erwin, C. R., Rauch, D. S., Lucas, D. S., and Rubingh, D. N. (1994) Site-specific variants of subtilisin BPN' with enhanced surface stability. *J. Biol. Chem.* **269,** 23,538–23,543.
5. Katz, B. A. and Kossiakoff, A. (1986) The crystallographically determined structures of atypical strained disulfides engineered into subtilisin. *J. Biol. Chem.* **261,** 15,480–15,485.
6. Mitchinson, C. and Wells, J. A. (1989) Protein engineering disulfide bonds in subtilisin BPN'. *Biochemistry,* **28,** 4807–4815.
7. Falke, J. J. and Koshland, D., Jr. (1987) Global flexibility in a sensory receptor: a site-directed cross-linking approach. *Science* **237,** 1596–1600.
8. Flitsch, S. L. and Khorana, H. G. (1989) Structure studies on transmembrane proteins. 1. Model study using bacteriorhodopsin mutants containing single cysteine residues. *Biochemistry* **28,** 7800–7805.
9. Sarsawat, L. D., Pastra-Landis, C., and Lowey, S. (1992) Mapping single cysteine mutants of light chain 2 in chicken skeletal myosin. *J. Biol. Chem.* **29,** 21,112–21,118.
10. Ling, R. and Luckey, M. (1994) Use of single-cysteine mutants to probe the location of the disulfide bond in LamB protein from *Escherichia coli. Biochem. Biophys. Res. Commun.* **201,** 242–247.
11. Beavis, R. C. and Chait, B. T. (1990) Rapid, sensitive analysis of protein mixtures by mass spectrometry. *Proc. Natl. Acad. Sci. USA* **87,** 6873–6877.
12. Hillenkamp, F., Karas, M., Beavis, R. C., and Chait, B. T. (1991) Matrix-assisted laser desorption/ionization mass spectrometry of biopolymers. *Anal. Chem.* **63,** 1193A–1203A.
13. Carr, S. A., Hemling, M. E., Bean, M. F., and Roberts, G. D. (1991) Integration of mass spectrometry in analytical biotechnology. *Anal. Chem.* **63,** 2802–2824.
14. Briggs, R. G. and Fee, J. A. (1978) Sulfhydryl reactivity of human erythrocyte superoxide dismutase: on the origin of the unsual spectral properties of the protein when prepared by a procedure utilizing chloroform and ethanol for the precipitation of hemoglobin. *Biochim. Biophys. Acta* **537,** 100–109.
15. Beavis, R. C. and Chait, B. T. (1990) High accuracy mass determination of proteins using matrix-assisted laser desorption mass spectrometry. *Anal. Chem.* **62,** 1836–1840.

16. Markland, F. S. and Smith, E. L. (1967) Subtilisin BPN' VIII. Isolation of CNBr peptides and the complete amino acid sequence. *J. Biol. Chem.* **242,** 5198–5211.
17. Wells, J. A., Ferrari, E., Henner, D. J., Estell, D. A., and Chen, E. Y. (1983) Cloning sequencing and secretion of *Bacillus amyloliquefaciens* subtilisin in *Bacillus subtilis. Nucleic Acids Res.* **11,** 7911–7925.
18. Lee, B. and Richards, F. M. (1971) The interpretation of protein structures: estimation of static accessibility. *J. Mol. Biol.* **55,** 379–400.
19. Bott, R., Ultsch, M., Kossiakoff, A., Graycar, T., Katz, B., and Power, S. (1988) The three dimensional structure of *Bacillus amyloliquefaciens* subtilisin at 1.8 Å and an analysis of the structural consequences of peroxide inactivation. *J. Biol. Chem.* **263,** 7895–7906.

14

Disulfide Bond Location in Proteins

Yiping Sun, Mark D. Bauer, Thomas W. Keough, and Martin P. Lacey

1. Introduction

The sulfhydryl groups of cysteine residues are among the most reactive side chains in proteins. Free sulfhydryl groups of cysteines are often involved in the active sites for enzymatic catalysis, as in cysteine proteases (1,2). Common reactions of free sulflhydryl groups include oxidation to form cysteic acids and disulfide bonds. Disulfide bonds in proteins play an important role in the folding and refolding process, and in stabilizing the three-dimensional structure of the molecule (3). Although the amino acid sequences of proteins can be deduced from the corresponding cDNA sequences, the state and connectivity of the cysteine residues after translation cannot be predicted accurately. Characterization of disulfide linkages in proteins has become an essential part of the analysis of recombinant DNA products, because molecules having incorrect disulfide linkages may have much lower activity than that of the desired product. Therefore, it is important to verify that recombinant materials have the correct disulfide linkages.

The general strategy for locating disulfide bonds in proteins by conventional methods involves several steps (4,5). The first step is to determine the number of free sulfhydryl groups and disulfide bonds. Second, the protein is cleaved by chemical reagents and/or enzymes between the half-cysteinyl residues. Third, the cysteine-containing peptides are separated using chromatographic techniques. Finally, the disulfide-bonded peptides are identified by amino acid analysis and sequence analysis, and related to specific segments of the protein. Although this approach is well established and has been used with much success, it is limited in many respects. For example, isolation and identification of cysteine-containing peptides present in complex mixtures is tedious and time consuming.

From: *Methods in Molecular Biology, Vol. 61: Protein and Peptide Analysis by Mass Spectrometry*
Edited by: J. R. Chapman Humana Press Inc., Totowa, NJ

Mass spectrometry (MS), which is now used extensively for structural characterization of peptides and proteins, has proven useful for locating disulfide bonds in proteins. MS-based strategies for the assignment of disulfide crosslinkages in proteins have been reviewed *(6–8)*. In this chapter, methods for locating disulfide bonds in proteins using fast-atom bombardment mass spectrometry (FABMS), on-line microbore/capillary LC/ionspray mass spectrometry (mLC/ISMS, cLC/ISMS), and matrix-assisted laser desorption/ionization time-of-flight mass spectrometry (MALDI-TOFMS) are described.

2. Materials

2.1. Enzymes Used for Digestions

1. *N*-Tosyl-L-phenyl-alanine chloromethyl ketone (TPCK)-treated trypsin (Sigma, St. Louis, MO).
2. Pepsin (Sigma).
3. Chymotrypsin (Sigma).
4. Carboxypeptidase B (Worthington, Freehold, NJ).
5. Carboxypeptidase Y (Sigma)

2.2. Peptides and Proteins

1. Echistatin (Bachem, Torrance, CA).
2. Oxytocin.
3. Glutathione (oxidized).
4. Bovine insulin.
5. Bovine pancreatic ribonuclease A.
6. Hen egg white lysozyme.
7. β-Lactoglobulin B.
8. Bovine trypsinogen.
9. Horse-heart cytochrome c.
10. Horse-heart myoglobin.
11. Papain.

Unless indicated, all materials were obtained from Sigma and stored at –20°C.

2.3. Chemical Reagents

1. Cyanogen bromide (CNBr) (Aldrich, Milwaukee, WI).
2. Dithiothreitol (DTT) (Sigma).
3. Iodoacetic acid (Sigma).
4. Trifluoroacetic acid (TFA) (Aldrich).
5. Oxalic acid (Sigma).
6. Sinapinic acid (Aldrich).
7. Dihydroxybenzoic acid (Aldrich).
8. α-Cyano-4-hydroxycinnamic acid (Sigma).

9. EDTA (Sigma).
10. 88% Formic acid (Baker, Phillipsburg, NJ).
11. 22% Formic acid; prepared by appropriate dilution of item 10.
12. 30% Hydrogen peroxide (Baker).
13. Cesium iodide (Aldrich).
14. Glycerol (Fisher Scientific, Fair Lawn, NJ).
15. Thioglycerol (Aldrich).

2.4. Solvents and Buffers

1. $0.1N$ HCl (Fisher).
2. Alkylation buffer: $0.5M$ Tris-HCl, $6M$ guanidine:HCl, $0.02M$ EDTA (pH 8.5).
3. $0.1M$ Ammonium acetate (pH 7.0).
4. $0.1M$ Ammonium bicarbonate (pH 8.4).
5. Acetonitrile (Baker, HPLC grade).
6. Deionized water (18 MΩ) was prepared on a Millipore water purification system.
7. HPLC solvent A: 0.02% TFA in water.
8. HPLC solvent B: 0.02% TFA in acetonitrile.

2.5. Instrumentation

2.5.1. FABMS

FABMS analyses were performed with a VG ZABSPEC (EBE) high-field mass spectrometer (VG Analytical Ltd., Manchester, UK). The instrument was operated at 8 kV ion acceleration and equipped with a high-voltage (35 kV) cesium ion gun, static FAB source, and standard electron multiplier detectors. Magnet scans from m/z 4000–100 at 5 s/decade were used with a mass resolution of 1500. Cesium iodide was used for mass calibration with an accuracy of 0.5 Da. All data were acquired with a VAX-based VG Opus data system. Chromatographic fractions from enzymatic digestions were collected, freeze-dried, dissolved in 7–10 μL of 22% formic acid, and analyzed by FABMS. A 50/50 (v/v) mixture of glycerol and thioglycerol or 50/50 glycerol and 3-nitrobenzoic acid with 1% TFA was used as the matrix.

For peptide sequencing using high-energy collision-induced dissociation (CID) by FABMS, the same instrument with an orthogonal acceleration (oa) TOF mass analyzer was used. The magnetic sector instrument (MS1) was tuned to transmit ions of interest at 8 kV. The ions were decelerated to 800 eV before entering the collision cell. The precursor and product ions were analyzed using the TOF instrument (MS2). The collision gas was Xe at $3-6 \times 10^{-6}$ mbar. CsI was used to calibrate the TOF mass analyzer. Mass spectrometric data from enzymatic digests of papain were obtained using a Kratos MS-50 high-field mass spectrometer (Kratos Scientific Instruments, Manchester, UK) with an FAB gun (Iontech, Middlesex, UK) at the School of Pharmacy, Purdue University (West Lafayette, IN).

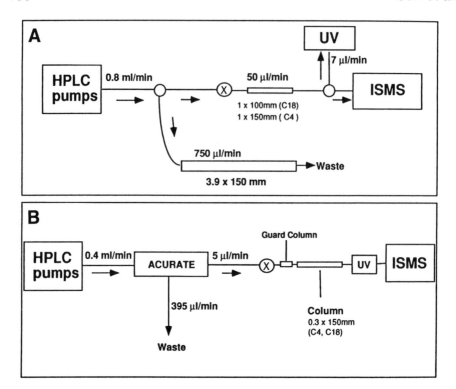

Fig. 1. Schematic diagrams of on-line mLC/ISMS system **(A)** and on-line cLC/ISMS system **(B)**.

2.5.2. ISMS

All ISMS analyses were performed on a PE Sciex API-III triple quadrupole mass spectrometer (PE Sciex, Thornhill, Canada). A delivery solvent of 50/50 methanol/water with 0.1% formic acid at a flow rate of 5 μL/min was used for flow injection experiments (Micro Flow Pump, ISCO). Samples were introduced with a Rheodyne 7125 loop injector having a 50-μL loop. For the tandem MS (MS/MS) sequencing of peptides, pure Ar was used as the collision gas at a gas thickness of $70–150 \times 10^{-14}$ atoms/cm^2.

2.5.3. LC/MS Systems

On-line mLC/ISMS and cLC/ISMS were used for the analysis of protein digests. The schematic diagrams describing these two systems are shown in Fig. 1. The microbore system (Fig. 1A) employed a Waters MS-600-high-performance liquid chromatography (HPLC) unit (Waters, Millipore), a Valco tee, a C18 (3.9 × 150 mm) "dummy" column, an Aquapore-C18 (1 × 100 mm, 5 μ, 300 Å) or a Vydac-C4 (1 × 150 mm, 5 μ, 300 Å) column, and an HPLC-UV detector with a capillary flow cell (LC-Packings, San Francisco, CA). The capillary LC system (Fig. 1B) used for LC-ISMS analysis is similar to the microbore system. An ACURATE flow splitter (LC-Packings) was used to provide a flow rate of 5 μL/min to the capillary column (0.3 × 150 mm, C4, 5 μ, 300 Å). The UV detector was set at 220 nm. Typical mobile phases used were 0.02% TFA in water (solvent A) and 0.02% TFA in acetonitrile (solvent B) (*see* Note 1). Mass spectra were recorded from *m/z* 2400 to 300 with a step size of 0.4 and a dwell time of 1.0 ms.

2.5.4. MALDI-TOFMS

MALDI experiments were performed on a Vestec VT2000 linear TOF mass spectrometer (PerSeptive Biosystems, Vestec Division, Houston, TX). Peptides and proteins were irradiated with the frequency-tripled output of a Q-switched Nd:YAG laser (Lumonics HY-400, 355 nm, 10-ns pulse width) operated at a repetition rate of 10 Hz. The instrumentation is described more fully in Chapter 13.

2.5.5. HPLC

Protein and peptide separations were carried out using reversed-phase columns (C4 and C18, 3.9 × 150 mm, 2.0 × 150 mm, Waters Delta-Pak 300 Å, 5 μ) under the control of a Waters 600-MS system. The effluent was detected with a Waters 490E programmable wavelength detector at 216 and 280 nm. Typical mobile phases used were 0.05% TFA in water (solvent A) and 0.05% TFA in acetonitrile (solvent B). The gradients varied depending on the sample. HPLC fractions were collected in polypropylene tubes and dried in a SpeedVac.

3. Methods

The general procedure to locate disulfide bonds is given in Fig. 2.

3.1. Determination of the Number of Cysteines, Free Sulfhydryl Groups, and Disulfide Bonds Using ISMS and MALDI-TOFMS

The first step in disulfide bond location is to determine the number of cysteines, free sulfhydryl groups, and disulfide bonds. Feng et al. *(9)* demonstrated a method that combines reduction/alkylation and ISMS for rapidly counting cysteines, free sulfhydryl groups, and disulfide bonds in proteins. The method involves three steps:

1. Reducing all disulfide bonds and alkylating the free sulfhydryl groups;
2. Alkylating only the original free sulfhydryl groups without reduction of disulfide bonds; and
3. Determining the molecular weight of the native protein and of the proteins formed in steps 1 and 2.

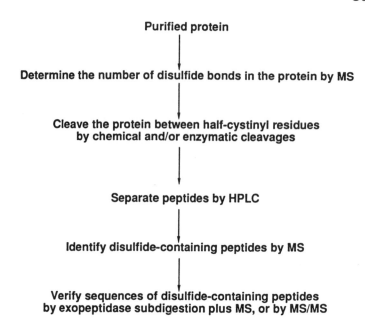

Fig. 2. General strategy for locating disulfide bonds in proteins by MS.

Several protein samples, including ribonuclease A, lysozyme, insulin, and β-lactoglobulin-B were analyzed using a similar procedure in our laboratory. The experimental conditions for the reduction and alkylation of proteins were as follows:

1. Dissolve 100–250 µg of protein in 100–200 µL of the Tris buffer (Section 2.4., item 2) in a microcentrifuge tube. Purge the sample solution with N_2.
2. Add approx 10 µL of DTT solution ($50M$ excess over disulfide bonds in protein) to the sample solution. DTT was predissolved in the Tris buffer. Incubate at 50°C for 4 h.
3. Let sample cool down, and then add approx 10 µL of iodoacetic acid solution (two times amount over DTT added). The iodoacetic acid was predissolved in the Tris buffer, and the pH adjusted to 8.0 with 5 µL of 5% ammonium hydroxide. Incubate the solution at 37°C for 1 h.
4. Repeat steps 1 and 3 with a second batch of sample for alkylation of only the original free sulfhydryl groups.
5. Purify the proteins obtained from steps 3 and 4 by reversed-phase HPLC with a C4 column (Waters, 2 × 150 mm, 300 Å, 5 µ) (*see* Note 2).

Purified proteins were analyzed by flow injection ISMS on the Sciex AP-IIII instrument. The results are summarized in Table 1. From the measured molecular mass of native proteins (M_{nat}), reduced and alkylated proteins (M_{r+a}),

Table 1
Determination of the Number of Cysteines, Free Sulfhydryl Groups, and Disulfide Bonds in Proteins by ISMS^a

	Lysozyme	Ribonuclease A	β-Lactoglobulin B	Insulin
M_{nat}	14,304.5	13,682.3	18,276.7	5733.4
M_{r+a}	14,779.5	14,156.8	18,572.8	2571.2 (A chain)
				3516.3 (B chain)
M_a	14,304.2	13,680.8	18,333.8	5733.4
N_{Cys}	8	8	5	6
N_{-SH}	0	0	1	0
N_{-S-S-}	4	4	2	3

$^a N_{Cys} = (M_{r+a} - M_{nat})/59$; $N_{-SH} = (M_a - M_{nat})/(59 - 1)$; $N_{-S-S-} = (N_{Cys} - N_{-SH})/2$.

alkylated proteins (M_a), and the known mass of the alkylating reagent (m = 59 for –CH$_2$COOH), the number of cysteines, free sulfhydryl groups, and disulfide bonds was easily determined.

Ionspray mass spectra of the native ribonuclease A and the fully reduced and alkylated protein are shown in Fig. 3. The mol-wt difference between M_{r+a} and M_{nat} revealed that the total number of cysteines in the protein is 8 (i.e., [14,156.8 – 13,682.3]/59 = 8). Similarly, the mol-wt difference between M_a and M_{nat} revealed that the number of the free sulfhydryl groups is 0 (i.e., [13,680.8 – 13,682.3]/ [59 – 1] = 0). Therefore, the number of disulfide bonds in this protein is 4.

Insulin represents a protein with two peptide chains, which are linked together through intermolecular disulfide bonds. Two peptides (A chain, mol wt = 2571.2 and B chain, mol wt = 3516.3) were derived from the intact protein (mol wt = 5733.4) after reduction and alkylation. The sum of the mol wt of the product peptides is 6 × 59 greater than the mol wt of insulin itself (Table 1).

The number of cysteines, free sulfhydryl groups, and disulfide bonds in a protein can be also determined using MALDI-TOFMS. Mass spectra of the native ribonuclease A (MH$^+$ = 13,680.0), the reduced and alkylated protein (MH$^+$ = 14,158.5), and the alkylated protein (MH$^+$ = 13,687.4) are shown in Fig. 4. From the measured molecular masses, the number of disulfide bonds in the protein was easily determined.

3.2. Generation of Disulfide-Linked Peptides Using Chemical and Enzymatic Cleavage

Generation of disulfide-linked peptides is an important step toward assigning disulfide linkages in proteins. The goal is to obtain, by chemical and enzymatic methods, a mixture of peptides, each containing only one disulfide bond. One should keep in mind that digestion conditions like high pH and long diges-

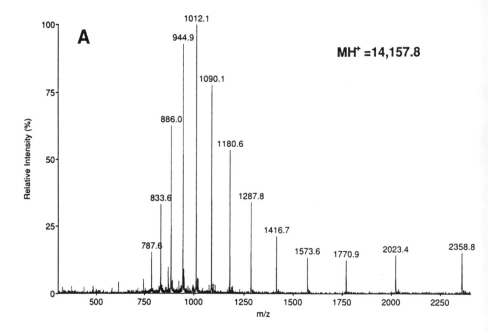

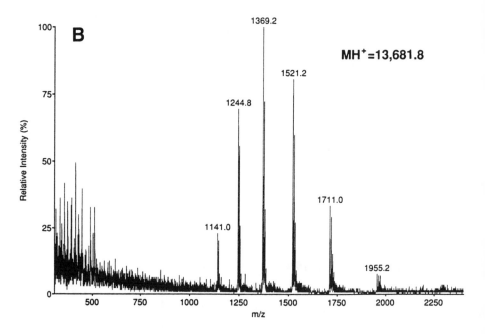

Fig. 3. IS mass spectra of the fully reduced and alkylated ribonuclease A **(A)** and alkylated ribonuclease A **(B)**.

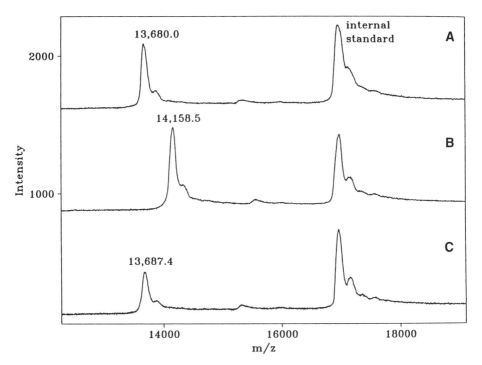

Fig. 4. MALDI mass spectra of ribonuclease A: native protein **(A)** after reduction and alkylation **(B)**, and after alkylation only **(C)**.

tion time may promote disulfide scrambling, and should be avoided *(3)*. Specific cleavage reagents, like CNBr, trypsin, and V8 protease, are commonly used. It is always easy to identify the peptides generated by these reagents because the N-terminal or C-terminal residues of the peptides are known. If a protein cannot be cleaved between half-cysteinyl residues to produce the desired disulfide-linked peptides, nonspecific enzymes like pepsin or partial acid hydrolysis can be used. The major advantage of using pepsin and partial acid hydrolysis to generate peptides is that disulfide bonds in proteins are fairly stable under mild acid conditions *(10,11)*. However, the protein digests using such nonspecific cleavages are complex mixtures, which make the separation and identification much more difficult.

3.3. Rapid Identification of Disulfide-Containing Peptides in Whole Protein Digests by On-Line mLC-ISMS, cLC-ISMS, and MALDI-TOFMS

With the advances of recombinant DNA technology, it is possible to produce proteins having any desired amino acid sequence. Many of the proteins of interest contain cysteine residues. Their biological activity is often affected by correct disulfide bond formation. It is therefore important to verify rapidly that recombinant materials have the correct disulfide bond arrangement. Ionspray ionization and MALDI are two new analytical techniques that allow rapid characterization of complex mixtures like protein digests *(12–15)*. Applications of these two methods to identify disulfide-containing peptides in protein digests were demonstrated with a tryptic digest of CNBr-treated hen egg white lysozyme. The tryptic digest of CNBr-treated lysozyme was prepared as follows:

1. Dissolve 1 mg of lysozyme in 200 μL of 0.1*N* HCl (*see* Note 3) in a microcentrifuge tube.
2. Add one small crystal of CNBr to the solution.
3. Incubate the solution at 37°C for 2 h.
4. Add 800 μL of water (*see* Note 3) to the solution, and dry the sample in a SpeedVac.
5. Dissolve the sample in 1 mL of 0.1*M* ammonium acetate (pH 7.0).
6. Add 20 μg of trypsin (dissolved in 1 μg/μL ammonium acetate buffer [see Note 4]).
7. Incubate the solution at 37°C for 3 h.
8. Analyze the whole lysozyme digest by MALDI (*see* Note 5 and footnote to Table 2) using several matrices, including sinapinic acid, 2,5-dihydrobenzoic acid, and α-cyano-4-hydroxycinnamic acid.

Portions of the MALDI spectra obtained with sinapinic acid as the matrix before and after DTT reduction are shown in Fig. 5 (for DTT reduction procedure for MALDI analysis, *see* Chapter 13). The disulfide-containing peptides detected in the whole mixture are listed in Table 2. Six disulfide-containing peptides that account for three of four disulfide bonds in the protein were identified using only a few picomoles of material.

The same tryptic digest was analyzed by flow injection ISMS. Many peptides with and without disulfide bonds were detected. It should be pointed out that the ionspray mass spectrum (not shown) of the whole mixture is much more complicated than the MALDI spectrum owing to the formation of multiply charged ions for individual peptides. Furthermore, peaks from some of the multiply charged ions of the peptides were very weak or suppressed completely. The developments of on-line mLC-ISMS and cLC-ISMS make it possible for the rapid characterization of peptides in protein digests. The mLC-ISMS system (Fig. 1) can be used for analyzing 50–1000 pmol of material, whereas the capillary system can be used for analyzing 1–50 pmol of material (data not shown). Disulfide-containing peptides detected using on-line mLC-ISMS are also summarized in Table 2. It is obvious that the on-line mLC-ISMS system provided a more complete identification of the peptides having disulfide bonds in the tryptic digest. For comparision, an aliquot of the digest was also analyzed by FABMS. Only a few peptides were detected (data not shown). None of the predicted disulfide-containing peptides were observed.

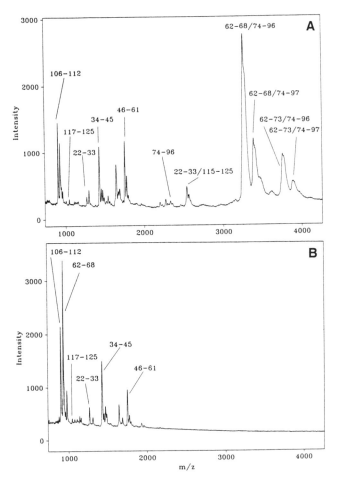

Fig. 5. Portions of the MALDI mass spectra of a tryptic digest of CNBr-treated lysozyme obtained with sinapinic acid as the matrix before **(A)** and after DTT reduction **(B)**.

On-line mLC-ISMS, in combination with DTT reduction, is also useful for rapidly identifying HPLC fractions that contain disulfide-linked peptides in a protein digest. Peaks corresponding to intermolecular disulfide-bonded peptides on the UV trace and on the total-ion current profile disappear once the digest is treated with DTT, and new peaks corresponding to their constituent peptides occur in the early eluting fractions. Therefore, by comparing two chromatographic profiles, the disulfide-containing fractions can be easily identified. The usefulness of this approach is demonstrated again with the same tryptic lysozyme sample. About 1 nmol of material was analyzed using the

Table 2
Disulfide-Containing Peptides Identified
in the Tryptic Digest of CNBr-Treated Lysozyme
Using MALDI-TOFMS and On-Line mLC-ISMS

		$MH^+_{Meas.}$	
Residues	$MH^+_{Calc.}$	MALDI-TOFMS[a]	mLC-ISMS
62–68/74–96	3270.7	3272.5	3270.8
62–68/74–97	3398.7	3401.0	3398.7
62–73/74–96	3769.2	3772.8	3769.6
62–73/74–97	3897.2	3897.1	3897.2
22–33/115–125	2543.9	2543.7	2543.5
22–33/115–116	1516.8	1516.0	1516.0
6–12/126–129	1105.8	ND	1105.8
6–12/126–128	922.8	ND	992.2

[a]Three different matrices (sinapinic acid, dihydrobenzoic acid, and α-cyano-4-hydroxycinnamic acid) were used. None of the disulfide-containing peptides were detected in dihydroxybenzoic acid, whereas similar results were obtained with sinapinic acid and α-cyano-4-hydroxycinnamic acid. ND, not detected.

mLC-MS system. The results before and after reduction are given in Fig. 6A and B, respectively. The peak eluting at 22 min contained the disulfide-linked peptide (62–68/74–96). Therefore, the corresponding fractions can be collected from off-line HPLC, and the disulfide-containing peptides can be further verified by other methods (*see* Section 3.4.). Sun et al. *(16,17)* reported the use of an HPLC electrochemical detector for the selective and simultaneous detection of thiol- and disulfide-containing peptides in protein digests with identification by MS. This approach is particularly useful for rapid identification of disulfide-containing peptides in large proteins having only a few disulfide bonds.

3.4. Detection and Sequence Verification of Disulfide-Containing Peptides by FABMS and MS/MS

3.4.1. Detection of Disulfide-Containing Peptides by FABMS with On-Probe Reduction and Oxidation

After separation, the purified peptides can be analyzed directly by FABMS. On-probe reduction and oxidation of peptides can be employed to detect rapidly fractions that have disulfide-containing peptides *(18–20)*. If a peptide contains an intramolecular disulfide bond, a new peak will appear 2 Da higher once reduced on the probe tip with the glycerol/thioglycerol matrix. If a peptide contains an intermolecular disulfide bond, new peaks corresponding to the two constituent peptides may appear in the lower mass range, and the peak

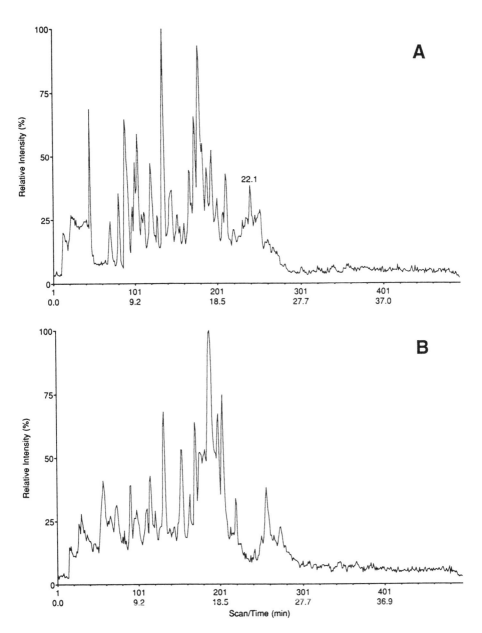

Fig. 6. Total-ion current-profiles of a tryptic digest of CNBr-treated lysozyme obtained using on-line mLC/ISMS before **(A)** and after DTT reduction **(B)**. The peak eluting at 22 min contained the disulfide-linked peptide (62–68/74–96).

197

corresponding to the intact disulfide-containing peptide will disappear. The procedure for on-probe reduction and FABMS analysis is as follows:

1. Dissolve each dried fraction in 7–10 μL of 22% formic acid.
2. Load 1 μL of peptide solution and 1 μL of FAB matrix (glycerol/thioglycerol, 1:1) on the probe tip.
3. Acquire the FAB mass spectrum.
4. Load 1 μL of the sample solution onto another probe tip.
5. Dry the sample in the air.
6. Add 1 μL of 30% ammonium hydroxide, and 1 μL of the FAB matrix.
7. Acquire the spectrum again.
8. Analyze the FAB data obtained before and after reduction.

Although this reduction approach is simple and useful, it can not be used to identify thiol-containing peptides. It also requires good instrument resolution to identify intramolecular disulfide-containing peptides with relatively high molecular masses. Furthermore, the constituent peptides are not always observed in the FAB mass spectra. This is particularly true when the sample is a mixture of several peptides. On-probe oxidation is another simple approach to identify both thiol- and disulfide-containing peptides rapidly *(18)*. Performic acid can convert a cysteine residue to cysteic acid, which increases the molecular mass by 48 Da. Intramolecular disulfide-linked peptides will increase in mol wt by 98 Da, whereas the intermolecular disulfide-linked peptides will disappear and form two constituent peptides containing cysteic acid residues. The oxidation procedure is as follows:

1. Dissolve each dried fraction in 7–10 μL of 22% formic acid.
2. Load 1 μL of the sample solution and 1 μL of FAB matrix onto the probe tip.
3. Acquire the FAB mass spectrum.
4. Load 1 μL of the sample solution onto another probe tip.
5. Dry the sample in the air.
6. Add 1 μL of performic acid reagent (33% hydrogen peroxide and 88% formic acid [1:19, v/v]), allow the solution dry on the tip, and then add 1 μL of FAB matrix (*see* Note 6).
7. Acquire the spectrum again.
8. Analyze the FAB data obtained before and after oxidation.

Figure 7 shows FAB mass spectra of an HPLC fraction from a peptic digest of papain before (A) and after (B) oxidation with performic acid. The peak at *m/z* 1875 in Fig. 7A corresponds to two peptides, segments Val150-Ala162 and Gly198Tyr203, joined by a disulfide bond, and peaks at *m/z* 611 and 1267 correspond to the peptides formed by cleavage of the disulfide bond. A similar series of ions (*m/z* 1946, 1338, and 611) is formed by cleavage after Ala163. Once oxidized (Fig. 7B), peaks for the peptides with disulfide bonds (*m/z* 1875

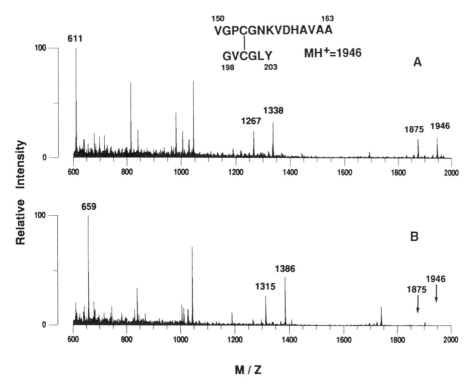

Fig. 7. FAB mass spectra of an HPLC fraction from a peptic digest of papain before (**A**) and after (**B**) oxidation with performic acid.

and 1946) were not present, and the molecular masses for the constituent peptides increased by 48 Da. These results demonstrate that Cys153 and Cys200 in papain were joined by a disulfide bond.

3.4.2. Sequence Verification by Subdigestion of Peptides with Carboxypeptidases Followed by FABMS Analysis

FABMS can be used to tentatively identify peptides derived from enzymatic digests of proteins. One of the methods for rapidly verifying sequences is to remove one or more N-terminal residues by Edman degradation or aminopeptidase, or to remove one or more C-terminal residues with carboxypeptidase B or Y. Carboxypeptidase B can easily remove many different C-terminal residues, especially Lys and Arg, which is very suitable for peptides from a tryptic digestion. The procedures are outlined as follows *(21)*:

1. Dissolve carboxypeptidase B (Y) in 0.1*M* ammonium bicarbonate (see Note 4) (pH 8.4).

2. Add buffered carboxypeptidase B solution (about 1–2 μg enzyme) to the dried HPLC fractions and vortex.
3. Incubate the sample solution at 37°C for 20 min.
4. Dry the solution in a SpeedVac.
5. Analyze the sample by FABMS.

One example is given in Fig. 8 (pp. 202,203). The mass spectrum of a fraction from the chymotryptic digest of papain had major peaks at m/z 611, 1657, and 2265 (Fig. 8A). On-probe oxidation and reduction were used to identify these peaks. The peaks at m/z 611 and 1657 were from the peptides 198–203 and 150–166, and the the peak at m/z 2265 was the disulfide-bonded peptide 150–166/198–203. This assignment was further verified by FABMS analysis of the subdigestion with carboxypeptidase B, which removed two C-terminal tyrosines at the same time to give the peaks at m/z 448, 1494, and 1939 (Fig. 8B).

3.4.3. Sequence Verification by CID MS/MS

3.4.3.1. Peptide Sequencing by Low-Energy CID Using a Triple Quadrupole Mass Spectrometer

Peptide sequences can be verified by low-energy CID using a triple quadrupole mass spectrometer *(22)*. The precursor ion of a peptide is selected in the first quadrupole and is collided with argon gas in the second quadrupole to generate fragment ions. The fragment ions generated are then mass-separated in the third quadrupole and detected. Peptides derived from a tryptic digestion can be easily sequenced by ionspray MS/MS of the doubly charged species. Disulfide-containing peptides can also be sequenced using this technique (for examples, *see* Section 3.5.) *(23)*.

3.4.3.2. Peptide Sequencing by High-Energy CID Using a Hybrid Magnetic Sector-oa-TOF Instrument

Stults et al. *(24)* demonstrated the use of high-energy CID with FABMS for the identification of disulfide-bonded peptides from recombinant relaxin. Bean and Carr *(25)* also reported the application of a BEEB four-sector mass spectrometer for the characterization of disulfide-linked peptides from an enzymatic digest of des-Val-Val-TGF-α. With the novel combination of a magnetic sector instrument and oa-TOF instrument, high-energy CID mass spectra of peptides can be obtained using both FAB and electrospray ionization *(26,27)*. A magnetic sector instrument as MS1 of a tandem mass spectrometer provides unit mass selection of a precursor ion with high ion transmission, and an oa-TOF instrument as MS2 offers high sensitivity and detection of all product ions. Mass spectra obtained with this type of instrument are very similar to those observed with four-sector-type MS/MS equipment. However, the hybrid instrument is cheaper and much easier to use. An example of the sequencing of an intramolecular disulfide-containing peptide, oxytocin, using this new technique, is shown in Fig. 9 (pp. 204,205). Figure 9A is the MS/MS spectrum of the molecular ion (MH)$^+$, whereas Fig. 9B is that of the reduced form of the peptide (MH + 2)$^+$, which was derived during the FAB ionization *(20)*. It is obvious that Fig. 9B has more sequence information than Fig. 9A.

3.5. Assignment of Disulfide Crosslinkages in Small Cysteine-Rich Proteins

Many naturally occurring toxins contain a high-density core of cysteine residues. The disulfide bridges in these molecules play a key role in maintaining their three-dimensional structures. Echistatin is a 49 amino acid protein from *Echis carinatus* venom, which contains four disulfide bonds. It is a member of a class of proteins called disintegrins. Echistatin represents a challenging case for disulfide assignment, since the cysteine residues are very close to each other and its structure is very compact. Proton NMR has demonstrated that Cys2-Cys11 and Cys20-Cys39 are disulfide linked. However, the spatial proximity of Cys7 and Cys8 makes it difficult to resolve their disulfide pattern using this technique. As mentioned, for the disulfide-bonding assignment to be successful, proteins have to be cleaved between the cysteinyl residues by chemical or enzymatic methods without disulfide exchange. Several approaches, which combined CNBr cleavage and enzymatic digestion, tried by this group and another lab *(28)* failed to produce disulfide-containing peptides without scrambling. Partial acid hydrolysis is an alternative means of producing peptides that are diagnostic for the location of disulfide bonds in proteins *(10)*. This method is particularly attractive because the disulfide bonds are unlikely to undergo exchanges in dilute acid. Zhou and Smith *(11)* reported the successful assignment of the disulfide linkages in hen egg white lysozyme by partial acid hydrolysis followed by FABMS analysis. A similar strategy, which combined mild acid hydrolysis and ionspray MS/MS, was used for locating disulfide bonds in echistatin *(23)*. The procedure used is as follows:

1. Dissolve 100 μg of the protein in 100 μL of 250 m*M* oxalic acid that has been purged with dry nitrogen (*see* Note 7).
2. Transfer the sample solution into two glass capillary tubes (1.5–1.8 × 90 mm, prewashed with nitric acid and deionized water), flush the tubes with nitrogen, and seal them.
3. Incubate the solution in a vacuum oven at 100°C for 4 h.
4. Isolate hydrolysis products by HPLC.
5. Analyze each HPLC fraction by ionspray MS and MS/MS.

Under these conditions, many overlapping disulfide-containing peptides were identified by ISMS. ISMS/MS data indicate that the four disulfide

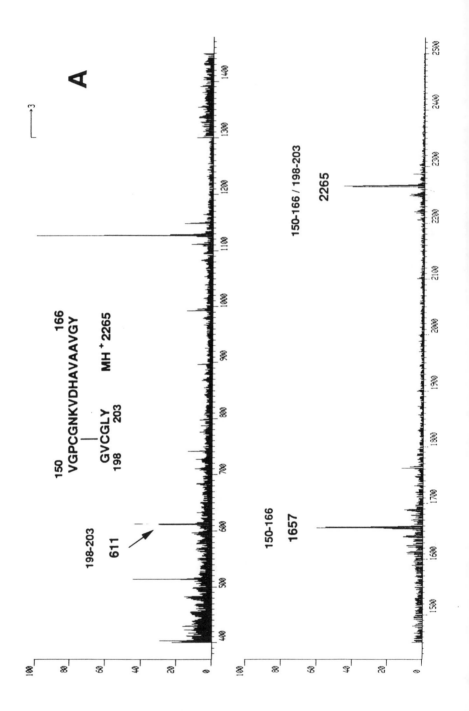

A

150
VGPCGNKVDHAVAAVGY
166

GVCGLY
198 203

MH⁺ 2265

198-203
611

150-166 / 198-203
2265

150-166
1657

202

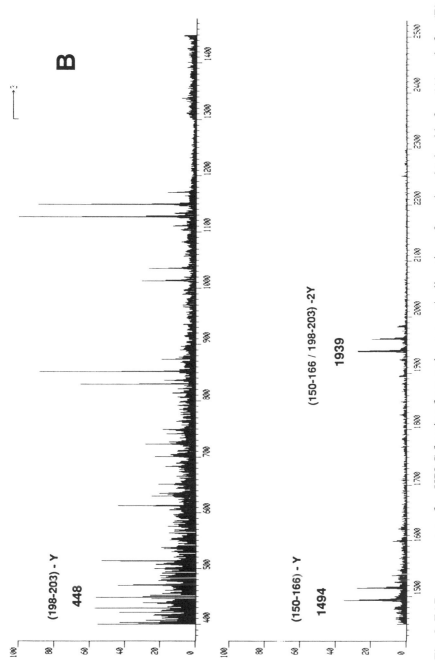

Fig. 8. FAB mass spectra of an HPLC fraction from a chymotryptic digestion of papain, obtained before (**A**) and after (**B**) subdigestion with carboxypeptidase B.

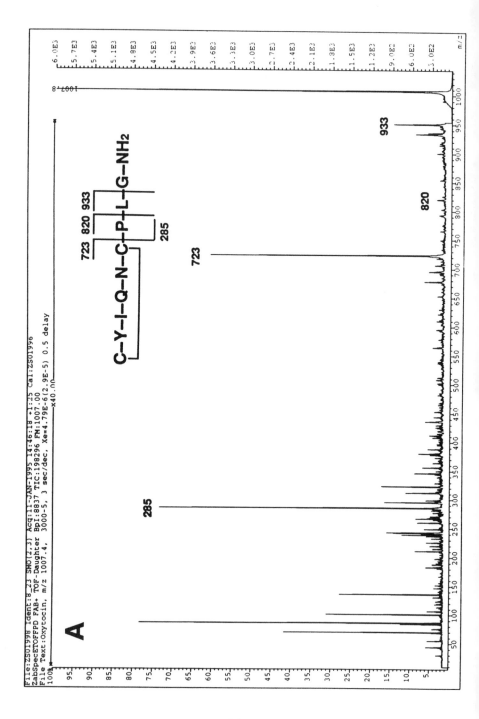

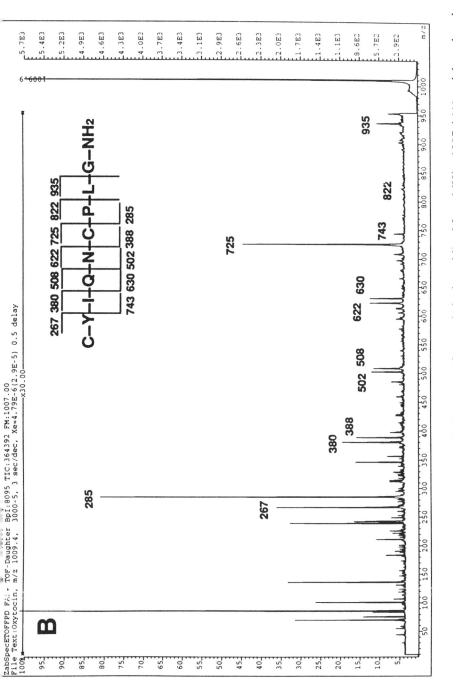

Fig. 9. Collision-induced FAB-oa-TOF-MS/MS spectra of oxytocin in the oxidized form MH$^+$ = 1007.4 (**A**) and the reduced form MH$^+$ + 2 = 1009.4 (**B**), using glycerol:thioglycerol (1:1) as the matrix.

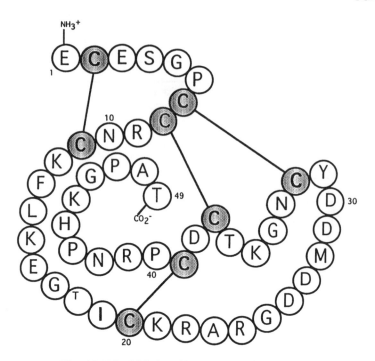

Fig. 10. Disulfide bonding pattern of echistatin.

bonds are Cys2-Cys11, Cys7-Cys32, Cys8-Cys37, and Cys20-Cys39. The disulfide bonding pattern for this protein is shown in Fig. 10. The MS/MS spectrum of the doubly charged ion at m/z 574 (Cys7-Cys32/Cys8-Cys37) is shown in Fig. 11.

4. Notes

1. TFA is a common additive to the LC mobile phase. It has been reported that TFA (0.1% for a typical LC analysis) affects IS ionization and suppresses the peptide signals *(29)*. For better sensitivity, LC solvents with 0.02% TFA can be used for on-line LC/ISMS analysis.

2. It should be pointed out that minor components, corresponding to the attachment of one or two extra alkylating groups, were detected in the carboxymethylated samples. The modification sites might be at residues Lys and His. It is recommended that proteins be purified immediately after carboxymethylation to minimize these side reactions. For the purification of proteins from carboxymethylation, off-line HPLC with a column (2 mm id or above) should be used. It is difficult to analyze directly the sample using on-line LC/MS with a small column like a microbore (1 mm id) or a capillary column (0.3 mm id), because the column might be clogged owing to the high concentration of guanidine-HCl and Tris

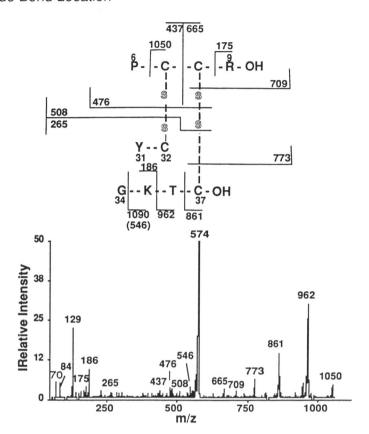

Fig. 11. ISMS/MS spectrum of the doubly charged ion at *m/z* 574 (Cys7-Cys32/Cys8-Cys37).

buffer. MALDI-TOFMS is more tolerant to salts and buffers. It can even be used for analyzing carboxymethylated proteins without removal of guanidine-HCl and Tris buffer. Both the carboxymethylated A-chain and B-chain of insulin were successfully detected in buffer solution after carboxymethylation (data not shown).

3. CNBr can be used for specific cleavage at the C-terminal methionine of a protein to give the corresponding homoserine lactone using either 88% formic acid or 0.1*N* HCl as a solvent. According to our experience, 0.1*N* HCl is a better solvent because in 88% formic acid, peptides can undergo formylation, which may complicate the MS analyses. The sample solution should be diluted with water (1:4) before drying in a SpeedVac.

4. Whenever possible, volatile solvent and buffer solutions should be used for sample preparations. Ammonium acetate buffer and ammonium bicarbonate buffers are most commonly used, and are compatible with most types of mass spectrometric analyses.

5. The solutions of the UV-absorbing MALDI matrix were prepared fresh daily at a concentration of 10 μg/μL These solutions were stored at room temperature in the dark when not in use.

6. Performic acid is a powerful oxidation reagent. With the combination of FABMS analysis, it can be used for rapidly identifying disulfide-containing peptides. However, under the experimental conditions used here, two other residues, methionine and tryptophan, can also be oxidized. Changes in molecular weights for Cys, Met, and Trp are diagnostic for oxidation at each of these residues. Two oxygens can be added to methionine to form methionine sulfone, whereas one, two, or three oxygens can be added to tryptophan. Oxidized tryptophan typically shows three peaks in a group, with the most intense peak from the addition of two oxygens. Although the presence of three methionine residues would give the same mass shift in molecular weight as two cysteine residues, this is not normally a problem because CNBr converts methionine to the corresponding homoserine lactone if proteins are first cleaved with this reagent prior to subsequent enzymatic digestion. It is important to point out that the performic acid reagent should be dried in air on the probe-tip before the FAB matrix is loaded, because the reagent can react with the matrix, which then affects FAB ionization of the samples.

7. Because of the stability of disulfide bonds in mild acid, partial acid hydrolysis is very attractive for the determination of disulfide linkages in proteins, especially cysteine-rich small proteins like those from snake venom. However, the hydrolysis products are complex mixtures, which makes separation and analysis more difficult. Furthermore, it requires more material than enzymatic approaches. Dehydrated peptides can form during acid hydrolysis; therefore, one should be careful when using computer-assisted mass searching against protein sequences.

Acknowledgment

The authors thank David L. Smith for use of his instrumentation to obtain data on papain analysis.

References

1. Brocklehurst, K., Willenbrock, F., and Salih, E. (1987) Cysteine proteases, in *Hydrolytic Enzymes, vol. 16* (Neuberger, A. and Brocklehurst, K., eds.), Elsevier, Amsterdam, pp. 39–145.
2. Varughese, K. I., Ahmed, F. R., Grey, P. R., Hasnain, S., Huber, C. P., and Storer, A. C. (1989) Crystal structure of a papain-E-64 complex. *Biochem.* **28,** 1330–1332.
3. Creighton, T. E. (1984) Disulfide bond formation in proteins, in *Methods in Enzymology, vol. 107* (Wold, F. and Moldave, K., eds.), Academic, San Diego, CA, pp. 305–329.
4. Wetlaufer, D. B., Parente, L., and Haeffner-Gormley, L. (1985) Use of proline-specific endopeptidases in the isolation of all four "native" disulfides of hen egg white lysozyme. *Int. J. Pept. Protein Res.* **26,** 83–91.
5. Staswick, P. E., Hermodson, M. A., and Nielsen, N. C. (1984) Identification of the cysteines which link the acidic and basic components of the glycinin subunit. *J. Biol. Chem.* **259,** 13,431–13,435.

6. Smith, D. L. and Sun, Y. (1991) Detection and location of disulfide bonds in proteins by mass spectrometry, in *Mass Spectrometry of Peptides* (Desiderio, M. D., ed.), CRC, Boca Raton, FL, pp. 275–287.

7. Smith, D. L. and Zhou, Z. (1990) Stragegies for locating disulfide bonds in proteins, in *Methods in Enzymology, vol. 193* (McCloskey, J. A., ed.), Academic, New York, pp. 374–389.

8. Hirayama, K. and Akashi, S. (1994) Assignment of disulfide bonds in proteins, in *Biological Mass Spectrometry: Present and Future* (Matsuo, T., Caprioli, R. M., Gross, M. L., and Seyama, Y., eds), Wiley, New York, pp. 299–312.

9. Feng, F., Bell, A., Dumas, F., and Konishi, Y. (1990) Reduction/Alkylation Plus Mass Spectrometry: A Fast and Simple Method for Accurate Counting of Cysteines, Disulfide Bridges and Free SH Groups in Proteins. *Proceedings of the 38th Annual Conference on Mass Spectrometry and Allied Topics*, Tucson, AZ, pp. 273,274.

10. Sun, Y., Zhou, Z., and Smith, D. L. (1989) Location of disulfide bonds in proteins by partial acid hydrolysis and mass spectrometry, in *Techniques in Protein Chemistry* (Hugli, T. E., ed.), Academic, New York, pp. 176–185.

11. Zhou, Z. and Smith, D. L. (1990) Assignment of disulfide bonds in proteins by partial acid hydrolysis and mass spectrometry. *J. Protein Chem.* **9,** 523–532.

12. Beavis, R. C. and Chait, B. T. (1990) Rapid, sensitive analysis of protein mixtures by mass spectrometry. *Proc. Natl. Acad. Sci USA* **87,** 6873–6877.

13. Hillenkamp, F., Karas, M., Beavis, R. C., and Chait, B. T. (1991) Matrix-assisted laser desorption mass spectrometry of biopolymers. *Anal. Chem.* **63,** 1193A–1203A.

14. Covey, T. R., Huang, E. C., and Henion, J. D. (1991) Structural characterization of protein tryptic peptides via liquid chromatography/mass spectrometry and collision-induced dissociation of their doubly charged molecular ions. *Anal. Chem.* **63,** 1193–1200.

15. Carr, S. A., Hemling, M. E., Bean, M. F., and Roberts, G. D. (1991) Integration of mass spectrometry in analytical biotechnology. *Anal. Chem.* **63,** 2802–2824.

16. Sun, Y., Smith, D. L., and Shoup, R. E. (1991) Simultaneous detection of thiol- and disulfide-containing peptides by electrochemical high performance liquid chromatography with identification of mass spectrometry. *Anal. Biochem.* **197,** 69–76.

17. Sun, Y., Andrews, P. C., and Smith, D. L. (1990) Identification of disulfide-containing peptides in endocrine tissue extracts by HPLC-electrochemical detection and mass spectrometry. *J. Protein. Chem.* **9,** 151–157.

18. Sun, Y. and Smith, D. L. (1988) Identification of disulfide-containing peptides by performic acid oxidation and mass spectrometry. *Anal. Biochem.* **172,** 130–138.

19. Yazdanparast, R., Andrews, P. C., Smith, D. L., and Dixon, J. E. (1987) Assignment of disulfide bonds in proteins by fast atom bombardment mass spectrometry. *J. Biol. Chem.* **262,** 2507–2513.

20. Yazdanparast, R., Andrews, P. C., Smith, D. L., and Dixon, J. E. (1986) A new approach for detection and assignment of disulfide bonds in proteins. *Anal. Biochem.* **153,** 348–353.

21. Qin, W., Smith, J. B., Smith, D. L., and Edmonds, C. G. (1991) Mass spectrometric analysis of the structure of γII bovine lens crystallin. *Exp. Eye Res.* **54,** 23–32.

22. Hunt, D. F., Yates, J. R., Shabanowitz, J., Winston, S., and Hauer, C. R. (1986) Protein sequencing by tandem mass spectrometry. *Proc. Natl. Acad. Sci. USA* **83,** 6233–6237.
23. Bauer, M., Sun, Y., Degenhardt, C., and Kozikoski, B. (1993) Assignment of all four disulfide bridges in echistatin. *J. Protein Chem.* **12,** 759–764.
24. Stults, J. T., Bourell, J. H., Canova-Davis, E., Ling, V. T., Laramee, G. R., Winslow, J. W., Griffin, P. R., Rinderknecht, E., and Vandlen, R. L. (1990) Structural characterization by mass spectrometry of native and recombinant human relaxin. *Biomed. Environ. Mass Spectrom.* **19,** 655–664.
25. Bean, M. F. and Carr, S. A. (1992) Characterization of disulfide bond position in proteins and sequence analysis of cystine-bridged peptides by tandem mass spectrometry. *Anal. Biochem.* **201,** 216–226.
26. Clayton, E. and Bateman, R. H. (1992) Time-of-flight mass analysis of high energy collision-induced-dissociation fragment ions. *Rapid Commun. Mass Spectrom.* **6,** 719,720.
27. Bateman, R. H., Bordoli, R., Braybrook, C., and Clayton, E. (1994) A Comparative Study of MS/MS Data Produced by Tandem Magnetic and Hybrid MS/MS Instruments and the Implications of the Conclusions on the Design of a Hybrid Magnetic Sector-Orthogonal Acceleration Time-of-Flight MS/MS Instrument. *Proceedings of the 41st Annual Conference on Mass Spectrometry and Allied Topics,* pp. 928a,928b.
28. Calvete, J. J., Wang, Y., Mann, K., Schafer, W., Niewiarowski, S., and Stewart, G. J. (1992) The disulfide bridge pattern of snake venom disintegrins, flavoridin and echistatin. *Fed. Eur. Biochem. Soc.* **309,** 316–320.
29. Apffel, A., Fischer, S., Goodley, P. C., and Sahakian, J. A. (1994) Enhanced Sensitivity for Mapping with Electrospray LC/MS. *Proceedings of the 42nd Annual Conference on Mass Spectromtry and Allied Topics,* Chicago, IL, p. 642.

15

Determination of Loading Values and Distributions for Drugs Conjugated to Proteins and Antibodies by MALDI-MS and ESI-MS

Marshall M. Siegel

1. Introduction

The covalent attachment of drugs to proteins and antibodies has been a subject of numerous investigations (1). These conjugated proteins are being evaluated as a means for delivering anticancer drugs to tumor sites (2) and are being used to elicit antibody responses from drugs in animals (3–5). The general analytical problem posed by these materials is the determination of the numbers of moles of materials of lower molecular weight (generally <2000 Da) relative to a mole of the materials of higher molecular weight (>10,000 Da), referred to as carriers, to which they are covalently bound. The molar ratio of the material of lower molecular weight to the carrier is referred to as the loading value. A related analytical problem is the determination of the distribution of the numbers of molecules of lower molecular weight covalently bound to carrier molecules. The relationship between the distribution and loading value is that the number average of the individual components making up the distribution is the average loading value.

The loading value has usually been determined either by UV absorption spectrometry, radiolabeling, immunoassay, or wet chemical measurements. In a number of cases, the loading values cannot be determined by any of these techniques either because the technique may not be applicable or is very time consuming to develop. For example, loading values cannot be determined by UV-absorption spectrometry when the chromophores for the drug and carrier overlap or when the drug has no UV chromophore. Again, the synthesis of a radioactive label for the drug and the development of an immunoassay or wet-

From: *Methods in Molecular Biology, Vol. 61: Protein and Peptide Analysis by Mass Spectrometry*
Edited by: J. R. Chapman Humana Press Inc., Totowa, NJ

Table 1
Comparison of Analysis Methods for Protein–Drug Conjugates[a]

Category	MALDI-MS and ESI-MS	UV absorption spectrometry	Immunoassay	Radiolabeli•
Sensitivity	pM-fM	mM	pM	fM
Accuracy	+	±	+	+
Selectivity for covalent vs noncovalent bound drug	++	−	+	−
Loading distribution	Yes	No	No	No
Assay development time	−	+	++	++
Assay time	~1 h	~1 h	2 d	~1h
Assay cost	+	−	−	±
Operator skill	+	−	++	+

[a]−, low; ±, medium; +, high; ++, very high.

chemical method may be technically complex and very time consuming. None of these traditional methods are capable of determining the distribution of the numbers of drug molecules in the carriers.

Because of the limitations posed by traditional techniques for determining loading values, matrix-assisted UV- and IR-laser desorption/ionization mass spectrometry (UV- and IR-MALDI-MS) *(6,7)* and electrospray ionization mass spectrometry (ESI-MS) *(8)*, which have both recently been developed for the analysis of high-mass molecules, have been evaluated by a number of mass spectrometrists for the determination of loading values and the distributions of drugs conjugated to carriers. In general, UV- and IR-MALDI-MS and ESI-MS have been found to be routine, reliable, and general methods for the determination of loading and distribution values. Furthermore, the MALDI and ESI mass spectral methods can be utilized to produce responses selectively from only covalently bound drug in samples where covalently and noncovalently bound drug may be present and difficult or impossible to differentiate using the more traditional methods. A comparison of the MALDI-MS, ESI-MS, UV absorption spectrometry, immunoassay, and radiolabeling methods for the determination of loading and distribution values of drugs conjugated to carriers is summarized in Table 1.

2. Materials

The chemical manipulations necessary to produce protein conjugates as well as the mass spectrometric instrumentation used to perform the measurements for determining loading and distribution values are fully described in the relevant references to the studies described herein. All the MALDI-MS and ESI-MS work which is described utilized time-of-flight (TOF) and quadrupole

Table 2
Examples of Protein and Antibody Drug Conjugate Structures[a]

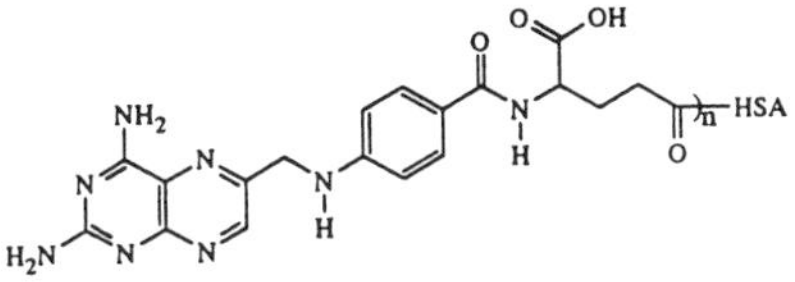

1 HSA-aminopterin

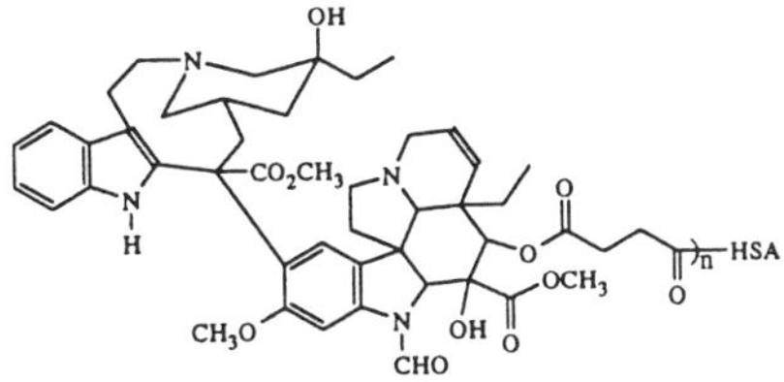

2 HSA-DAVCR

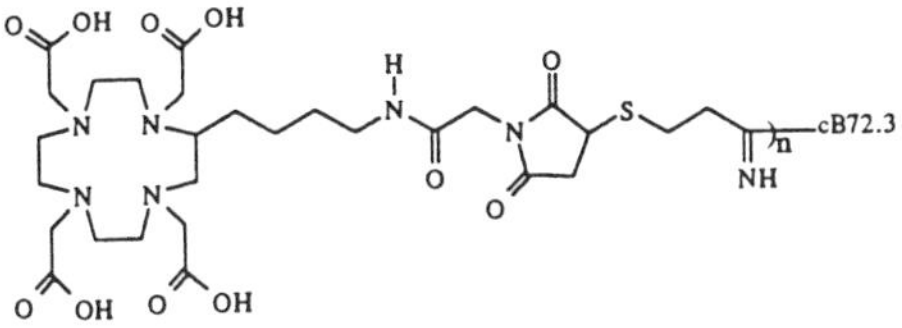

3 cB72.3 MAb-macrocycle 12N4-maleimide

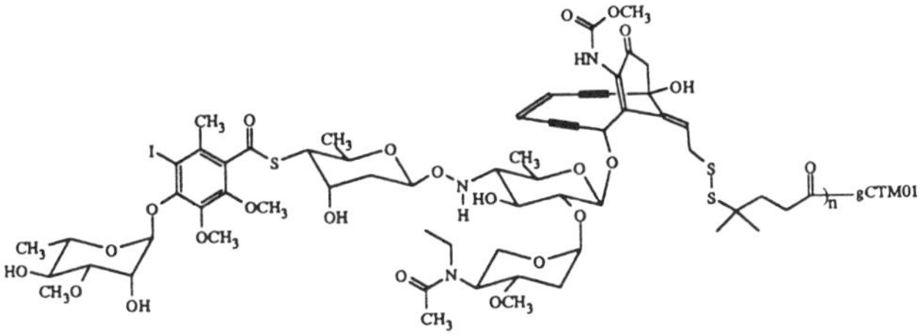

4 gCTM01 MAb-calicheamicin DMA

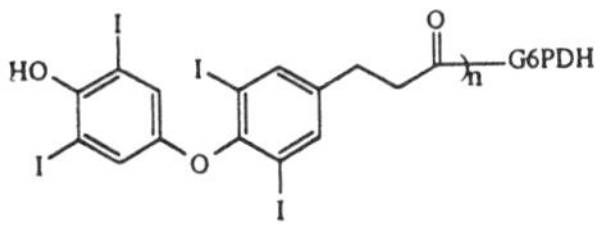

5 G6PDH-desaminothyroxine

[a]The drugs are covalently bound to the γ-amino group of the lysine residues of the protein.

mass analyzers, respectively. Table 2 lists the chemical structures of the conjugated drugs whose MALDI or ESI mass spectra are reviewed.

3. Methods and Discussion

3.1. Determination of Average Loading Values from MALDI-MS Data

The MALDI-MS average loading value is calculated (Eq. [1]) from the difference between the measured, chemically averaged, mass values for the conjugated (mol wt_{av} [conj.]) and pure proteins (mol wt_{av} [prot.]) divided by the calculated change in chemical mass of the protein when conjugated with one drug molecule (Δ mol wt [conj.]). The most prominent ions observed in MALDI mass spectra (e.g., Fig. 1A) are the molecular ions M^{1+}, M^{2+}, and the less abundant M^{3+}. Other, often observed, ions of low abundance are the dimer ions $2M^{1+}$ and $2M^{3+}$ and the trimer ion $3M^{2+}$. The mass centroid is computed for each of these molecular ions. The averaged chemical masses for the conjugated and pure proteins are computed from the product of the respective charge and the centroided *m/z* value for the M^{1+}, M^{2+} and M^{3+}, ions (Eq. [2]).

Average loading value = (mol wt_{av} [conj.] − mol wt_{av} [prot.]) / Δ mol wt [conj.] (1)

mol $wt_{av}[x]$ = 1/3 {*m/z* (M^+) + 2 [*m/z* (M^{2+})] +3 [*m/z* (M^{3+})]} (2)

where x = conj. or prot.

Figure 1A illustrates a typical UV-MALDI mass spectrum of human serum albumin (HSA) with an average mol wt of 66,440 ± 50 Da, and Fig. 1B and C illustrates the UV-MALDI mass spectra of the HSA protein conjugated with the anticancer agents aminopterin, **1**, and DAVCR, **2**, respectively, having average mol wt values of 74,020 ± 40 Da (HSA-aminopterin) and 87,540 ± 270 Da (HSA-DAVCR). The mass differences between the conjugated and pure proteins (7590 ± 90 and 21,100 ± 280 Da, respectively) when divided by the mass change of the protein on conjugation with 1 mol of drug (422 and 865 Da, respectively) correspond to loadings of 18.0 ± 0.2 and 24.4 ± 0.3, respectively. These values compare favorably with loading values of 15.9 and 18.6, obtained by UV absorption spectrometry methods.

It should be noted that the molecular ion peaks are quite broad. The M^+ peak for HSA has a full width at half-maximum intensity (FWHM) of 1700 Da, which originates from the heterogeneity of the sample *(9–11)* and the low resolution of the mass spectrometer (*see* Fig. 1A'). The peak widths of the M^+ peaks for the HSA–drug conjugates are even larger (HSA-aminopterin FWHM: 3600 Da; HSA-DAVCR: FWHM 12,000 Da), as illustrated in Fig. 1B' and 1C', and are a result of the broad distribution in the number of drug molecules conjugated to HSA. Because the resolution achieved at high masses with the UV-MALDI-TOFMS was only ~40, which is insufficient to resolve the individual anticancer drug components, the distribution of the anticancer drug molecules cannot be directly measured. Additional HSA and bovine serum albumin (BSA) protein conjugates studied in ref. *12* are HSA-methotrexate (anticancer agent), HSA-lactose (sugar), HSA-helicin (sugar), BSA-FLT (anti-AIDS drug), and BSA-CL184005 (platelet-activating factor antagonist). Reliable loading values were obtained for all these materials by UV-MALDI-MS. No routine analytical method other than MALDI-MS is presently available for determining the average drug-loading values for the BSA–FLT and BSA-CL184005 conjugates.

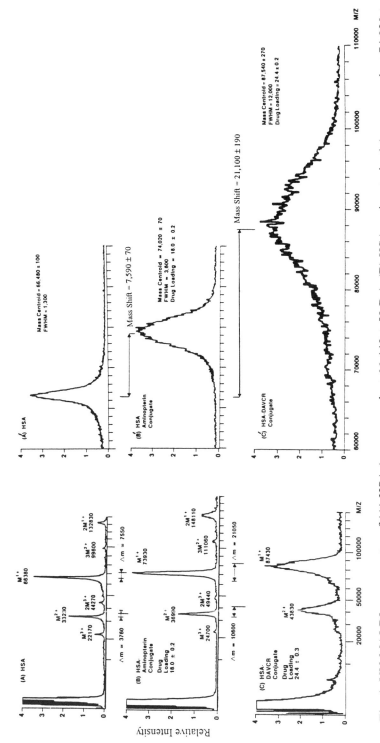

Fig. 1. UV-MALDI mass spectra of **(A)** HSA (average mol wt 66.440 ± 50 Da), **(B)** HSA-aminopterin, **1** (average mol wt 74,020 ± 40 Da) with an average loading of 18.0 ± 0.2, and **(C)** HSA-DAVCR, **2** (average mol wt 87,540 ± 270 Da), with an average loading of 24.4 ± 0.3. Exploded views of the respective M[1+] peaks are illustrated in figures **(A')**, **(B')**, and **(C')**. Reprinted with permission from ref. *12*.

A number of researchers have studied similar conjugate systems by UV-MALDI-MS to determine average loading values for protein conjugates with molecular weights ranging from 40–70 kDa. Protein–pesticide conjugates as well as protein–drug conjugates were prepared for the purpose of producing antibodies against the respective pesticide or drug. Wengatz et al. *(13)* measured the loading values for a variety of protein–pesticide conjugates. The protein carriers used included BSA, ovalbumin (OVA), horseradish peroxidase (POD), and alkaline phosphatase (AP) and the pesticides and related materials included pyrethroids, 3-phenoxybenzoic acid, methabenzthiazurons, and atrazine-aminohexanoic acid. The UV-MALDI-MS loading results were considerably lower than the loading values obtained by UV absorption spectrometry. This was rationalized as probably the result of the measurement of only covalent conjugates by UV-MALDI-MS and the measurement of both covalent and noncovalent conjugates by UV absorption spectrometry. No additional experimental corroboration of the proposed noncovalent binding effects was presented. Shoyama and coworkers also used UV-MALDI-MS to determine the loading values of BSA-forskolin *(14)* and BSA-opium alkaloids *(15)*, including morphine, codeine, and thebaine, which do not have specific UV absorbances. Adamczyk et al. *(16)* obtained the loading values of a variety of BSA–drug conjugates (BSA–tricyclic antidepressant drug conjugates, BSA–steroid related conjugates, a BSA–homovanillic acid conjugate, and a BSA–3-methoxy-4-hydroxyphenylglycol conjugate) by UV-MALDI-MS, trinitrobenzenesulfonic acid titration, and by UV absorption spectrometry. The MALDI-MS loading values were generally consistent with the values obtained by titration, but were generally inconsistent with the loading values obtained by UV absorption spectrometry. Unusual features observed in two samples analyzed by UV-MALDI-MS were bimodal distributions of the drug conjugates where the higher mass distribution corresponded to the maximum possible number of drugs that could conjugate with BSA. The bimodal distributions suggested that the conjugates were produced under conditions where the protein became partially denatured.

In a similar fashion, the loading values of chelating agents and anticancer drugs, covalently bound to monoclonal antibodies (MAbs) with molecular weights of ~150 kDa, were determined by UV-MALDI-MS *(17)*. The pharmaceutical use of these antibody conjugates is for targeting tumor sites with chelated radioactive metal ions for imaging or therapy and with anticancer drugs for chemotherapy *(1,2)*. Figure 2A illustrates a typical UV-MALDI mass spectrum of a purified MAb, chimeric B72.3, with an average mol wt of 147,430 ± 50 Da, and Fig. 2B illustrates the UV-MALDI mass spectrum of cB72.3 conjugated with chelator 12N4-maleimide, **3,** with an average mol wt of 152,180 ± 50 Da. The mass difference between the conjugated and pure

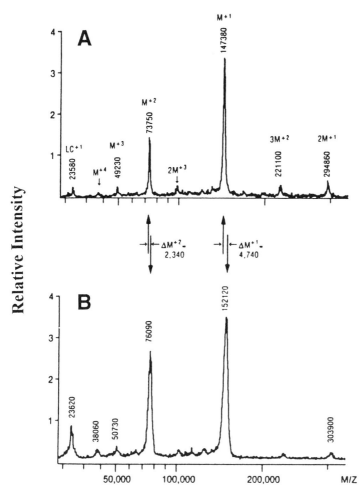

Fig. 2. **(A)** UV-MALDI mass spectrum of chimeric B72.3 MAb with an average mol wt of 147,430 ± 50 Da. **(B)** UV-MALDI mass spectrum of chimeric B72.3 conjugated with chelator 12N4-maleimide, **3**, with an average mol wt of 152,180 ± 50 Da. Note the respective mass differences between the singly and doubly charged conjugated and pure antibody molecules correspond to an average value of 4750 ± 50 Da, which is equivalent to an average drug loading of 6.5 ± 0.1. Reprinted with permission from ref. *17*.

antibody molecules (4,750 ± 50 Da) when divided by the mass change equivalent to conjugation with 1 mol of chelator (727.8 Da) corresponds to an average loading value of 6.5 ± 0.1. This value compares favorably with an average loading value of 4.7 obtained by a titration assay. Extensive studies of MAbs conjugated with stable molecules produced reliable UV-MALDI-MS loading values when compared with loading values obtained by UV absortion spectrometry and radioactivity measurements. Examples of other chelating agents conjugated to antibodies studied by UV-MALDI-MS were derivatives of diethylenetriaminepentaacetic acid (DTPA), macrocycle 12N4 (1,4,7,11-tetraazacyclododecane tetraacetic acid), and macrocycle 12N4P4 (1,4,7,11-tetraazacyclododecane tetraphosphinic acid). Stable anticancer agents conjugated to antibodies successfully studied by UV-MALDI-MS included mitoxantrone analogs and methotrexate hydrazide.

Mabs conjugated with the chelating agents described in the above paragraph were chemically treated with dithiothreitol to produce light-chain and heavy-chain conjugates. Each of the mixtures was analyzed by UV-MALDI-MS without isolation and purification and the distribution of the chelating agent between the light and heavy chains was determined *(17)*. In a related study, Alexander et al. *(18)* determined the loading value for the anticancer agent doxorubicin conjugated to the purified heavy-chain fragment of a chimeric MAb by UV-MALDI-MS. Similar UV-MALDI-MS measurements for the purified light chain showed no loading of doxobubicin, whereas UV absorption spectrometry indicated the presence of doxorubicin. This inconsistency of the MALDI-MS and UV absortion spectrometry results has not been resolved by the authors.

3.2. Determination of Distributions and Average Loading Values from Partially Resolved MALDI-MS and ESI-MS Data

UV-MALDI-MS loading values of MAbs conjugated with derivatives of the potent anticancer agent calicheamicin, **4**, were not measureable or were considerably less than the values obtained by immunoassay, UV absorption spectrometry, and biological assays *(19,20)*. Recent studies have demonstrated that, whereas calicheamicin and calicheamicin conjugates are labile under UV-laser irradiation, reliable calicheamicin loading values, as well as drug distributions, can be obtained by MALDI-MS when using IR-lasers *(21,22)*. Figure 3A-1, -2, and -3 illustrates the peak profiles of the singly, doubly, and triply charged molecular ions of pure gCTM01 MAb, respectively, obtained by

---

Fig. 3. *(opposite page)* **(A-1, -2, and -3)** The IR-MALDI mass spectra of singly, doubly, and triply charged pure gCTM01 MAb and mathematically fitted peak profiles, respectively. **(B-1, -2, and -3)** The IR-MALDI mass spectra of singly, doubly, and triply charged calicheamicin-DMA conjugated to gCTM01 MAb, **4**, respectively. Note the modulation of the mass spectral peaks and the deconvoluted peak profiles for each of the drug conjugates contributing to the drug distribution. Also, included with each figure is the relative difference (percent error) between the experimental and fitted data. Reprinted with permission from ref. *21*.

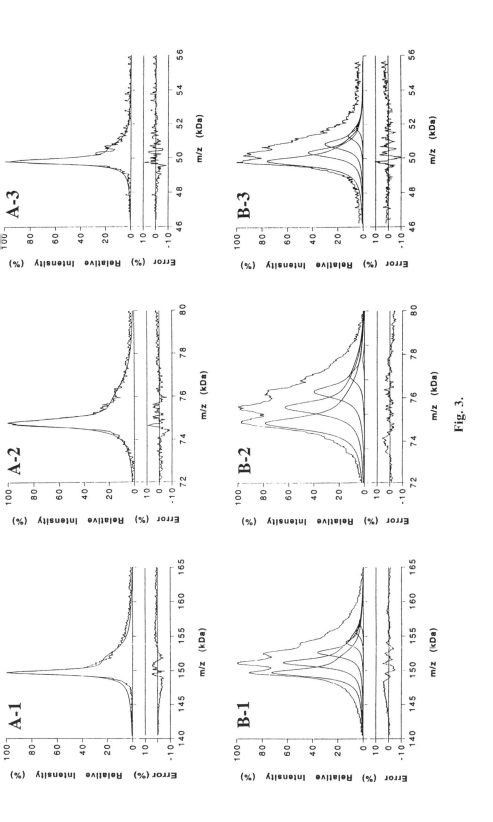

Fig. 3.

IR-MALDI-MS. Likewise, Fig. 3B-1, -2, and -3 illustrates the corresponding peak profiles of the singly, doubly, and triply charged molecular ions of calicheamicin-DMA (calicheamicin *N*-acetyl γ_1^I-dimethyl butyric acid) conjugated to gCTM01 MAb, **4**, respectively. Note the modulation of the peak profiles of the calicheamicin-DMA conjugates. The mass difference between adjacent maxima in the peak modulation is ~1460 Da corresponding to the conjugation of a calicheamicin-DMA molecule to the antibody, i.e., the peak maximum at lowest mass corresponds to no conjugated drug and each maximum with increasing mass corresponds to an additional conjugated drug molecule. Included with each Fig. 3A is the fitted peak profile and with each Fig. 3B is the deconvoluted spectrum for each of the drug conjugates, which corresponds to the drug distribution in the MAb. (Also included with each figure is the relative difference [percent error] between the experimental and fitted data.) The computed distribution of the conjugated drugs is as follows: no conjugated drug molecule 41%, one conjugated drug molecule 30%, two conjugated drugs 20%, three conjugated drugs 7%, and four conjugated drugs 2%. The average loading value obtained from the IR-MALDI-MS data is 1.1, which compares favorably to 1.9 obtained from UV absorption spectrometry data. The ability to determine the distribution of conjugated drugs from the IR-MALDI mass spectrum was possible owing to the achievement of a resolution of ~140, which is about twice the resolution observed in comparable UV-MALDI measurements at 150 kDa.

Recent ESI-MS results with gCTM01-calicheamicin-DMA conjugate, **4**, under higher resolution conditions with a quadrupole analyzer *(22)* produced a loading of 1.3 with a distribution for 0, 1, 2, 3, and 4 conjugated drugs of 38, 22, 21, 13, and 6%, respectively. These ESI-MS results are less reliable than the IR-MALDI-MS results owing to the difficulty in resolving the overlapping multiply charged peaks that arise from the heterogeneity of the covalently bound calicheamicin-DMA drug and the polysaccharide in the MAb.

3.3. Determination of Distributions and Average Loading Values from Fully Resolved MALDI-MS and ESI-MS Data

When the molecular weight of the conjugating molecule is large enough and the mass spectrometer resolution is high enough, the distribution of the conjugates can be directly measured without resorting to mathematical deconvolution, since each component is fully resolved. The contribution of each component in the distribution is computed from the corresponding integrated peak intensity. The underlying assumption is that the instrumental response is independent of the number of conjugates attached to the carrier molecule. The average loading value is then simply the number average for the distribution.

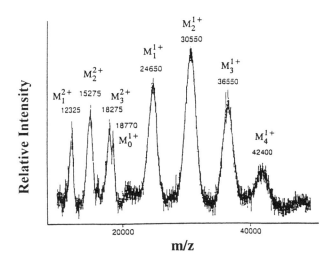

Fig. 4. UV-MALDI mass spectrum illustrating the distribution of PEG polymers (average mol wt 6000 Da) conjugated to recombinant SCF protein (mol wt 18,796 Da). Note the separation of the M^{1+} ions by ~6000 Da and the M^{2+} ions by ~3000 Da consistent with loadings of increasing numbers of conjugated PEG polymers. The n subscripts for the charged molecular ions, e.g., M^{z+}_{n}, correspond to the number of PEG polymers conjugated to each SCF protein molecule. Reprinted with permission from ref. *25*.

Proteins with polyethylene glycol (PEG) polymers covalently attached have been shown to have longer circulating half-lives and reduced immunogenicity than the unmodified protein *(23,24)*. Using UV-MALDI-MS, Watson et al. *(25)* demonstrated that the distribution of the number of PEG units (each with an average mol wt of ~6000 Da) covalently attached to recombinant stem-cell factor (SCF) protein (mol wt 18,796 Da) can be routinely determined. Figure 4 illustrates the UV-MALDI spectrum of SCF-PEG and clearly shows M^{1+} and M^{2+} ions corresponding to the attachment of 0 to 4 PEG units to the SCF protein. ESI-MS utilizing a quadrupole analyzer is unlikely to generate successfully a resolved spectrum because of the large number of overlapping peaks owing to multiple charging and heterogeneity of the PEG oligomers.

On the other hand, if the protein and the covalently attached molecules are initially available as pure monomeric materials and the protein is covalently bound to a low number of molecules having a reasonably sharp distribution, ESI-MS with a quadrupole analyzer can be used to determine the average loading and distribution of drugs covalently attached to a protein. Straub and Levy *(26)* utilized ESI-MS to determine the distribution and average loading values of drugs (desaminothyroxine, digoxigenin, valproic acid, carbamazepine) covalently bound to glucose 6-phosphate dehydrogenase (G6PDH, mol wt

54.5 kDa). Figure 5A,B illustrates the ESI mass spectra for pure G6PDH and G6PDH–thyroxine conjugate, **5**, respectively, and, Fig. 5A',B' illustrates the respective spectra transformed to a mol wt mass axis. Note the distribution of 0 to 4 conjugated drugs with the predicted mass difference for the drug of 744 Da between the peaks in the transformed spectrum (Fig. 5B'). The average ESI-MS loading value corresponds to 1.2, which is consistent the loading value of 1.5 obtained by radioactive labeling experiments. Similar consistent results were obtained for the other three G6PDH–drug conjugates.

In a related work, Le Blanc and Siu *(27)* applied ESI-MS to the analysis of the reaction products of proteins with ketones. The reactions of bovine ubiquitin and equine myoglobin with acetone to form Schiff bases were monitored. The relative ion abundances of the Schiff bases were found to follow expected statistical distributions. Using nonenzymatic methods to produce glycated proteins, UV-MALDI-MS was used to determine the average number of glucose molecules ("loading values") bound to proteins *(28,29)*. Other protein conjugates studied by ESI-MS, where resolved distributions were observed, are hemoglobin–acrylamide conjugates *(30)*, hemoglobin–acrylonitrile conjugates *(31)*, myoglobin–quinone methide conjugates *(32)*, and β–lactoglobulin B–4-hydroxy-2-nonenal conjugates *(33)*.

4. Conclusions

MALDI-MS and ESI-MS have been successfully used to measure rapidly, routinely, and reliably the average loading values and distributions of small molecules (drugs, sugars, pesticides, chelators, polymers, and pollutants) covalently bound to large molecules (proteins). It is anticipated that, in the near future, these mass spectrometric methods will also be applied to the measurement of loading values and distributions for protein–oligonucleotide conjugates *(34)*, protein–peptide conjugates, protein–polysaccharide conjugates, and protein–protein conjugates. The reliability of the MALDI and ESI mass spectral techniques for determining average loading values has been confirmed in many cases by correlation with UV absorption spectrometry, radiochemical, wet chemical, and chromatographic techniques for determining average loading values. The determination of the distributions of small molecules covalently attached to large molecules can be obtained from MALDI and ESI mass spec-

---

Fig. 5. *(opposite page)* ESI mass spectra of **(A)** pure G6PDH and **(B)** G6PDH-desaminothyroxine conjugate, **5**. The respective spectra, transformed to a mol-wt mass axis, are illustrated as **(A')** and **(B')**. Note that the measured mol wt of pure G6PDH is 54,477 ± 5 Da (A') and the distribution of the desaminothyroxine conjugates range from 0 to 4 molecules (B') as exhibited by the mass differences of 744 Da for each desaminothyroxine molecule. Reprinted with permission from ref. *26*.

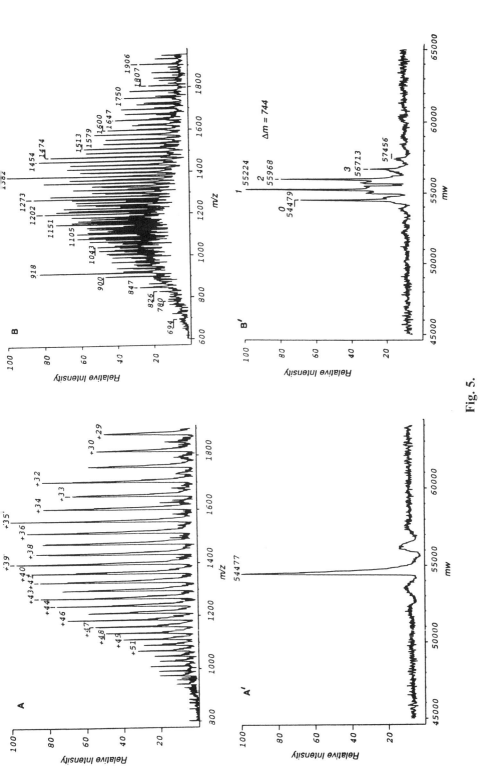

Fig. 5.

tral data, provided the resolution needed to achieve mass separation of the components is available. Of all the analytical methods, MALDI and ESI mass spectrometry are the most rapid and reliable methods for distribution analysis. It is anticipated that with further technical improvements in UV- and IR-MALDI-MS and ESI-MS, the mass spectrometric method for the determination of loading values and distributions may become the analytical method of choice.

Acknowledgments

The author greatfully appreciates the assistance of Irwin Hollander in critically reviewing the manuscript and Keiko Tabei in preparing the figures for publication.

References

1. Collier, R. J. and Kaplan, D. A. (1984) Immunotoxins. *Sci. Am.* **251,** 56–64.
2. Koppel, G. A. (1990) Recent advances with monoclonal antibody drug targeting for the treatment of human cancer. *Bioconj. Chem.* **1,** 13–23.
3. Butler, V. P., Jr. (1977) The immunological assay of drugs. *Pharmacological Rev.* **29,** 103–184.
4. Landon, J. and Moffat, A. C. (1976) The radioimmunoassay of drugs: a review. *Analyst* **101,** 225–243.
5. Park, B. K. and Kitteringham, N. R. (1990) Drug–protein conjugation and its immunological consequences. *Drug Metab. Rev.* **22,** 87–144.
6. Karas, M., Bahr, U., and Giessmann, U. (1991) Matrix-assisted laser desorption ionization mass spectrometry. *Mass Spectrom. Rev.* **10,** 335–357.
7. Hillenkamp, F., Karas, M., Bevis, R. C., and Chait, B. T. (1991) MALDI/MS of biopolymers. *Anal. Chem.* **63,** 1193A–1203A.
8. Smith, R. D., Loo, J. A., Ogorzalek Loo, R. R., Busman, M., and Udseth, H. R. (1991) Principles and practice of electrospray ionization-mass spectrometry for large polypeptides and proteins. *Mass Spectrom. Rev.* **10,** 359–452.
9. Geisow, M. J. and Green, B. N. (1991) An introduction to protein electrospray mass spectrometry. *BioSpec* (VG Biotech Electrospray Application Notes).
10. Geisow, M. J., Harris, R., Dodsworth, N., Green, B. N., and Hutton, T. (1991) Characterization and quality assurance of recombinant human serum albumin by ESI/MS, in *Techniques in Protein Chemistry, II* (Villafranca, J. J., ed.), Academic, San Diego, CA, pp. 567–572.
11. Loo, J. A., Edmonds, C. G., and Smith, R. D. (1991) Tandem mass spectrometry of very large molecules: serum albumin sequence information from multiply charged ions formed by electrospray ionization. *Anal. Chem.* **63,** 2488–2499.
12. Siegel, M. M., Tsou, H.-R., Lin, B., Hollander, I. J., Wissner, A., Karas, M., Ingendoh, A., and Hillenkamp, F. (1993) Determination of the loading values for high levels of drugs and sugars conjugated to proteins by MALDI/MS. *Biol. Mass Spectrom.* **22,** 369–376.

13. Wengatz, I., Schmid, R. D., Kreissig, S., Wittmann, C., Hock, B., Ingendoh, A., and Hillenkamp, F. (1992) Determination of the hapten density of immuno-conjugates by UV-MALDI/MS. *Anal. Lett.* **25**, 1983–1997.

14. Shoyama, Y., Sakata, R., and Murakami, H. (1993) Direct determination of forskolin-bovine serum albumin conjugate by UV-MALDI/MS. *Org. Mass Spectrom.* **28**, 987,988.

15. Shoyama, Y., Fukada, T., Tanaka, T., Kusai, A., and Nojima, K. (1993) Direct determination of alkaloid-bovine serum albumin conjugate by UV-MALDI/MS. *Biol. Pharm. Bull.* **16**, 1051–1053.

16. Adamczyk, M., Buko, A., Chen, Y.-Y., Fishpaugh, J. R., Gebler, J. C., and Johnson, D. D. (1994) Characterization of protein-hapten conjugates. 1. UV-MALDI/MS of immuno BSA-hapten conjugates and comparison with other characterization methods. *Bioconj. Chem.* **5**, 631–635.

17. Siegel, M. M., Hollander, I. J., Hamann, P. H., James, J. P., Hinman, L., Smith, B. J., Farnsworth, A. P. H., Phipps, A., King, D. J., Karas, M., Ingendoh, A., and Hillenkamp, F. (1991) UV-MALDI/MS analysis of monoclonal antibodies for the determination of carbohydrate, conjugated chelator, and conjugated drug content. *Anal. Chem.* **63**, 2470–2481.

18. Alexander, A. J., Dodsworth, D. W., Kirkley, D. H., Root, B., and Tous, G. I. (1992) Characterization of the Doxorubicin Conjugate of a Chimeric Monoclonal Antibody by UV-MALDI/MS. *Proceedings of the 40th ASMS Conference on Mass Spectrometry and Allied Topics*, Washington, DC, May 31–June 5, pp. 941,942.

19. Hinman, L. M., Wallace, R. E., Hamann, P. R., Durr, F. E., and Upeslacis, J. (1990) Calicheamicin immunoconjugates: influence of analog and linker modification on activity *in vivo. Antibody, Immunoconj., Radiopharm.* **3(Abstr. 84)**, 59.

20. Wallace, R. E., Hinman, L. M., Hamann, P. R., Lee, M., Upeslacis, J., and Durr, F. E. (1990) Evaluation of calicheamicin immunoconjugates in human tumor xenografts. *J. Cancer Res. Clin. Oncol.* **116 (Suppl. Part I)**, 323.

21. Siegel, M. M., Kunz, A., Bell, D., Berkenkamp, S., Karas, M., Ingendoh, A., and Hillenkamp, F. (1994) Determination of Distribution and Loading Values for Calicheamicin Derivatives Conjugated to Antibodies by IR-MALDI/MS. *Proceedings of the 42nd ASMS Conference on Mass Spectrometry and Allied Topics*, Chicago, IL, May 29–June 3, p. 967.

22. Siegel, M. M., Tabei, K., Kunz, A., Berkenkamp, S., and Hillenkamp, F. (1995) Distribution and Loading of Calicheamicin Derivatives Conjugated to Monoclonal Antibodies (MW ~150 kDa) by IR-MALDI/MS and ESI/MS. *Proceedings of the 43rd ASMS Conference on Mass Spectrometry and Allied Topics*, Atlanta, GA, May 21–26, p. 307.

23. Katre, N. V., Knauf, M. J., and Laird, W. J. (1987) Chemical modification of recombinant interlukin-2 by polyethylene glycol increases its potency in the murine meth A sarcoma model. *Proc. Natl. Acad. Sci. USA* **84**, 1487–1491.

24. Katre, N. V. (1990) Immunogenicity of recombinant IL-2 modified by covalent attachment of polyethylene glycol. *J. Immunol.* **144**, 209–213.

25. Watson, E., Shah, B., DePrince, R., Hendren, R. W., and Nelson, R. (1994) UV-MALDI mass spectrometric analysis of a pegylated recombinant protein. *BioTechniques* **16,** 278–281.

26. Straub, K. M. and Levy, M. J. (1994) Characterization of hapten binding to immunoconjugates by ESI/MS. *Bioconj. Chem.* **5,** 194–198.

27. Le Blanc, J. C. Y. and Siu, K. W. M. (1994) Electrospray mass spectrometric study of protein-ketone equilibria in solution. *Anal. Chem.* **66,** 3289–3296.

28. Lapolla, A., Gerhardinger, C., Baldo, L., Fedele, D., Keane, A., Seraglia, R., Catinella, S., and Traldi, P. (1993) A study on in vitro glycation processes by MALDI/MS. *Biochim. Biophys. Acta* **1225,** 33–38.

29. Lapolla, A., Fedele, D., Seraglia, R., Catinella, S., and Traldi, P. (1994) MALDI capabilities in the study of non-enzymatic protein glycation. *Rapid. Commun. Mass Spectrom.* **8,** 645–652.

30. Springer, D. L., Bull, R. J., Goheen, S. C., Sylvester, D. M., and Edmonds, C. G. (1993) Electrospray ionization mass spectrometric characterization of acrylamide adducts to hemoglobin. *J. Toxicol. Environ. Health* **40,** 161–176.

31. Stevens, R. D., Bonaventura, J., Bonaventura, C., Fennel, T. R., and Millington, D. S. (1994) Application of ESI/MS for analysis of haemoglobin adducts with acrylonitrile. *Biochem. Soc. Trans.* **22,** 543–547.

32. Bolton, J. L., Le Blanc, J. C. Y., and Siu, K. W. M. (1993) Reaction of quinone methides with proteins: analysis of myoglobin adduct formation by ESI/MS. *Biol. Mass Spectrom.* **22,** 666–669.

33. Bruenner, B. A., Jones, A. D., and German, J. B. (1994) Maximum entropy deconvolution of heterogeneity in protein modification: protein adducts of 4–hydroxy-2-nonenal. *Rapid Commun. Mass Spectrom.* **8,** 509–512.

34. Bunnell, B. A., Askari, F. K., and Wilson, J. M. (1992) Targeted delivery of antisense oligonucleotides by molecular conjugates. *Somatic Cell and Mol. Genet.* **18,** 559–569.

16

Hydrophobic Proteins and Peptides Analyzed by Matrix-Assisted Laser Desorption/Ionization

Kevin L. Schey

1. Introduction

Hydrophobic proteins are vital components of every living cell. Often located in or associated with cell membranes, their functions vary from ion or molecule transport to cell recognition to signal transduction. Owing to their limited solubility in aqueous solvents, structural analysis by conventional techniques has been problematic. Moreover, studies of hydrophobic peptides from these and other proteins have found these materials to be equally difficult to analyze.

Mass spectrometry (MS) has become a valuable tool for peptide and protein analysis; however, to date, most successful analyses have involved water-soluble species. The versatility of matrix-assisted laser desorption/ionization (MALDI) MS makes it a viable alternative for the analysis of water-insoluble proteins. More specifically, the attractive features of MALDI for the analysis of hydrophobic species include its insensitivity to contaminants, including salts, lipids, and some detergents, as well as the hydrophobic character of the matrices employed. Although the mechanism of protein selectivity in MALDI analysis remains unclear, the cocrystallization of matrix and protein (and therefore sample preparation) clearly plays an important role in successful analyses (1).

Sample preparation protocols are presented in this chapter for the analysis of hydrophobic peptides and proteins by MALDI-MS. The strategy employed involves solubilization of both analyte and matrix in an appropriate solvent (2,3). This approach provides two alternatives for sample preparation: solubilization in strong organic solvents, such as formic acid, or solubilization in detergent solution (cf Chapter 11). Once solubilization of matrix and protein has been achieved, standard methods of MALDI analysis can be employed. However, particular attention must be paid to deleterious aspects of using

From: *Methods in Molecular Biology, Vol. 61: Protein and Peptide Analysis by Mass Spectrometry*
Edited by: J. R. Chapman Humana Press Inc., Totowa, NJ

strong organic solvents (*see* Notes 1 and 2). The development of MS for the analysis of hydrophobic peptides and proteins provides an extremely valuable tool in structural studies of this important class of proteins.

2. Materials

1. 50 m*M* Sinapinic acid (3,5-dimethoxy-4-hydroxycinnamic acid, Aldrich, Milwaukee, WI) solution in 70% formic acid.
2. A 7:3 (v/v) formic acid:hexafluoroisopropanol (Brand-Nu Laboratories, Meriden, CT) solution.
3. Protein/peptides:
   a. Integral membrane proteins typically in the form of washed membranes suspended in deionized water (≥ 1 mg/mL) (*see* Note 3).
   b. Lyophilized or precipitated protein or peptide

3. Method

1. Solubilize 2 μL of washed membrane preparation in 5 μL formic acid/hexafluoroisopropanol solution or solubilize lyophilized or precipitated protein samples in formic acid/hexafluoroisopropanol solution to a concentration of ≥ 4 pmol/μL (*see* Notes 1 and 2).
2. Place the sealed container in an ultrasonic water bath for 2–5 min. This procedure significantly assists solubilization of the membrane preparation.
3. Mix 1 μL of protein solution with 3 μL sinapinic acid solution (*see* Note 4).
4. Place 1 μL of protein/matrix solution on the MALDI probe, and allow to air-dry (Fig. 1).
5. During MALDI analysis, focus laser on ring of dried matrix (*see* Note 5).

4. Notes

1. A potential problem with using formic acid as a solvent is its reactivity toward serine and threonine residues in proteins. For small peptides, the author has observed formyl esterification of these amino acid residues in the form of satellite peaks at 28-Da intervals of higher molecular weight. Typically, the intensities of these products are <50% of the $(M + H)^+$ ion signal. Although such additions to large proteins cannot be resolved, the author has not observed any significant shifts in measured molecular weights in the time frame required to prepare and analyze a sample. Typical errors in mol-wt determination are 0.08% for proteins in the 20,000–40,000 Da range. If the sample remains in formic acid solution for several hours, significant shifts in molecular weight to higher mass and tailing to higher masses appears, most likely because of extensive formylation. Attempts at eliminating formic acid from the protocol by substituting isopropanol for formic acid have produced less intense and variable MALDI signals. A protocol has been reported for deformylation of formylated peptides generated by cyanogen bromide (CNBr) cleavage in formic acid by treatment with ethanolamine *(4)*. This procedure has not been attempted in the author's MALDI experiments with CNBr digests.

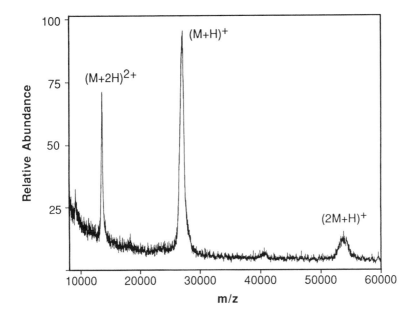

Fig. 1. MALDI spectrum of bacteriorhodopsin ($M_r \approx 27{,}000$).

2. A further complication of using strongly acidic solvents is cleavage of acid-labile peptide bonds, such as the aspartic acid–proline bond. Cleavage of this bond in small and large proteins has been observed within 1 h after sample preparation, and cleavage products increase in intensity with time.

3. For membrane preparations, it is helpful to start with a concentrated protein preparation and dilute in formic acid/hexafluoroisopropanol solution. Diluting the protein solution very often improves the MALDI signal, perhaps by diluting contaminating lipids while the matrix concentrates the protein. Detection limits in the femtomole range have been achieved by this method.

4. The sample preparation method of Vorm et al. *(5)*, which involves depositing sinapinic acid from a saturated acetone solution prior to spotting the protein solution, has given variable results. It is worthwhile attempting, because in some cases this method provides dramatic improvements in signals for hydrophobic proteins.

5. The surface tension of the protein/matrix solution is such that a ring of matrix/ protein crystals forms on the sample surface in contrast to the uniform crystal surface observed in standard MALDI sample preparations. The formation of this ring effectively concentrates the protein on the sample surface, and it is critical that the MALDI laser be focused onto the ring of matrix crystals. In the case of integral membrane proteins, the ring of crystals appears to be "wax-coated" owing to the large amount of lipid present in the membrane preparation. In practice, higher laser powers are often required to produce signals from integral mem-

brane protein samples. Attempts to wash the dried sample with hexane or chloroform:methanol (1:1, v/v) to remove the lipid coating have produced, at best, only modest improvements in MALDI signals. Use of other solvents, such as hexafluoroisopropanol alone or trifluoroacetic acid, produces lower surface tension solutions resulting in sample flowing off the sample probe. In addition, no signals were obtained with the use of α-cyano-4-hydroxycinnamic acid or 2,5-dihydroxybenzoic acid matrices. This protocol has been used successfully in the analysis of hydrophobic peptides and proteins as large as 40,000 Da *(2)*. The amount of protein on the sample probe tip is usually in the 1–10 pmol range. A typical MALDI spectrum from this class of compound is illustrated in Fig. 1.

6. Calibration of the mass scale is accomplished by using a standard MALDI sample preparation protocol with protein or peptide standards chosen to bracket the molecular weight of the analyte. The calibrant solution can be placed on top of the hydrophobic sample or, alternatively, spotted on another portion of the sample surface.

References

1. Strupat, K., Karas, M., and Hillenkamp, F. (1991) 2,5-Dihydroxybenzoic acid: a new matrix for laser desorption-ionization mass spectrometry. *Int. J. Mass Spectrom. Ion Processes* **111,** 89–102.
2. Schey, K. L., Papac, D. I., Knapp, D. R., and Crouch, R. K. (1992) Matrix-assisted laser desorption mass spectrometry of rhodopsin and bacteriorhodopsin. *Biophys. J.* **63,** 1240–1243.
3. Salehpour, M., Perera, I. K., Kjellberg, J., Hedin, A., Islamian, M. A., Hakansson, P., Sundqvist, B. U. R., Roepstorff, P., Mann, M., and Klarskov, K. (1990) UV-laser induced desorption mass spectrometry of hydrophobic- and glyco-proteins, in *Ion Formation from Organic Solids (IFOS V), Proc. 5th. Int. Conf., Lovanger, Sweden* (Hedin, A., Sundqvist, B. U. R., and Benninghoven, A., eds.), Wiley, Chichester, UK, pp. 119–124.
4. Tarr, G. E. and Crabb, J. W. (1983) Reversed-phase high-performance liquid chromatography of hydrophobic proteins and fragments thereof. *Anal. Biochem.* **131,** 99–107.
5. Vorm, O., Roepstorff, P., and Mann, M. (1994) Improved resolution and very high sensitivity in MALDI TOF of matrix surfaces made by fast evaporation. *Anal. Chem.* **66,** 3281–3287.

17

Analysis of Glycoproteins and Glycopeptides Using Fast-Atom Bombardment

Geert Jan Rademaker and Jane Thomas-Oates

1. Introduction

The choice of approach for analyzing glycoprotein samples using fast-atom bombardment-mass spectrometry (FAB-MS) is dependent on the information required (*see* Scheme 1). If only the carbohydrate portion of the molecule is of interest, then *N*-linked glycans may be released directly from the intact glycoprotein using peptide-N^4-(*N*-acetyl-β-glucosaminyl) asparagine amidase F (PNGase F) (Section 3.1.), while *O*-linked glycans can be cleaved on reductive β-elimination (Section 3.2.). The released glycan chains and the residual (glyco)protein can be separated from each other using a C18 reversed-phase cartridge (Section 3.3.), on which released glycans are not retained. Glycans are then recovered from the unretained fraction, or directly from the reaction mixture, following acid-catalyzed acetylation of the dried fraction and extraction of the acetylated glycans into dichloromethane (Section 3.4. and *see* Note 1).

If information is required on both the glycans and the peptides to which they are attached, then the protein should first be reduced and carboxymethylated (Section 3.5. and *see* Note 2) to remove disulfide bridges, and subsequently digested with trypsin (Section 3.6.6. and *see* Note 3) or another specific protease. A rapid indication of the site(s) of *N*-linked glycosylation can then be obtained from FAB mapping *(1,2)* (*see* Note 4). This technique involves direct FAB-MS analysis (Section 3.7.) of the mixture of peptides (*see* Note 5) and glycopeptides before and after release of the *N*-linked glycans with PNGase F. Since peptides, but not glycopeptides, ionize from such mixtures, any new peptide signal that appears following PNGase F treatment corresponds to a peptide that contains an *N*-glycosylation site.

From: *Methods in Molecular Biology, Vol. 61: Protein and Peptide Analysis by Mass Spectrometry*
Edited by: J. R. Chapman Humana Press Inc., Totowa, NJ

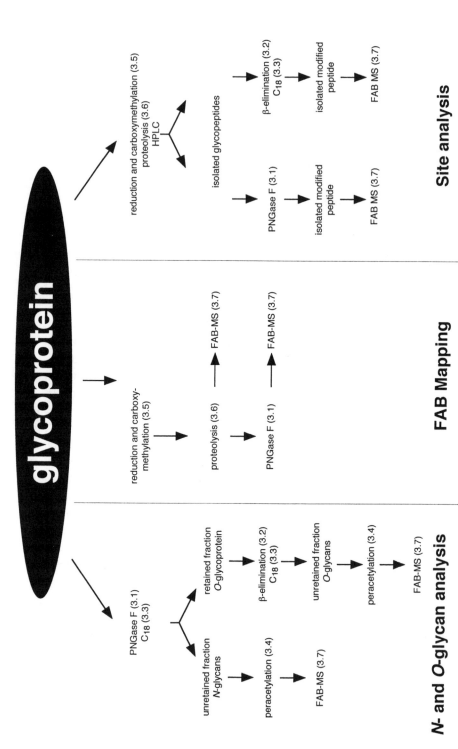

Scheme 1. General strategy for the FAB mass spectrometric analysis of glycoproteins.

If a detailed analysis of *N*- and *O*-glycosylation sites is required, the mixture of tryptic peptides and glycopeptides (Section 3.6.) may be fractionated using reversed-phase high-performance liquid chromatography (HPLC) to isolate individual glycopeptides. Although these may then be analyzed directly by FAB-MS (Section 3.7. and *see* Note 6), in order to obtain information on the site of glycan attachment it is almost always necessary first to release *N*-linked glycans with PNGase F (Section 3.1.). Direct FAB-MS analysis of the PNGase F reaction mixture is possible, since the enzyme treatment is carried out in a volatile buffer. Thus, the FAB mass spectrum of the released peptide affords the amino acid sequence in which a formerly *N*-glycosylated asparagine residue is present as aspartic acid (*see* Note 7). The *N*-glycans released with PNGase F may be purified in the same way as those released from whole glycoproteins, either on a C18 cartridge (Section 3.3.) or following acid-catalyzed acetylation and extraction of the dried reaction mixtures (Section 3.4.).

Sites of *O*-glycosylation can be determined following reductive β-elimination (Section 3.2.) of an *O*-linked glycopeptide, which simultaneously releases the carbohydrate chain and generates a modified amino acid residue in an otherwise undegraded peptide *(3)* (*see* Note 8). The modified peptide is separated from the released carbohydrate using a C18 cartridge (Section 3.3.) and the carbohydrate can be isolated from the unretained fraction. Analysis of the modified peptide, using FAB-MS, yields sequence information and allows the position of the modified amino acid, and thus the site of *O*-glycosylation, to be identified (e.g., Fig. 1).

2. Materials (*see* Note 9)

2.1. PNGase F Digestion

1. Round-bottomed glass tubes (13 × 100 mm) with Teflon™-lined screw caps.
2. 50 m*M* Ammonium bicarbonate in distilled water, pH adjusted to 8.4 using 1*M* NH_4OH.
3. PNGase F (e.g., Boehringer-Mannheim, Mannheim, Germany, cat. no. 1365-169).
4. Toluene.
5. Water bath or heating block at 37°C.
6. Freeze dryer (or vacuum pump and cold trap).

2.2. β-Elimination of O-Glycans

1. Round-bottomed glass tubes (13 × 100 mm) with Teflon-lined screw caps.
2. β-Elimination reagent: a freshly made aqueous solution of 1*M* $NaBD_4$ in 0.1*M* NaOH.
3. Vortex mixer.
4. Glacial acetic acid.
5. Oven or water bath at 45°C.
6. Vacuum centrifuge (e.g., Savant SpeedVac, Savant, Farmingdale, NY).

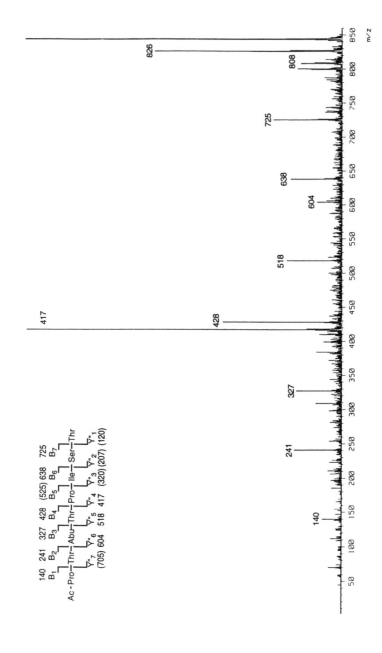

Fig. 1. CID mass spectrum of [M + H]⁺ of peptide obtained on reductive β-elimination of compound 2 (*m/z* 844) (*3*).

234

2.3. C18 Separation of Peptide and Carbohydrate

1. C18 cartridge (e.g., Alltech, Deerfield, FL, 205350 or Baker, Phillipsburg, NJ, 7020-06).
2. Methanol.
3. Round-bottomed glass tubes (13 × 100 mm) with Teflon-lined screw caps.
4. 5% Aqueous acetic acid solution.
5. 1-Propanol.
6. Vacuum centrifuge (e.g., Savant SpeedVac).

2.4. Acid-Catalyzed Acetylation

1. Round-bottomed glass tubes (13 × 100 mm) with Teflon-lined screw caps.
2. Acetylation reagent:trifluoroacetic anhydride (2 vols), glacial acetic acid (1 vol) mixed in a glass round-bottomed tube, capped, and allowed to cool to room temperature.
3. Vacuum centrifuge (e.g., Savant SpeedVac) or source of air or nitrogen.
4. Vortex mixer.
5. Dichloromethane.

2.5. Reduction and Carboxymethylation

1. Source of oxygen-free nitrogen
2. $0.6M$ Tris-HCl buffer, pH 8.5 (adjusted with glacial acetic acid) and degassed by bubbling a stream of N_2 through the solution for approx 15 min.
3. 50 mL Pear-shaped flask.
4. Dithiothreitol (DTT).
5. Water bath or heating block at 37°C.
6. Iodoacetic acid solution in Tris buffer: molarity to be approx the same as that of the disulfides.
7. Glacial acetic acid.
8. 5% (v/v) Aqueous acetic acid solution.

2.6. Trypsin Digestion

1. Round-bottomed glass tubes (13 × 100 mm) with Teflon-lined screw caps.
2. 50 mM Ammonium bicarbonate in distilled water, pH adjusted to 8.4 using 1M NH_4OH.
3. Trypsin (e.g., bovine pancreatic trypsin, type XIII; Sigma, St. Louis, MO, cat. no. T8642).
4. Water bath or heating block at 37°C.
5. Freeze dryer (or vacuum pump and cold trap).

2.7. Obtaining a FAB Spectrum in the Positive- or Negative-Ion Mode

1. Sample (0.1–10 μg) dried in a glass vial or tube.
2. Solvent for dissolving sample, e.g., water, methanol, or dimethyl sulfoxide, which must be miscible with the matrix used.

3. Monothioglycerol or other suitable matrix.
4. Disposable graduated glass capillaries (1–5 µL) or glass syringe (1–10 µL).
5. Methanolic 1M HCl or 50% aqueous trifluoroacetic acid.
6. Mass spectrometer with FAB source tuned and optimized, and resolution set to 1000–1500.

3. Methods

3.1. PNGase F Digestion

1. Dissolve 5–10 nmol of glycoprotein or glycopeptide in 100–200 µL of ammonium bicarbonate solution, pH 8.4, in a screw-capped tube.
2. Add 1 U of PNGase F (e.g., 5 µL of the Boehringer-Mannheim preparation).
3. Add 1 drop of toluene, and cap the tube.
4. Incubate overnight at 37°C.
5. Stop the reaction by freeze-drying.

3.2. β-Elimination of O-Glycans

1. Dry ~10 nmol of glycopeptide or glycoprotein in a glass tube.
2. Add 200 µL of β-elimination reagent to the dry sample, mix, centrifuge to bring sample and reagents to the bottom of the tube, and incubate at 45°C for 18 h.
3. Quench the reaction by adding, dropwise, glacial acetic acid until the evolution of deuterium gas ceases (typically 2–4 drops).
4. Dry the reaction mixture using a SpeedVac. Alternatively, load the mixture directly onto a C18 cartridge, and separate glycan and peptide fractions (see Section 3.3.).

3.3. C18 Separation of Peptide and Carbohydrate

1. Prewet a C18 cartridge with 10 mL of methanol, followed by 10 mL of 5% (v/v) aqueous acetic acid without allowing the cartridge to dry out at any stage.
2. Load the (quenched) reaction/digest mixture slowly onto the cartridge without letting the cartridge dry out. Collect the eluate.
3. Wash the cartridge with 10 mL of 5% (v/v) aqueous acetic acid to elute freed carbohydrate chains and salts. Collect the eluate and pool with the first eluate.
4. Elute the peptide or protein with 5 mL of 20% (v/v) 1-propanol in 5% (v/v) aqueous acetic acid, followed by 5 mL of 40% (v/v) 1-propanol in 5% (v/v) aqueous acetic acid. Pool the two propanol-containing fractions.
5. Evaporate the aqueous and the propanol-containing fractions separately to dryness using a SpeedVac.

3.4. Acid-Catalyzed Acetylation

1. Evaporate the reaction mixture or aqueous fraction from the C18 cartridge to dryness in a round-bottomed, screw-capped glass tube, using a vacuum centrifuge.
2. Add 250–500 µL acetylation reagent, cap tube, and mix well.

3. Allow to react at room temperature for 15–20 min.
4. Evaporate to dryness in vacuum centrifuge or under a stream of air or nitrogen.
5. Redissolve the residue in 1 mL dichloromethane, cap the tube, and vortex.
6. Add ~1 mL of water, vortex, centrifuge at low speed to separate the phases, and remove the upper aqueous phase with a Pasteur pipet. Repeat twice.
7. Evaporate the dichloromethane using a SpeedVac or stream of air or nitrogen.

3.5. Reduction and Carboxymethylation

1. Dissolve the protein in ~0.5 mL 0.1M Tris-HCl buffer in the pear-shaped flask, and blow nitrogen gently over the solution for 15–30 min.
2. Add DTT to give a final concentration of 4 mol DTT/mol of disulfide.
3. Incubate at 37°C for 30 min.
4. Add sufficient iodoacetic acid solution to give a final ratio of 5 mol iodoacetic acid/mol DTT.
5. Incubate at room temperature in the dark for 30 min. Maintain the pH of the reaction mixture at 8.0–8.5 by addition of solid Tris.
6. Stop the reaction by adjusting the pH to 3.5–4.0 using glacial acetic acid.
7. Dilute the reaction mixture by adding 1 mL 5% aqueous acetic acid solution.
8. Isolate the reduced and carboxymethylated product using the C18 separation procedure (*see* Section 3.3.), loading the diluted reaction mixture onto the primed cartridge, and recovering the protein from the pooled 20 and 40% propanol washes.

3.6. Trypsin Digestion

1. Dissolve ~500 µg of (glyco)protein in 100–200 µL of ammonium bicarbonate, pH 8.4, in a screw-capped tube.
2. Dissolve trypsin in the same buffer to make a 0.5–1.0 mg/mL solution.
3. Add enough trypsin solution to the sample solution to make the trypsin:(glyco)-protein ratio somewhere between 1:100 and 2:100 (w/w).
4. Cap the tube and incubate for 5 h at 37°C.
5. Stop the digestion by freeze-drying.

3.7. Obtaining a FAB Spectrum in the Positive- or Negative-Ion Mode

1. Dissolve the sample in 10 µL of solvent.
2. Transfer 2–3 µL of matrix (usually monothioglycerol) to the target of the FAB probe, using a glass capillary, glass rod, or stainless-steel spatula.
3. Load 1–2 µL of sample solution into the matrix using a glass capillary or syringe. Addition of 1 µL of acid (methanolic 1M HCl or 50% aqueous trifluoro-acetic acid) to the matrix at this point will displace cations from pseudo-molecular ions.
4. Introduce probe into source and acquire 30-s scans using the highest accelerating voltage available for the mass range required.

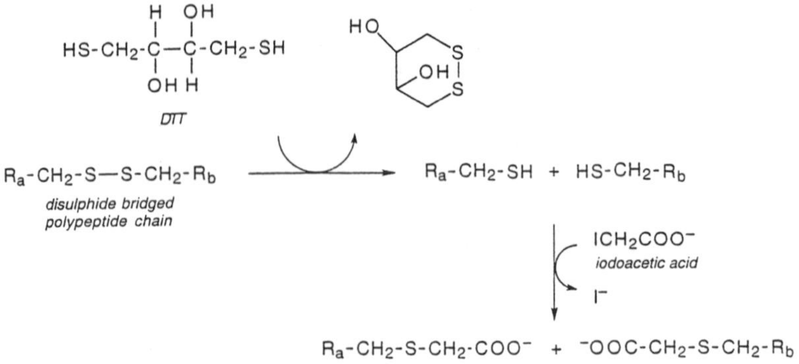

Fig. 2. Reduction and carboxymethylation of proteins containing disulfide bridges.

4. Notes

1. Although the isolation of the free *N*- and *O*-glycans is described herein, their subsequent handling and mass spectrometric analysis are beyond the scope of this chapter. For a detailed description of the analysis of free glycans and their derivatives, *see* ref. *4*.

2. Proteins that contain disulfide bridges should first be reduced and carboxymethylated (Fig. 2) in order to remove these linkages irreversibly and improve the accessibility of the polypeptide backbone to the proteolytic enzyme.

3. Proteolysis is most conveniently achieved on treatment of the polypeptide with trypsin, which cleaves Arg-*X* and Lys-*X* bonds (where *X* is any amino acid except proline). The average frequency with which Arg and Lys residues occur within proteins makes trypsin a good choice of enzyme, since it typically releases peptides having around 10 amino acid residues, and thus a molecular mass appropriate for FAB-MS analysis.

4. Knowing the amino acid sequence of a glycoprotein (usually from the cDNA sequence) and the specificity of trypsin (*see* Note 3), it is possible to predict the molecular weight (and sequence) of the peptides and glycopeptides that should be generated when a glycoprotein is treated with trypsin. Positive-ion FAB-MS analysis of a small aliquot of such a tryptic digest generates a "FAB map" of the digest. A FAB map contains pseudomolecular ions for many or all of the peptides in the mixture, whereas glycopeptides fail to ionize, so that their signals are absent from the map. Subsequent treatment of the mixture with PNGase F converts *N*-glycopeptides to new, modified peptides in which the glycosylated asparagine is converted to aspartic acid. The FAB map of the PNGase F digest will thus contain additional pseudomolecular ions corresponding to the new, modified peptides. From the *m/z* value for each "new" peptide signal, the identity of the peptide and therefore its location in the original protein sequence can be identified (remembering that asparagine and aspartic acid differ in mass by 1 Da). The knowledge that *N*-glycosylation requires the consensus sequence -Asn-*X*-Ser/Thr-

allows the precise location of the glycosylation site to be determined, since only asparagine residues occurring within these sequences may be *N*-glycosylated.

5. In a mixture of peptides, fragmentation of individual peptides is uncommon and, if present, difficult to interpret. Although this absence of fragmentation is essential for FAB mapping, it precludes the sequencing of peptides directly from the mixture. Selection of peptides of interest can be achieved by selecting the precursor $[M + H]^+$ ion. Many mass spectrometers then offer options to monitor fragmentation of a selected precursor ion, using some form of tandem mass spectrometry, be it MS/MS, B/E, MIKES, or MIKES/CID (on reversed geometry sector instruments), or using the third quadrupole in triple-quadrupole instruments. Resolution in ion selection is highest on four-sector tandem mass spectrometers, but in most cases, this resolution is not needed. Peptides exhibit specific fragmentation at the peptide bond, allowing sequencing of the peptide from both termini (*see* Chapter 3 and Appendix).

6. FAB mass spectra (or CID spectra) obtained from underivatized glycopeptides are generally rather disappointing, given the usually good mass spectrometric characteristics of peptides. Since *N*-glycans are generally quite large, obtaining a pseudomolecular ion from their glycopeptides is not always easy, although *O*-glycopeptides should yield $[M + H]^+$ ions. If fragmentation occurs at all, it tends to be rather uninformative, since it most commonly arises by cleavage of the peptide–carbohydrate bond accompanied by hydrogen transfer, rather than by simple cleavage of the peptide backbone to yield information on the site of glycosylation. Since glycosidic linkages are relatively weak, they also fragment very readily, so that fragmentation may also occur by β-cleavage of the glycosidic linkages *(5)*. The fact that a consensus sequence for *N*-glycosylation exists may allow the site of *N*-glycosylation to be determined solely from the peptide fragmentation, which occurs subsequent to the mass spectrometric cleavage of the carbohydrate–peptide bond.

7. In order to identify the site of attachment of the glycan chain to an *N*-glycosylated peptide, it is almost always necessary to remove the carbohydrate using PNGase F, and then to analyze the peptide using FAB-MS (this analysis may be carried out directly on an aliquot of the PNGase F digestion mixture, since this contains a volatile buffer that evaporates in the vacuum lock of the mass spectrometer). The fragment ions generated from both the amino- and carboxy-termini of the peptide allow its sequence, and thus the former site of *N*-glycosylation, to be deduced (*see* Note 4).

8. Routine identification of the site of *O*-glycosylation is more problematic, since no pan-specific enzyme for the removal of *O*-linked glycans exists, no consensus sequence for *O*-glycosylation has been identified, and sites of *O*-glycosylation tend to be found in the neighborhood of several other serine and threonine residues. Reductive β-elimination may be used to release the *O*-glycan chain and generate an unsaturated amino acid residue in an otherwise undegraded peptide. Both the unsaturated amino acid and the carbohydrate are reduced using a deuterated reagent, which specifically modifies the amino acid residue to which the

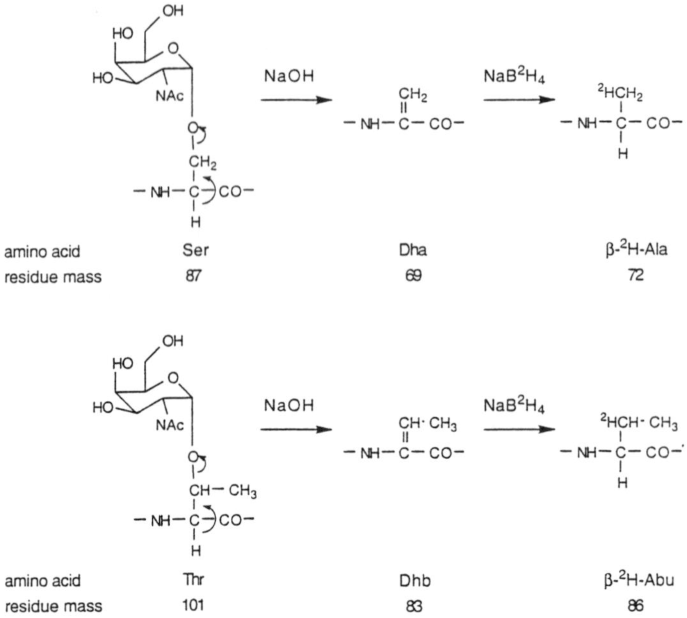

Fig. 3. Use of β-elimination to identify the attachment site of an *O*-glycan chain.

glycan was attached, giving it a unique mass, and also prevents peeling of the carbohydrate chain. The use of deuterated reductant is particularly important if the glycan is attached to a serine residue, since this, on β-elimination and reduction, is converted to alanine. The use of a deuterated reductant allows a chemically generated alanine to be distinguished from an endogenous alanine residue (*see* Fig. 3). Since some peptides are labile in the presence of sodium borodeuteride, β-elimination and recovery of a modified peptide should then be carried out in the absence of reductant, while maintaining all other conditions as given. Such conditions, may however, compromise recovery of intact carbohydrate chains, owing to peeling.

Isolation and analysis of the peptide fraction using positive ion FAB-MS generates a spectrum containing both a pseudomolecular ion for the modified peptide, and also fragment ions that allow the location of the modified amino acid (and thus the site of glycosylation) to be determined. The residue mass for threonine is 101 Da, whereas its dehydration product, dehydro-2-aminobutyric acid, and its deuterium-reduced counterpart, β-deutero-2-aminobutyric acid (β-^{2}H-Abu), have residue masses of 83 and 86 Da, respectively. The analogous masses for serine and its derivatives, dehydroalanine and β-deutero-alanine (β-^{2}H-Ala), are 87, 69, and 72 Da, respectively.

9. Unless extensive HPLC separations are necessary, 5–10 nmol of sample are sufficient to carry out any of the glycopeptide or glycoprotein manipulations

described. All reagents should be of the highest quality available, water should be double glass distilled, and all liquids should be stored and chemical reactions carried out in glass vessels.

Acknowledgments

We are very grateful to our colleagues Koen van der Drift, Leonore Blok-Tip, and Jerry Thomas for critically reading the manuscript.

References

1. Dell, A., Thomas-Oates, J. E., Rogers, M. E., and Tiller, P. R. (1988) Novel fast atom bombardment mass spectrometric procedures for glycoprotein analysis. *Biochimie* **70,** 1435–1444.
2. Carr, S. A., Roberts, G. D., Jurewicz, A., and Frederick, B. (1988) Structural fingerprinting of Asn-linked carbohydrates from specific attachment sites in glycoproteins by mass spectrometry: application to tissue plasminogen activator. *Biochimie* **70,** 1445–1454.
3. Rademaker, G. J., Haverkamp, J., and Thomas-Oates, J. E. (1993) Determination of glycosylation sites in *O*-linked glycopeptides: a sensitive mass spectrometric protocol. *Org. Mass Spectrom.* **28,** 1536–1541.
4. Dell, A. (1990) Preparation and desorption mass spectrometry of permethyl and peracetyl derivatives of oligosaccharides. *Methods Enzymol.* **193,** 647–660.
5. Dell, A. (1987) FAB-mass spectrometry of carbohydrates. *Adv. Carbohydr. Chem. Biochem.* **45,** 19–72.

18

Identification of Cleaved Oligosaccharides by Matrix-Assisted Laser Desorption/Ionization

David J. Harvey

1. Introduction

Matrix-assisted laser desorption/ionization (MALDI) mass spectrometry (MS) was introduced by Karas and Hillenkamp in 1988 *(1)* for the ionization of proteins and peptides, but was soon shown to be capable of ionizing many other types of biomolecules, such as oligosaccharides *(2)*, glycolipids *(3)*, and nucleotides *(4)*. The technique involves cocrystallization of the sample with a UV-absorbing matrix and bombarding the mixture with a beam of laser radiation, usually in the UV range. Radiation of 337 nm from a nitrogen laser is most commonly used, although the frequency-quadrupled output at 266 nm from a Nd:YAG laser is sometimes used. The functions of the matrix are to absorb the laser energy, to dilute the sample to inhibit cluster formation, and to produce ions, most probably through reaction of the sample with active photochemical degradation products. The most common matrices are currently sinapinic acid (3,5-dimethoxy-4-hydroxycinnamic acid) *(5)*, which is most often used for proteins, α-cyano-4-hydroxycinnamic acid *(6)* used for smaller peptides and glycolipids, and 2,5-dihydroxybenzoic acid (2,5-DHB, also known as gentisic acid) *(7)*, which produces signals from a wide range of compounds.

Since the MALDI process involves pulses of radiation from the laser and is capable of ionizing large molecules (several thousand Daltons), it is ideally coupled to a time-of-flight (TOF) mass spectrometer. Linear TOF instruments, unfortunately, have low mass resolution (typically 200–500), but are, nevertheless, common, as bench-top instruments, and capable of giving a mass measurement accuracy of ±0.1%, particularly if the measurement is made in the presence of a suitable internal standard. It should be noted, however, that

From: *Methods in Molecular Biology, Vol. 61: Protein and Peptide Analysis by Mass Spectrometry*
Edited by: J. R. Chapman Humana Press Inc., Totowa, NJ

because the peaks produced by natural isotopes are not resolved, the mass produced will approximate the chemical (or isotopically averaged) mass rather than the monoisotopic mass that is most commonly encountered in the MS of molecules of lower molecular weight.

Resolution can be improved by use of a TOF instrument fitted with a reflectron, which compensates for the small velocity spread of ions of the same m/z value, or by use of a magnetic mass spectrometer fitted with an array detector (8), which can accommodate the pulsed nature of the ion beam. Both types of instrument are capable of producing mass resolutions of up to 5000.

1.1. MALDI of Oligosaccharides

The first matrix to be described for the analysis of oligosaccharides was 3-amino-4-hydroxybenzoic acid (2). Only $(M + Na)^+$ ions are produced, in contrast to the ionization of peptides where $(M + H)^+$ ions are usually the major species. The spectra are simple and lack dimeric and multiply charged ions and fragments, all of which are seen with other matrices, in varying proportions, from other types of compounds. Furthermore, no derivatization of the oligosaccharides is necessary, although some improvement in sensitivity can be achieved by permethylation. Thus, MALDI provides an excellent technique for profiling oligosaccharides of the type released from glycoproteins. In 1991, Stahl et al. (9) reported that 2,5-DHB gave improved signals from oligosaccharides, and this matrix is now used routinely for such MALDI analysis. Further improvements in sensitivity have been achieved by Harvey et al. (10) and involve recrystallization of the dried matrix solution from ethanol. This procedure achieves a more even distribution of crystals over the target surface and is thought to improve uptake of the sample molecules into the crystal lattice. In contrast to 3-amino-4-hydroxybenzoic acid, which appears to become saturated with low concentrations of sample, the use of 2,5-DHB as a matrix gives a response that directly reflects the concentration of the sample (11).

1.1.1. Release of Oligosaccharides from Glycoproteins

Oligosaccharides attached to glycoproteins are generally classified as O-linked (attached to serine or threonine) or N-linked (attached to asparagine in an Asn-X-[Ser or Thr]) motif. They may be released from the glycoproteins either chemically or enzymatically.

1.1.1.1. CHEMICAL RELEASE

O-Linked glycans may be released with base (sodium hydroxide) in the presence of a reducing agent, such as sodium borohydride, to minimize decomposition. The process is known as β-elimination and produces an oligosaccharide with a reduced reducing terminus, thus preventing the use of radio- or fluorescent labeling at this point. By far the most important chemical release method is hydrazinolysis *(12,13)*. Both *N*- and *O*-linked oligosaccharides can be released by heating the glycopeptide with anhydrous hydrazine at 95°C for 4 h and *O*-linked oligosaccharides can be released fairly specifically by heating at only 60°C for 5 h. The reducing terminus remains intact and can be labeled. The method has been automated, and a machine capable of producing a glycan pool directly from a glycoprotein (GlycoPrep-1000) is available from Oxford GlycoSystems (Abingdon, Oxfordshire, UK).

Hydrazinolysis cleaves all peptide–oligosaccharide bonds, including the oligosaccharide–asparagine bond and the *N*-acetyl bonds of constituent acylamino-sugars, so that oligosaccharides are usually reacetylated with acetic anhydride under mild conditions after release to restore the original oligosaccharide structure. This technique, unfortunately, presupposes the existence of only *N*-acetyl groups in the parent molecule, and the information on the possible presence of other acyl groups is lost. In addition, some oligosaccharides with *N*-acetylamino groups at the reducing terminus are produced as normal byproducts of the hydrazinolysis reaction. Overreaction with hydrazine can result in loss of the reducing-terminal GlcNAc residue from *N*-linked glycans and possible substitution of an acetylamino group by a hydroxy group.

1.1.1.2. ENZYMATIC RELEASE

All of the problems associated with hydrazine release are avoided by the use of enzymes, but unfortunately, no enzyme is currently known that will remove all oligosaccharide structures. Peptide-*N*-glycosidase-F (PNGase-F or *N*-glycanase) is the most widely used of these enzymes, and cleaves the amide bond between the oligosaccharide and the peptide in *N*-linked glycans. The enzyme is relatively nonspecific, but oligosaccharides with 1,3-linked fucose at the reducing terminus are resistant. Since cleavage removes the nitrogen atom from the asparagine to produce aspartic acid, the one mass unit increment in the mass of this amino acid can be used diagnostically to determine if a potential glycosylation site is occupied. The other common endoglycosidase in general use is endo-*N*-acetylglucosaminidase-H (endo-H). This enzyme cleaves between the GlcNAc residues of the chitobiose core of high mannose and hybrid *N*-linked oligosaccharides, but does not release complex oligosaccharides. Lack of the reducing-terminal GlcNAc residue in the released oligosaccharide results in loss of information about its substituents. *O*-Glycanase may be used to release *O*-linked oligosaccharides, but the enzyme is relatively nonspecific, and results are often poor.

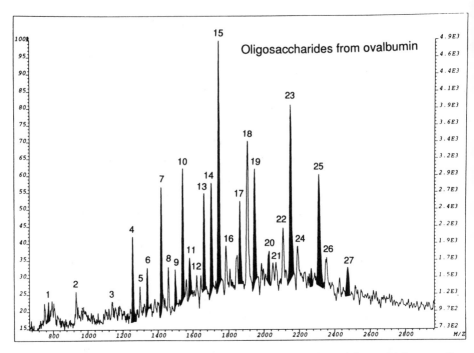

Fig. 1. MALDI mass spectrum of oligosaccharides released from chicken ovalbumin by hydrazinolysis and recorded on a TOF instrument using 2,5-DHB as the matrix. Compositions of the peaks are given in Table 1. Filled peaks are those whose compositions have been confirmed by exoglycosidase digestion. Peaks 5, 8, 12, 16, 24, and 26 appear to represent by-products of hydrazinolysis. Reproduced with permission from ref. *10.*

1.1.2. Examination of Released Oligosaccharides by MALDI

Oligosaccharides released by hydrazinolysis, followed by reacetylation and purification by descending paper chromatography, are suitable for direct examination by MALDI. Oligosaccharides obtained from the GlycoPrep-1000 should be further purified by passing the sample through a mixed-bed resin, consisting of 20 μL each of Chelex 100 (Na⁺ form) over Dowex AG50-X12 over AG3-X4 (OH⁻ form) over Sephadex A25 (OH⁻ form), which has been washed with water (3 mL).

By way of illustration, Fig. 1 shows a typical profile of *N*-linked oligosaccharides (from chicken ovalbumin) recorded with a TOF instrument (*see* Section 3.1.) whereas Fig. 2 shows the increased resolution that can be obtained when the same mixture is analyzed using a magnetic sector instrument (*see* Section 3.2.). The composition of each oligosaccharide of known structure in this mixture is detailed in Table 1.

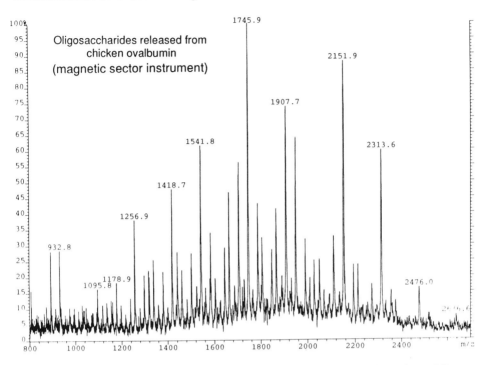

Fig. 2. High-resolution MALDI mass spectrum of oligosaccharides released from chicken ovalbumin and recorded on a magnetic sector instrument using 2,5-DHB as the matrix. Reproduced with permission from ref. *10*.

2. Materials

2.1. MALDI on a TOF Mass Spectrometer

1. Matrix solution: 2,5-DHB obtained from Aldrich (Gillingham, Dorset, UK) (*see* Note 1). The solution is prepared by dissolving the matrix (7 mg, 50 nmol) in acetonitrile (0.7 mL) and water (0.3 mL) (*see* Notes 2 and 3).
2. TOF mass spectrometer and targets.
3. Microliter syringes or pipet with disposable tips for loading target.
4. Absolute ethanol.

2.2. MALDI on a Magnetic Sector Mass Spectrometer

1. Matrix solution: 2,5-DHB as in Section 2.1., item 1. A saturated solution of the matrix in acetonitrile is used (*see* Notes 2 and 4).
2. Magnetic sector mass spectrometer fitted with an array detector and combination fast-atom bombardment (FAB)/MALDI ion source (Fisons VG Autospec-QFPD).
3. Microliter syringes or pipet with disposable tips for loading target.
4. Absolute ethanol.

Table 1
Compositions of N-Linked Oligosaccharides
Released from Chicken Ovalbumin by Hydrazinolysis[a]

Peak No.	[M + Na]$^+$	Calculated mass	Hexose	HexNAc
1	771.2	771.7	2	2
2	933.7	933.8	3	2
3	1138.3	1137.0	3	3
4	1257.8	1258.1	5	2
6	1340.9	1340.2	3	4
7	1419.6	1420.3	6	2
9	1501.6	1502.4	4	4
10	1543.0	1543.4	3	5
11	1582.3	1582.4	7	2
13	1664.1	1664.5	5	4
14	1704.7	1705.6	4	5
15	1745.9	1746.6	3	6
17	1866.1	1867.7	5	5
18	1907.6	1908.7	4	6
19	1949.3	1949.8	3	7
20	2030.1	2029.8	6	5
21	2070.1	2070.9	5	6
22	2111.1	2111.9	4	7
23	2152.3	2153.0	3	8
25	2314.3	2315.1	4	8
27	2476.8	2477.3	5	8

[a]Reproduced with permission from ref. 10.

3. Methods

3.1. Mass Measurement on a TOF Mass Spectrometer

1. Prepare an aqueous solution of the sample to a concentration of about 10 pmol/μL or greater.
2. Place the matrix solution (1 μL) onto the target followed quickly by the sample solution (0.5 μL), and allow the two to mix. Leave at room temperature until all of the solvent has evaporated (see Notes 5 and 6).
3. Redissolve the crystals in the minimum amount of absolute ethanol (about 0.2–0.5 μL) with stirring, and allow the solvent to evaporate (see Notes 7 and 8).
4. Place the target in the mass spectrometer, fire the laser, and adjust the laser power to just above that needed to obtain a signal (see Note 9).
5. Average the signal from about 10 laser shots in order to achieve a good signal-to-noise ratio.

6. For the most accurate mass measurement (±0.1%), it is necessary to incorporate an internal standard (*see* Note 10). The target is prepared as in steps 1–3, but with the addition of an amount of the internal standard that will give a peak of equivalent height (*see* Note 11).

7. Acquire several (5–10) spectra, calibrate each one by use of the internal standard peak, and report the result as a mean value with its standard deviation (*see* Notes 12 and 13).

8. For quantitative information, average the signal from several laser shots at one spot. Measure the peak height. Repeat the measurements from several other areas of the target. Take the mean and standard deviations of the peak heights (*see* Notes 14 and 15).

3.2. MALDI on a Magnetic Sector Mass Spectrometer

1. Prepare an aqueous solution of the sample to a concentration of about 10–100 pmol/μL.

2. Place the matrix solution (about 3 μL) onto the target followed quickly by the sample solution (1 μL), and allow the two to mix. Leave at room temperature until all of the solvent has evaporated.

3. Redissolve the crystals in the minimum amount of absolute ethanol (about 0.5–1.0 μL) with stirring, and allow the solvent to evaporate.

4. Place the target in the mass spectrometer with the laser set to full power (*see* Note 16).

5. Adjust the mass range of the instrument to cover the region desired, and while data acquisition is proceeding, scan the laser beam over the target area by means of the vertical adjustment (*see* Notes 17 and 18). The beam should be moved up and down at a speed of about 1.5 Hz and periodically should be moved sideways if the end insertion probe is used. Use of the side insertion probe will require movement of the probe in order to bring a fresh area of the target under the laser beam.

6. Obtain 10–12 complete spectra in this way, and average the result with the data system. Since no peak broadening occurs (in contrast to the use of a TOF mass spectrometer), it should not be necessary to include an internal standard. However, since the instrument with the array detector gives an accuracy of about ±0.3 mass units, inclusion of an internal standard should resolve any ambiguities (*see* Notes 19 and 20).

4. Notes

1. The commercial 2,5-DHB appears to be sufficiently pure for direct use, but some increase in performance may be achieved by recrystallization.

2. The matrix solution appears to be stable for several weeks at room temperature, but since it absorbs UV energy, it is advisable to keep it out of direct sunlight.

3. It is advantageous to keep the sample spot size as small as possible. The presence of water, with its high surface tension, helps to achieve this.

4. Because it is not so important to keep the sample size small on the magnetic sector instrument, the water can be eliminated from the matrix solution to aid rapid evaporation of the solvent.

5. Drying may be assisted with a stream of air.

6. Crystals of matrix start to form at the edge of the droplet of solution and grow toward the center. With very dilute solutions containing few impurities, the crystals often meet. With more concentrated solutions, crystals stop growing before they reach the center and some amorphous material is left.

7. The amount of ethanol should be sufficient to dissolve the sample and matrix, but not enough to cause excessive spreading of the sample over the target surface.

8. The ethanol appears to serve two purposes. First, it produces a much more even distribution of crystals over the target surface than the aqueous acetonitrile. Second, it is thought that as the original aqueous acetonitrile solution dries, the acetonitrile evaporates first, leaving the water. Since the oligosaccharides are much more soluble in the water than in acetonitrile, they remain in solution while the crystals are growing and are deposited in the crystal-free area in the center of the target. When recrystallized from absolute ethanol, the lower solubility of the oligosaccharides promotes sample uptake by the crystals. Since the oligosaccharides are invariably dissolved in water at the start of the procedure in order to aid maximum transfer to the target, the initial solvent always will contain both aqueous and organic phases. Thus, the two-stage recrystallization process should be used.

9. Increasing the laser power will cause the peak to broaden toward the high mass end of the spectrum as the result of a greater energy spread of the ions. Thus, the increased ion yield from strong laser pulses causes greater screening of some ions from the acceleration potential with the result that these ions are accelerated more slowly, giving them longer flight times and hence a higher measured mass. If the peaks become sufficiently broad, other adduct ions, such as $(M + K)^+$ are no longer resolved, thus further biasing the peak toward high mass.

10. The internal standard should be a compound that is as structurally similar to the sample as possible in order to ensure that the kinetics of ion formation are equivalent. Its mass should preferably be higher than that of the sample if a one-point calibration is made (*see* Note 12).

11. The process of balancing the peak heights can take considerable time and is best done by trial and error.

12. It could be argued that bracketing standards will give an even more accurate mass measurement (two-point calibration). However, considerable difficulty will be experienced in achieving equivalent signals from all three peaks. Also, there will be an error associated with the calculation of the arrival time of all three peaks. It is often assumed that, because the mass of the internal standard(s) is known, there is then no error associated with the mass measurement. This is not necessarily so. Thus, any instrumental error encountered in mass measurement of the calibration peak will be reflected in the measured mass of the sample. The relationship between ion flight time and m/z value is linear, and a graph of the two values should pass through the origin. Since there is no variable error associated with the measurement of the zero point, a calibration line plotted using a single internal standard and zero (single-point calibration) will

**Table 2
Residue Masses of Common Monosaccharides
Found as Constituents of Oligosaccharides from Glycoproteins[a]**

		Mass	
Monosaccharide	Abbreviation	Isotopic average	Monoisotopic
Deoxypentose	DeoxyPen	116.12	116.05
Pentose	Pen	132.12	132.04
Deoxyhexose	DeoxyHex	146.14	146.06
Hexose	Hex	162.14	162.05
Hexosamine	HexN	161.16	161.07
N-Acetylhexosamine	HexNAc	203.20	203.08
Hexuronic acid	HexA	176.13	176.03
N-Acetylneuraminic acid	NeuAc	291.26	291.10
N-Glycolylneuraminic acid	NeuGc	307.26	307.09

[a]Molecular masses are calculated by adding the appropriate residue masses together with the mass of water (average and monoisotopic = 18.01). For the MNa$^+$ ion, a further 22.99 should be added. Note: The acids may form sodium salts.

be subject to less error than one involving a line plotted between the values from two internal standards.

13. The measured mass will give a confirmation of monosaccharide composition. If the structures of the constituent monosaccharides have been determined by other methods, their individual masses (Table 2) can be combined to give the masses of possible oligosaccharides. These calculated masses can be compared with the measured mass to obtain a composition in terms of the isobaric monosaccharide composition. It should be noted that this comparison does not prove a given composition and gives no information on the sequence, branching, linkage, or anomericity of the oligosaccharide. These properties must be determined independently.

14. Peak heights or peak areas may be used, but in practice, it is often easier to measure peak heights.

15. Since the sample is not homogeneous, the intensity of the signal from a single spot will rarely be an accurate reflection of the sample concentration. Thus, the average measurement from many sample spots is needed. On the Finnigan Lasermat, this can conveniently be achieved by setting the "Laser aim" control to "Auto," which causes the laser beam to fire at four positions in rotation for a set number of shots. The signals from these laser shots are averaged. If the laser is set to fire 20 shots, the resulting signal will be the mean of five laser shots from each of four spots. On instruments not fitted with an autosampler, the target can then be removed, turned through 90°, and the reading repeated. Two more rotations gives a total of four readings, each of 20 shots from four spots. For even greater precision, readings from several separate targets can be combined—this will compensate for errors incurred in measuring sample volumes and so on.

16. Unlike a linear TOF mass spectrometer, the ion optics of the magnetic sector instrument can accommodate the energy spread of the ions produced by high-energy laser beams and, consequently, the laser can be run at full power to achieve a better signal-to-noise ratio.

17. As a consequence of the high laser power, the sample is soon depleted under the laser beam. Thus, the beam needs to be moved over the target surface in order to maintain the ion-beam intensity.

18. Since several exposures of the array detector are required to cover the whole mass range and because the laser beam is moved over the target surface during spectral acquisition, different areas of the spectrum will be acquired with the laser beam hitting different areas of the target. Usually, the signal intensity varies over the surface of the target, and this effect could lead to ion-intensity variations across the spectrum. To minimize this effect, about a dozen complete spectra should be acquired with the laser beam being rastered over the target surface at about 1.5 Hz. These spectra should then be averaged.

19. The higher resolution of a magnetic sector instrument results in separation of the isotope peaks. Thus, the measured mass will be the monoisotopic mass, unlike the chemical (isotopically averaged) mass that is measured with the TOF instrument. However, at the high-mass end of the spectrum, isotope peaks might not be resolved, and consequently, masses approaching the chemical mass are recorded. Care should be taken to note which type of mass is being measured for each component of the mixture. It sometimes may be advantageous to operate the instrument at relatively low resolution to ensure that all masses are chemical.

20. The acceleration voltage is lower on a magnetic sector mass spectrometer (8 kV) than on a TOF instrument (20–25 kV), and therefore, ions take longer to accelerate. Consequently, more fragmentation occurs in the ion-source region, and fragment ions will appear in the spectrum. These ions are mainly the products of glycosidic cleavage and, if they involve losses from the nonreducing terminus of the oligosaccharide, will give fragment ions with an integral number of monosaccharide residues whose masses will be indistinguishable from those of natural oligosaccharides. A possible way to identify such fragment ions is to permethylate the oligosaccharide prior to MALDI analysis. In this case, loss of a nonreducing terminal sugar will leave a hydroxy group at the terminus providing a diagnostic mass shift. Fragmentation from the reducing terminus will usually result in the linking oxygen being removed with the neutral fragment so that the ion may then be more readily identified as a fragment.

Acknowledgment

I thank Raymond A. Dwek, Director of the Glycobiology Institute for his interest and encouragement.

References

1. Karas, M. and Hillenkamp, F. (1988) Laser desorption ionisation of proteins with molecular weights exceeding 10,000 Daltons. *Anal. Chem.* **60,** 2299–2301.

2. Mock, K. K., Davey, M., and Cottrell, J. S. (1991) The analysis of underivatised oligosaccharides by matrix-assisted laser desorption mass spectrometry. *Biochem. Biophys. Res. Commun.* **177,** 644–651.

3. Juhasz, P. and Costello, C. E. (1992) Matrix-assisted laser desorption ionization time-of-flight mass spectrometry of underivatized and permethylated ganglio-sides. *J. Am. Soc. Mass Spectrom.* **3,** 785–796.

4. Nordhoff, E., Karas, M., Kramer, R., Hahner, S., Hillenkamp, F., Kirpeker, F., Lezius, A., Muth, J., Meier C., and Engels, J. W. (1995) Direct mass spectrometric sequencing of low-picomole amounts of oligodeoxynucleotides with up to 21 bases by matrix-assisted laser desorption/ionization mass spectrometry. *J. Mass Spectrom.* **30,** 99–112.

5. Beavis, R. C. and Chait, B. T. (1989) Cinnamic acid derivatives as matrices for ultraviolet laser desorption mass spectrometry of proteins. *Rapid Commun. Mass Spectrom.* **3,** 432–435.

6. Beavis, R. C., Chaudhary, T., and Chait, B. T. (1992) Alpha-cyano-4-hydroxy-cinnamic acid as a matrix for matrix-assisted laser desorption mass spectrometry. *Org. Mass Spectrom.* **27,** 156–158.

7. Strupat, K., Karas, M., and Hillenkamp, F. (1991) 2,5-Dihydroxybenzoic acid: a new matrix for laser desorption-ionisation mass spectrometry. *Int. J. Mass Spectrom. Ion Proc.* **111,** 89–102.

8. Bordoli, R. S., Howes, K., Vickers, R. G., Bateman, R. H., and Harvey, D. J. (1994) Matrix-assisted laser desorption mass spectrometer on a magnetic sector instrument fitted with an array detector. *Rapid Commun. Mass Spectrom.* **8,** 585–589.

9. Stahl, B., Steup, M., Karas, M., and Hillenkamp, F. (1991) Analysis of neutral oligosaccharides by matrix-assisted laser desorption-ionisation mass spectrometry. *Anal. Chem.* **63,** 1463–1466.

10. Harvey, D. J., Rudd, P. M., Bateman, R. H., Bordoli, R. S., Howes, K., Hoyes, J. B., and Vickers, R. G. (1994) Examination of complex oligosaccharides by matrix-assisted laser desorption mass spectrometry on time-of-flight and magnetic sector instruments. *Org. Mass Spectrom.* **29,** 753–766.

11. Harvey, D. J. (1993) Quantitative aspects of the matrix-assisted laser desorption mass spectrometry of complex oligosaccharides. *Rapid Commun. Mass Spectrom.* **7,** 614–619.

12. Patel, T., Bruce, J., Merry, A., Bigge, C., Wormald, M., Jaques A., and Parekh, R. (1993) Use of hydrazine to release in intact and unreduced form both *N*-linked and *O*-linked oligosaccharides from glycoproteins. *Biochemistry* **32,** 679–693.

13. Patel, T. P. and Parekh, R. B. (1994) Release of oligosaccharides from glycoproteins by hydrazinolysis, in *Methods in Enzymology (Guide to Techniques in Glycobiology),* vol. 230 (Lennarz, W. J. and Hart, G. W., eds.), Academic, San Diego, CA, pp. 57–66.

19

Structural Characterization of Protein Glycosylation Using HPLC/ Electrospray Ionization Mass Spectrometry and Glycosidase Digestion

Christine A. Settineri and Alma L. Burlingame

1. Introduction

Glycosylation provides three common classes of co- and posttranslational modifications to proteins which are distinguished by the sites of attachment to the polypeptide and have distinct oligosaccharide structural motifs. These different structural types may be linked covalently to the side chains of:

1. Asparagine (*N*-linked Asn); or
2. Serine (*O*-linked Ser) and/or threonine (*O*-linked Thr) (reviewed in ref. *1*); or
3. To the protein C-terminal carboxyl group to form an amide through the amino function of a glycolipid-modified phosphoethanolamine moiety (reviewed in ref. *2*).

In addition, there are other less common modifications known that involve either a single residue, such as *N*-acetylglucosamine *(3,4)* or fucose *(5)*, or several residues, such as xylose-xylose-glucose *(6)*.

N-linked oligosaccharides are found on Asn residues primarily in the Asn-*Xxx*-Ser/Thr consensus sequence (where *Xxx* is any amino acid except proline), although not all Asn residues that are found in this sequence are glycosylated. There is no known consensus sequence for *O*-linked glycosylation, however, so the exact site(s) of attachment must be determined experimentally. Usually, different *N*- and *O*-linked oligosaccharide structures are attached to different sites on a glycoprotein (termed heterogeneity), and this heterogeneity often exists not only among these different sites, but also within the group of structures that occurs at each single site of glycosylation (termed microheterogeneity). This heterogeneity complicates the detailed characteriza-

From: *Methods in Molecular Biology, Vol. 61: Protein and Peptide Analysis by Mass Spectrometry*
Edited by: J. R. Chapman Humana Press Inc., Totowa, NJ

tion of carbohydrates in glycoproteins, because it is very difficult to separate the oligosaccharide components for individual analysis (reviewed in ref. 7).

Most earlier structural work on protein glycosylation has been carried out on the oligosaccharides isolated from a glycoprotein by chemical or enzymic degradation, which in many cases destroyed the information on the sites of attachment. For more than a decade, these studies have employed chemical derivatization and soft ionization mass spectrometry (reviewed in ref. 8). In addition, in some cases, advantage has been taken of polyanionic chromatography for the separation of mixtures of closely related oligosaccharides (9), yielding detailed structural information based on the elution time of known standards. Until recently, however, the structural characterization of such glycosylation remained a formidable challenge for many reasons. These issues and perspectives have been reviewed and discussed recently elsewhere (7).

The advent of electrospray ionization mass spectrometry (ESI-MS) (10,11) however, has provided the first truly practical technology for routine use with microbore or capillary high-performance liquid chromatography (HPLC) (12–14). It was quickly recognized that such a combination would provide the analytical power to ionize, selectively detect, and characterize covalently modified peptides among all the other peptides in a proteolytic digest of a protein. In the case of glycoprotein analysis, one can locate readily the components bearing glycosylation, and determine structural class and heterogeneity present. In addition, in some cases, measurement of the molecular weight of a glycoprotein can provide direct global assessment of the glycosylation present (15,16). The development of methods of performing collision-induced dissociation (CID) with HPLC/ESI-MS (13,17) has made routine the location and preparative fractionation of glycopeptides in complex digest mixtures by selectively monitoring for carbohydrate-specific fragment ions (selected-ion monitoring [SIM]). As the HPLC eluant enters the mass spectrometer, fragmentation can be induced either before the first quadrupole of a single or triple quadrupole mass analyzer (HPLC/ESI-CID-MS) or within the second quadrupole (HPLC/ESI-MS/MS) of a triple quadrupole. Monitoring for fragment ions specific for carbohydrate moieties (such as m/z 204 for N-acetylhexosamines [HexNAc] and m/z 163 for hexoses [Hex]) during the former of these CID methods, or for parent ions that yield these fragment ions for the latter of these CID methods, allows one to identify readily glycopeptide-containing HPLC fractions in even the most complex mixtures of peptides and glycopeptides (14,18–20). Interestingly, a recent comparison of the two CID methods concluded that HPLC/ESI-CID-MS (using only the first quadrupole) proved "far more sensitive, universal and versatile" than the counterpart CID method requiring the triple quadrupole instrument (19).

By combining the use of endo- and exoglycosidases with HPLC/ESI-MS of glycopeptides, oligosaccharide sequence and attachment site information can be obtained *(14,20)*. For example, if the same HPLC/ESI-MS conditions are used to analyze glycoprotein digests before and after treatment with specific endoglycosidases, such as peptide *N*-glycosidase F (*N*-glycanase) or *O*-glycosidase, potential glycosylation sites can be confirmed by the presence of newly generated deglycosylated peptides. Sequence and partial linkage information on oligosaccharides present also can be obtained by HPLC/ESI-MS analysis of glycopeptides before and after treatment with specific exoglycosidases. These include neuraminidase (sialidase), β-galactosidase, and β-*N*-acetyl-hexosaminidase, which, depending on the specific exoglycosidase employed, remove only nonreducing terminal monosaccharides linked in defined ways, or *O*-glycosidase, which cleaves only the disaccharide, galactosyl-β1→3*N*-acetyl-galactosamine (Galβ1→3GalNAc) linked to a Ser or Thr residue.

If more than one possible glycosylation site is present in a particular glyco-peptide, or if amino acid sequence information is required to establish the presence of a specific glycopeptide, tandem mass spectrometry (MS/MS) or Edman degradation analysis can be performed on the isolated glycopeptide fractions. For example, MS/MS of a subdigested tryptic glycopeptide was used successfully to determine the four exact sites of *O*-linked glycosylation on bovine fetuin *(21)*, whereas Edman degradation was used to determine the two sites (out of a possible three) of *O*-linked glycosylation on human lecithin:cholesterol acyltransferase (LCAT) by looking for blank cycles in the Edman data *(14)*.

In summary, determination of the structures, the attachment sites, and the structural heterogeneity of *N*- and *O*-linked carbohydrates attached to glyco-proteins may be achieved by combining the use of proteases and glycosidases with HPLC/ESI-MS and Edman degradation methods. Utilization of these methods is illustrated with two proteins derived from human plasma, which were analyzed in our laboratory, but which could not be completely separated, viz., LCAT and apolipoprotein D (Apo D) *(14)*. Additional examples are given from work performed by other laboratories using ESI-MS for the analysis of glycoproteins, including recombinant human erythropoietin (rhuEPO) *(15)*, and bovine fetuin *(18)*.

2. Materials

Materials listed in this section relate to the analysis of human LCAT (Apo D) *(14)*. Details relevant to the analysis of bovine fetuin and rhuEPO can be found in the appropriate publications *(15,18)*.

2.1. Reduction and Alkylation

1. 100 mM NH$_4$HCO$_3$, 6M guanidine-HCl (pH 8.5).
2. 3 mM Ethylenediaminetetraacetic acid (EDTA).
3. Dithiothreitol (DTT).
4. Iodoacetic acid sodium salt.

2.2. Tryptic Digest

Trypsin (sequencing-grade) (Boehringer-Mannheim, Indianapolis, IN).

2.3. Glycosidase Digests

See Table 1 for details.

2.4. HPLC Analysis

1. Dual-syringe pump for microbore-HPLC/Applied Biosystems 140A (Applied Biosystems, Foster City, CA).
2. Column: 1.0 mm id × 250 mm Aquapore 300 C-18 (Applied Biosystems).
3. HPLC solvents: water, acetonitrile (Fisher, Pittsburgh, PA); trifluoroacetic acid (TFA) (Pierce, Rockford, IL).
4. UV detector: variable-wavelength (Applied Biosystems 759A) with high sensitivity "U-Z View" capillary flow cell (LC Packings, San Francisco, CA).

2.5. Mass Spectrometry

1. Single quadrupole mass spectrometer with ESI source (VG Platform).
2. Microflow syringe pump for postcolumn solvent addition (Isco μLC-500).
3. Postcolumn solvents: 2-methoxyethanol and propan-2-ol (Aldrich).

2.6. Edman Microsequencing

1. Gas-phase protein sequencer (Model 470A, Applied Biosystems).
2. PTH analyzer (Model 120A, Applied Biosystems).

3. Methods

3.1. Reduction and Alkylation (Using LCAT as an Example)

1. Dissolve LCAT in 0.1 mM NH$_4$HCO$_3$, 6M guanidine-HCl (pH 8.5).
2. Add a 50-fold excess (over the number of cysteines) of DTT and incubate at 50°C for 1.5 h under N$_2$ or Ar.
3. Add a 20-fold excess of iodoacetic acid, and leave at room temperature in the dark for 1 h.
4. Dialyze overnight vs 50 mM NH$_4$HCO$_3$.

3.2 Tryptic Digest (Using LCAT as an Example)

1. Add trypsin, in an enzyme:substrate ratio of 1:50 (w/w), to the reduced and alkylated material from Section 3.1.
2. Incubate for 8 h at 37°C.

3.3. HPLC Analysis

1. Prepare solvents A (0.1% TFA in water) and B (0.08% TFA in CH$_3$CN).
2. Set up the solvent gradient: 2–70% B over 120 min, and then 70–98% B over 10 min for tryptic digest products; 2–40% B over 30 min or 10–50% B over 40 min for glycosidase products of the isolated glycopeptides.

Table 1
Exoglycosidase Reaction Mixtures Used for Sequencing Glycopeptides

Enzymes, in sequencing order	Enzyme source	Linkage specificity	Concentration, U/mL	Digestion buffer[a]	Digestion time, h
Sialidase	Newcastle Disease Virus	NeuNAcα2-3,8R NeuGAcα2-3,8R	0.02–0.05	A	3–7
β-Galactosidase	S. pneumoniae	Galβ1-4GlcNAc Galβ1-4GalNAc	0.03–0.07	A	3–7
β-GlcNAcase	Chicken liver	GlcNAcβ1-3,4R	1.0	A	3–7
α-Mannosidase	Jack bean	Manα1-2,3,6Man	3.5	A + 25 mM $ZnCl_2$	3–6
β-Mannosidase	Helix pomatia	Manβ1-4GlcNAc	2.0–4.0	A + 25 mM $ZnCl_2$	3–6
GlcNAcase	Chicken liver	GlcNAcβ1-3,4R	2.0	A	3–7
α-Fucosidase	Bovine epididymis	Fucα1-6(>2,3,4)R	0.2	A	3–6

[a]Buffer A: 30 mM sodium acetate, pH 5.0.

259

3. Set the HPLC pump flow at 50 µL/min.
4. Set the postcolumn flow at 35–40 µL/min of 2-methoxyethanol: propan-2-ol (1:1 [v/v]).
5. Set the effluent split, which is downstream from the postcolumn addition mixing tee, to give a split of approx 1:20, i.e., 4–5 µL/min to the mass spectrometer and the remainder to fraction collection (*see* Note 1).

3.4. Identification of Glycopeptides by HPLC/ESI-MS

The basic strategy used for identifying and sequencing glycopeptides is shown in Fig. 1. The reduced and alkylated glycoprotein is first proteolytically cleaved, and the resulting mixture of peptides and glycopeptides is then analyzed by HPLC/ESI-MS using either ESI-CID-MS with a single quadrupole and adjustment of the source-orifice potential to induce CID, or by ESI-MS/MS, using a triple quadrupole with selection of specific masses for parent-ion scans. HPLC/ESI-CID-MS, used to identify where in the chromatogram the glycopeptides elute, can be performed two different ways.

> Method 1: Two HPLC/ESI-MS analyses are carried out under identical chromatographic conditions. The first analysis is performed at low source-orifice potential, scanning the entire mass range to record the ions that result from ionization of the intact peptides and glycopeptides. The second analysis is performed at high source-orifice potential, recording only the oxonium ions of analytical interest using selected-ion monitoring (SIM) (*see* Note 2).
>
> Method 2: A single HPLC/ESI-MS analysis is performed where the source-orifice potential is stepped back and forth between each scan. At high voltage, the low mass region is scanned (e.g., m/z 150–400) to obtain the oxonium ions from the fragmented glycopeptides, whereas at low voltage the higher mass region is scanned (e.g., m/z 500–2200) to obtain the ions from the intact peptides and glycopeptides. Alternatively, in a similar manner, the voltage can be ramped from high to low during the low mass scan, followed by a constant low voltage during the higher mass scan.

As an example, an aliquot of a tryptic digest of reduced, carboxymethylated LCAT *(14)* was analyzed by microbore reversed-phase HPLC/ESI-MS as follows:

1. Set the mass spectrometer to scan the range m/z 300–2000 with a 5-s scan speed.
2. Set the mass spectrometer sampling cone (source-orifice) potential to approx 60 V (the precise voltage setting will depend on the particular instrument model being used).
3. Set up the appropriate HPLC conditions as detailed in Section 3.3.
4. Load a 300-pmol aliquot of the digest onto the column and begin the analysis.
5. After the analysis in step 4 is complete, set the mass spectrometer, in the selected-ion monitoring mode, to record several mass values, as described above, including m/z 204, 292, 366, and 657 ± 0.5, with a scan time of approx 0.2–0.25 s and an interscan delay of 0.02 s.

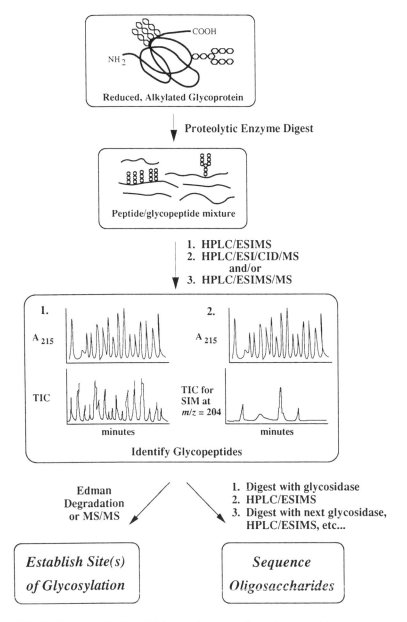

Fig. 1. Strategy for identifying and sequencing glycopeptides.

6. Raise the voltage of the sampling cone in the source of the mass spectrometer from 60–180 V (150–200 V) to induce fragmentation of the oligosaccharides attached to the peptides (the precise voltages for scanning will depend on the particular instrument model being used).

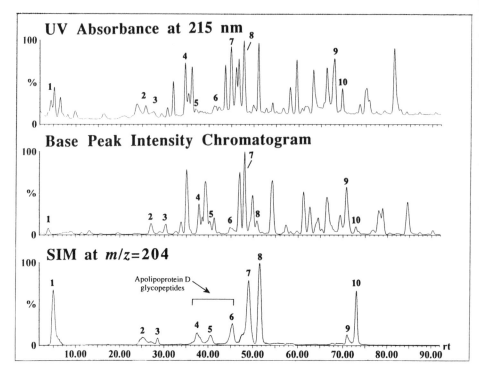

Fig. 2. HPLC/ESI-MS and HPLC/ESI-CID-MS data with selected-ion monitoring at $m/z = 204$ for the LCAT (Apo-D) tryptic digest. The identified glycopeptides are labeled 1–10.

7. Using the same HPLC conditions as in step 3, load another 300-pmol aliquot of the digest onto the column and begin the second analysis (Method 1).

The mass chromatogram representing base peak abundance, resulting from scanning the entire mass range, and an ultraviolet (UV) absorbance chromatogram, obtained from the HPLC/ESI-MS experiment (step 4), are shown in the top two panels of Fig. 2. Data from the HPLC/ESI-CID-MS experiment with SIM (step 7) are then used in order to identify which of the UV and MS peaks from this scanning HPLC/ESI-MS analysis contain glycopeptides.

Thus, the chromatogram representing SIM at m/z 204 is also illustrated in Fig. 2 (lower panel). The presence of a similar chromatogram for SIM at m/z 292 (not shown) indicates that complex type N-linked oligosaccharides are present on this protein (22). As shown in Fig. 2, 10 different glycopeptide-containing fractions were identified, and the corresponding UV, MS, and SIM peaks are labeled 1–10. Once the glycopeptide-containing peaks were identified, the electrospray mass spectra of the glycopeptides from the first analysis

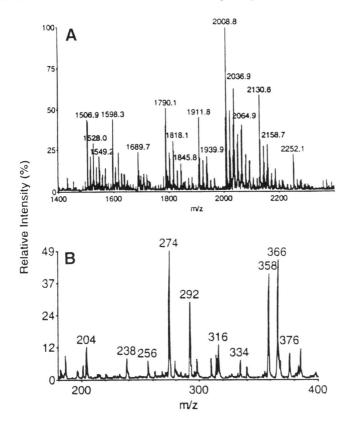

Fig. 3. HPLC/ESI-CID-MS data for *N*-83 glycopeptides from rhuEPO. **(A)** High *m/z* range (1500–2400) scan recorded at an orifice potential of 65 V. **(B)** Low *m/z* range (190–390) scan recorded at an orifice potential of 125 V. Reprinted from ref. *15* with permission from the American Chemical Society.

were identified by direct comparison of the UV chromatograms from the two experiments and the retention times of the SIM peaks. The molecular masses of the intact glycopeptides were then determined from the original HPLC/ESI-MS data. The further processing of these data is detailed in Section 3.5.

Alternatively, a digest of rhuEPO was analyzed in a single HPLC/ESI-CID-MS analysis where the orifice potential was stepped *(15)* from 125 V during a low mass scan (*m/z* 190–390) to 65 V during the high mass scan (*m/z* 1500–2400). Figure 3 shows an example of the data obtained for one group of glyco-peptides attached to Asn83 (termed *N*-83 glycopeptide) in the two different mass range scans. As shown in Fig. 3B, the low mass scan contains ions indicative of complex-type oligosaccharides containing HexNAc (*m/z* 204), NeuAc (*m/z* 274 and 292), and Hex-HexNAc (*m/z* 366). In addition, the low mass scan in Fig. 3B

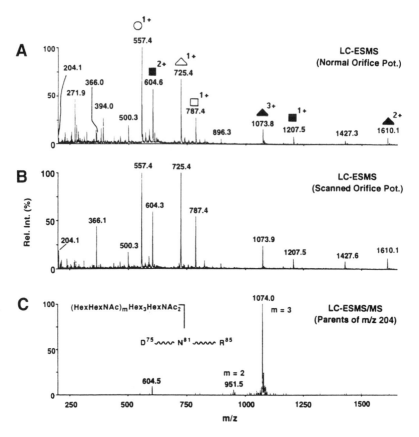

Fig. 4. HPLC/ESI-MS spectra of a group of glycopeptides from bovine fetuin. **(A)** HPLC/ESI-MS spectrum using "normal" orifice potiential (65 V). The different symbols, together with an indication of the charge state, identify five of the parent ions present. **(B)** HPLC/ESI-CID-MS spectrum resulting from ramping the orifice potential from 120 V, at *m/z* 150, to 65 V at *m/z* 500; from *m/z* 500 to 2000 the orifice potential was held constant at 65 V. Note the increased intensities at *m/z* 204 and 366 compared with A. **(C)** HPLC/ESI-MS/MS, parent-ion scan of *m/z* 204. Reprinted from ref. *18* with permission from the Protein Society.

also indicates the presence of mono- and di-*O*-acetyl NeuAc at *m/z* 334 and 376 and the corresponding dehydro forms at *m/z* 316 and 358.

As a comparison of the spectra obtained using HPLC/ESI-CID-MS vs HPLC/ESI-MS/MS, Fig. 4 shows the difference in the spectra of a group of glycopeptides from a digest of bovine fetuin *(18)* using:

1. "Normal" orifice potential (ESI-MS).
2. Scanned orifice potential (ESI-CID-MS).
3. Parent ions that fragment to form *m/z* 204 (ESI-MS/MS).

It is clear from Fig. 4 that selectivity is the advantage of using parent-ion scans compared with HPLC/ESI-CID-MS, since no other peptide species are present in the spectrum (*see* Fig. 4C). However, it also illustrates the lower sensitivity obtained using this method relative to ESI-CID-MS. Only one ion for each glycoform species, i.e., a single charge state, is present in the spectrum in Fig. 4C, compared to two ions for each glycoform in Fig. 4B. Only by using the data from Fig. 4A or B, is it possible to be sure of the charge state of the ions, and therefore the correct nominal molecular weight of the glycopeptides. In practice, scanning or stepping the orifice potential is often more useful. However, with the ability to do MS/MS on the triple quadrupole, the selectivity of the data can be used to confirm predictions from the varied orifice-potential experiments.

3.5. Interpretation of Electrospray Spectra of Glycopeptides

By subtracting the calculated molecular weight of the common complex-type oligosaccharide structures (listed in Table 2), it is possible to deduce the likely peptide components of the glycopeptides. This is illustrated with the data obtained from HPLC/ESI-MS analyses of human LCAT. For example, based on the molecular weights and the SIM data, the peaks labeled 1 and 7–10 in Fig. 2 represent the four predicted LCAT tryptic *N*-linked glycopeptides. In this case, peaks 9 and 10 contain the same glycopeptide, except that the peptide in fraction 10 contains an *N*-terminal pyroglutamic acid residue, rather than the expected *N*-terminal Gln, based on a mol-wt difference of 17 u between the two components. As an example, the ESI mass spectrum of fraction 9 is shown in Fig. 5. The major component in this spectrum has an average molecular mass of 6404.0. By subtracting the mass of a disialo biantennary structure (2206.0 Da) from 6404.0, a mass of 4198.0 is obtained, which corresponds closely to the molecular weight of LCAT tryptic peptide 363–399 (4199.0 Da).

Unexpectedly, the carbohydrate-containing components, labeled 2 and 3 in Fig. 2, revealed molecular weights that correspond to the sum of the expected C-terminal tryptic peptide, residues 400–416 (1655.8 Da), together with attached hexa- to undecasaccharides. Since this peptide contains no Asn, but one Thr and two Ser residues, it is likely that *O*-linked oligosaccharides are present in this compound. Again, a similar glycopeptide mol-wt distribution was observed in fraction 3, but 17 at mass units less than that in fraction 2, so this component was assumed to be a pyroglutamic acid-containing glycopeptide, since the peptide in fraction 2 also contained an *N*-terminal Gln. The ESI mass spectrum for fraction 3 is shown in Fig. 6A (*see* p. 268).

Based on the observed molecular weights for the components in fractions 4, 5, and 6, it was not clear that these components were LCAT glycopeptides, and further characterization was required to determine their identity. In particular,

Table 2
Average Molecular Weights
for Addition of Common N-linked Glycans to Peptides

Glycan	Composition	Asialo composition	Mol wt added to peptide	Number of sialic acids				
				0	1	2	3	4
High Mannose								
Man_5	$Man_5GlcNAc_2$		1217					
Man_6	$Man_6GlcNAc_2$		1380					
Man_7	$Man_7GlcNAc_2$		1542					
Man_8	$Man_8GlcNAc_2$		1704					
Man_9	$Man_9GlcNAc_2$		1866					
Hybrid								
Hybrid	$Man_5GlcNAc_3$		1421					
Bisecting hybrid	$Man_5GlcNAc_4$		1624					
Complex								
Biantennary		$Man_3Gal_2GlcNAc_4$		1624	1915	2206	2497	2789
Biant. fucosylated		$Man_3Gal_2GlcNAc_4Fuc_1$		1770	2061	2352	2643	2935
Bisecting biant.		$Man_3Gal_2GlcNAc_5$		1827	2118	2409	2700	2992
Bisecting biant. fuc.		$Man_3Gal_2GlcNAc_5Fuc_1$		1974	2265	2556	2847	3139
Triantennary		$Man_3Gal_3GlcNAc_5$		1989	2280	2571	2863	3154
Triant. fucosylated		$Man_3Gal_3GlcNAc_5Fuc_1$		2135	2426	2718	3009	3300
Tetraantennary		$Man_3Gal_4GlcNAc_6$		2354	2645	2937	3228	3519
Tetraant. fuc.		$Man_3Gal_4GlcNAc_6Fuc_1$		2500	2792	3083	3374	3665

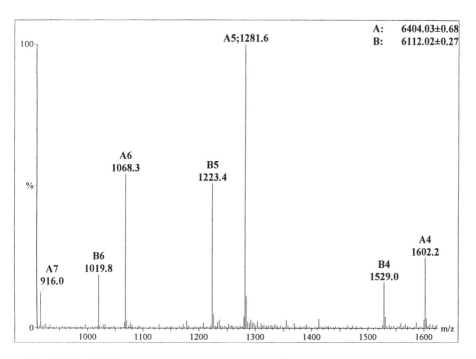

Fig. 5. HPLC/ESI mass spectrum of glycopeptide fraction 9, which represents LCAT tryptic peptide (363–399) with attached sialylated biantennary oligosaccharides.

a subtraction of the masses of the commonly found oligosaccharides from the glycopeptide molecular weights resulted in masses that did not closely match any of the predicted tryptic peptides of LCAT. The further analysis of these components is described in Section 3.6.

3.6. Edman Sequencing of Glycopeptides

Since several unexpected glycopeptide components were detected in the tryptic digest of LCAT, the 10 fractions (collected during the HPLC/ESI-MS experiments from the remaining 95% of the injected material), which were identified as containing glycopeptides from the SIM data, were sequenced by Edman degradation. Many of the fractions contained two or more components, so that the entire fractions from one of the 300-pmol HPLC/ESI-MS analyses were used for the Edman analyses. The initial yields from the analysis ranged from 2–200 pmol. As expected, blank cycles were obtained at the Asn residues, which were assumed to be glycosylated (except for fraction 9, where there was not enough material to sequence through the glycosylation site), confirming the presence of the proposed *N*-linked glycopeptides. By confirming the peptide sequence of the glycopeptides, the Edman data, combined with

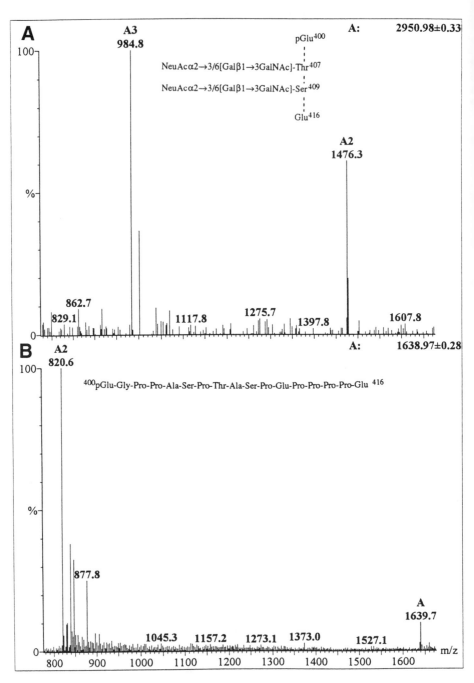

Fig. 6. (A) HPLC/ESI mass spectrum of glycopeptide fraction 3, which represents LCAT tryptic peptide pE(400–416) with attached $Hex_2HexNAc_2NeuAc_2$. **(B)** HPLC/ESI mass spectrum of glycopeptide fraction 3 after digestion with neuraminidase followed by O-glycosidase.

molecular weight and SIM data obtained from the corresponding mass spectra, established the overall size of the glycosylation unit for all of the expected LCAT *N*-linked glycopeptides and confirmed the presence of the unexpected *O*-linked glycopeptides. The Edman data obtained for fractions 1–10 is summarized in Table 3.

Edman degradation analysis of peaks 4, 5, and 6 indicated that these fractions did not contain LCAT glycopeptides, as mentioned in Section 3.5., but instead represented components with the following peptide sequences: (C*IQA-blank-YSLMENGK), (C*IQA-blank-YSLMENGKIK), and (ADGT-VNQIE-GEATPV-blank-LTEPAKKLEVK) (C* = carboxymethyl cysteine). From a search of the Swiss Protein database, the sequences obtained by Edman degradation were identified as peptides (41–53), (41–55), and (63–88) from a second glycoprotein, apolipoprotein D (Apo D), which apparently copurified with the LCAT glycoprotein. This protein, which contains two potential sites of *N*-linked glycosylation at Asn45 and Asn78, has, in fact, previously been shown to associate with LCAT *(23,24)*, although no data have been reported on its oligosaccharide structures. The molecular weights of the Apo D glycopeptides combined with the amino acid sequences indicated that glycopeptides (CIQANYSLMENGK) and (CIQANYSLMENGKIK), from peaks 4 and 5 respectively, bear primarily sialylated triantennary oligosaccharides, whereas glycopeptide (ADGTVNQIEGEATPVNLTEPAKKLEVK) from peak 6 must primarily bear fucosylated and sialylated biantennary and triantennary glycoforms. The ESI mass spectrum of the glycopeptide components in fraction 6 is shown in Fig. 7A.

3.7. Sequencing of Glycopeptides

3.7.1. N-*linked Glycopeptides*

Once glycopeptides have been detected and identified by the methods described, they can be preparatively isolated by microbore HPLC for further digestion with exoglycosidases *(14,25)* and analysis by HPLC/ESI-MS. In this experiment, depending on the exact specificities of the individual enzymes used, different linkages can either be identified or ruled out. The enzymes that may be used are listed in Table 1, together with the appropriate buffers and the final enzyme concentrations of the reactions recommended for exoglycosidase sequencing. The general procedure for this experiment is as follows:

1. Collect fractions containing glycopeptides from the analysis of tryptic digests detailed in Section 3.4.
2. Dry these fractions by vacuum centrifugation (*see* Note 3).
3. Solubilize each fraction in 10 μL 50 m*M* NaOAc (pH 5.0), or in an alternative buffer if noted in Table 1.

Table 3
Edman Sequence Analysis Results
from LCAT (Apo D) Glycopeptide Fractions 1–10 (Fig. 2) Identified by HPLC/ESI-MS[a]

Fraction	Glycosylation site	Sequence range	Edman sequence results[b]
1	LCAT Asn84	(81–85)	VVYXR
2	LCAT Thr407/Ser409	(400–416)	QGPPASPXAXPEPPPPE
3	LCAT Thr407/Ser409	pE(400–416)	No sequence
4	Apo D Asn45	(41–53)	C*IQAXYSL[MENGK]
5	Apo D Asn45	(41–55)	C*IQAXYSL[MENGKIK]
6	Apo D Asn78	(63–88)	ADGTVNQIEGEATPVXL[TEPAKKLEVK]
7	LCAT Asn20	(16–39)	AELSXHTRPVILVPG[C*LGNQLEAK]
8	LCAT Asn272	(257–276)	MAWPEDHVFISTPSFX[YTGR]
9	LCAT Asn384	(363–399)	QPQPVHLLPLHGIQ[HLLNMVFSXLTLEHINAILLGAYR]
10	LCAT Asn384	pE(363–399)	No sequence

[a]Abbreviations: C*, carboxymethyl cysteine; pE, pyroglutamic acid; X, blank cycle.
[b]The bracketed sequences were not identified by Edman analysis owing to insufficient sample.

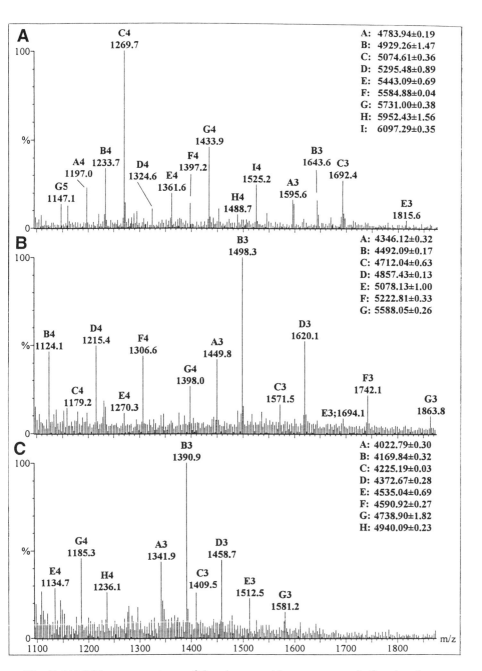

Fig. 7. (A) ESI mass spectrum of the glycopeptide components in fraction 6, representing Apo D peptide (63–88) with attached fucosylated and sialylated biantennary and triantennary oligosaccharides. (B) ESI mass spectrum of fraction 6 following neuraminidase digestion. (C) ESI mass spectrum of fraction 6 following neuraminidase and β-galactosidase digestions.

4. Add the appropriate glycosidase (*see* Note 4) so that a final enzyme concentration approximately equal to that recommended in Table 1 is reached (*see* Note 5).
5. Digest at 37°C for the time given in Table 1.
6. Analyze the digested fractions by HPLC/ESI-MS using the MS conditions itemized in Section 3.4., steps 1 and 2, and the appropriate HPLC conditions as detailed in Section 3.3.
7. Collect fractions from this analysis, and prepare these for further digestion and analysis by starting again at step 2.

Sequencing by exoglycosidase and HPLC/ESI-MS for product identification can be illustrated using the glycopeptides identified from Apo D in peaks 4, 5, and 6 in Fig. 2, which were collected in fractions during the previous HPLC/ESI-MS experiments. In this case, the Apo D protein was only present at approximately one-third of the concentration of the LCAT protein, so that at most only 50–200 pmol of the glycopeptides were present. Therefore, only two enzyme digestion steps were performed. Neuraminidase from *Arthrobacter ureafaciens* was employed first, since it removes terminal sialic acid residues linked $\alpha 2 \rightarrow 3$, 6, or 8 to Gal, NeuAc, GalNAc, or GlcNAc *(26)*, followed by β-galactosidase from *S. pneumoniae*, since this enzyme removes terminal galactose residues linked $\beta 1 \rightarrow 4$ to GlcNAc or GalNAc *(27,28)*.

Figure 7B,C shows the spectra obtained from HPLC/ESI-MS analysis after each of the digestions, using the glycopeptides in fraction 6 as starting material. The structural information determined from the masses recorded in the spectra is summarized in Table 4. As shown, the spectra contain a wealth of information, and each molecular mass can be followed after each digestion step, where the mass difference combined with the specificity of the enzyme used indicates the structure present. The molecular masses obtained from the HPLC/ESI-MS data after each digestion step indicate that primarily trisialo triantennary oligosaccharides are attached to Asn45 (from the glycopeptides in fractions 4 and 5), and that primarily fucosylated sialylated biantennary and triantennary oligosaccharides are attached to Asn78 (from the glycopeptides in fraction 6). These data are summarized in Table 4.

3.7.2. O-linked Glycopeptides

O-linked oligosaccharides from mammalian glycoproteins also can be identified using sequential glycosidase digestion. However, since these structures are usually smaller than the common *N*-linked oligosaccharides and often occur in clusters of glycans localized in one section of the protein sequence, the strategy for sequencing is different. For example, with LCAT, the HPLC/ESI-CID-MS data identified potential *O*-linked glycopeptides (Section 3.5.), but, unlike the *N*-linked glycopeptides, there are often several possible sites of glycosylation in a single glycopeptide. Therefore, in order to determine first which

Table 4
HPLC/Electrospray Mass Spectrometry Results Obtained from Sequential Glycosidase Digestion of Apo D _N_-linked Glycopeptides[a]

Fraction no.	Glycosylation site and sequence range	Intact glycopeptide mol wt, Da	Proposed carbohydrate composition	Mol wt after neuraminidase digestion, Da	Mol wt after β-galactosidase digestion, Da	Confirmed carbohydrate structure[b]
4	Asn45 (41–53)	3734.9	$Hex_5HexNAc_4NeuAc_2$	3167.9	2843.9	Disialo biantennary
		4100.3	$Hex_6HexNAc_5NeuAc_2$	3533.1	3047.1	Disialo triantennary
		4390.0	**$Hex_6HexNAc_5NeuAc_3$**	**3533.1**	**3047.1**	**Trisialo triantennary**
		4682.3	$Hex_6HexNAc_5NeuAc_4$	3533.1	3047.1	Tetrasialo triantennary
		4756.9	$Hex_7HexNAc_6NeuAc_3$	3899.2	3413.1	Trisialo triantennary Lac_1
		5044.4	$Hex_7HexNAc_6NeuAc_4$	3899.2	3413.1	Tetrasialo triantennary Lac_1
5	Asn45 (41–55)	3978.6	$Hex_5HexNAc_4NeuAc_2$	3409.8	3086.4	Disialo biantennary
		4268.6	$Hex_6HexNAc_5NeuAc_2$	3409.8	3086.4	Disialo triantennary
		4632.5	**$Hex_6HexNAc_5NeuAc_3$**	**3774.9**	**3287.9**	**Trisialo triantennary**
		4924.8	$Hex_6HexNAc_5NeuAc_4$	3774.9	3287.9	Tetrasialo triantennary
				4140.4	3655.7	Triantennary Lac_1
				4509.5	4025.1	Triantennary Lac_2
6	Asn78 (63–88)	4783.9	$Hex_5HexNAc_4NeuAc_1Fuc_1$	4492.1	4169.8	Fuc. monosialo biantennary
		4929.3	$Hex_5HexNAc_4NeuAc_2$	4346.4	4022.8	Disialo biantennary
		5074.6	**$Hex_5HexNAc_4NeuAc_2Fuc_1$**	**4492.1**	**4169.8**	**Fuc. disialo biantennary**
		5295.5	$Hex_6HexNAc_5NeuAc_2$	4711.7	4225.2	Disialo triantennary
		5443.1	$Hex_6HexNAc_5NeuAc_2Fuc_1$	4857.2	4372.7	Fuc. disialo triantennary
		5731.0	$Hex_6HexNAc_5NeuAc_3Fuc_1$	4857.2	4372.7	Fuc. trisialo triantennary
		5952.4	$Hex_7HexNAc_6NeuAc_3$	5077.0	4591.0	Trisialo triantennary Lac_1
		6097.3	$Hex_7HexNAc_6NeuAc_3Fuc_1$	5223.2	4738.8	Fuc. trisialo triantennary Lac_1
				5588.1	4940.0	Fuc. tetraantennary Lac_1

[a]Bold type indicates the major species identified based on the relative abundance of the molecular ions in the mass spectra.

[b]Abbreviations: Fuc, fucosylated; Lac, lactosylated; Gal-GlcNAc, indicating an additional lactosamine group extended on one of the branches.

sites are glycosylated, Edman degradation analysis is performed on the intact glycopeptides (*see* Note 6). As illustrated in Table 3, Edman degradation analysis of the *O*-linked glycopeptide in LCAT indicated that Thr407 and Ser409 (not Ser405) were glycosylated, based on the presence of dose-specific blank cycles during the analysis.

To characterize the oligosaccharides on the *O*-linked glycopeptide from LCAT, glycosidase sequencing followed by HPLC/ESI-MS (*see* the protocol in Section 3.7.1.) was again employed. In this case, however, the use of the exoglycosidase, neuraminidase, was combined with the endoglycosidase, *O*-glycosidase (endo-α-*N*-acetyl-D-galactosaminidase from *Diplococcus pneumoniae*), which specifically hydrolyzes unsubstituted Galβ1→3GalNAc disaccharides at the linkage to Ser or Thr *(30)* (*see* Note 7). Since mammalian *O*-linked oligosaccharides often contain this structure, it is best to try this enzyme first. However, any sialic acid residues present must be removed with neuraminidase before using the *O*-glycosidase. Figure 6A,B shows HPLC/ESI-MS spectra of the LCAT *O*-linked glycopeptide, identified as the *N*-terminal pyroglutamic acid form, before and after digestion with the two enzymes. As shown, the neuraminidase and *O*-glycosidase removed all the glycosylation from this glycopeptide, indicating that the structures present were NeuAc2→3 (or 2→6) or 6[Galβ1→3GalNAc]. If the *O*-glycosidase does not remove the oligosaccharides, a larger structure is likely to be present, and other exoglycosidases can be used to digest the glycopeptide as described in the case of *N*-linked glycopeptides.

3.8. Conclusions

The detection, identification, and sequencing of carbohydrates attached to glycoproteins are complex processes requiring a combination of several chromatographic and mass spectral analyses, together with various steps of enzyme degradation. The convenience and sensitivity that result from a combination of microbore HPLC with ESI-MS make HPLC/ESI-MS the method of choice for analyzing glycoproteins and glycopeptides. HPLC/ESI-MS combined with CID, Edman degradation, and glycosidase digestion are the primary tools required to establish the structural nature of the glycoforms present at each site.

The key factor that facilitates the characterization of the glycosylation on these glycoproteins is the use of HPLC/ESI-CID-MS and HPLC/ESI-MS/MS for the selective identification of glycopeptides in a complex mixture of peptides and glycopeptides. Using this strategy, the ESI mass spectrometer is used as a selective dectector for glycopeptides by monitoring diagnostic carbohydrate fragment ions generated by CID during the HPLC/ESI-MS experiment.

Sequencing of the oligosaccharides is performed using glycosidase digestion of glycopeptides followed by HPLC/ESI-MS. The availability of specific exo- and endoglycosidases, such as neuraminidase, β-galactosidase, and *O*-glycosidase, provides the means to perform extensive analyses on oligosaccharides attached to glycopeptides while retaining the attachment site information. Using the specificity of the glycosidases described, combined with the measured changes in glycopeptide mass following each digestion step, leads to the acquisition of detailed sequence and branching information on the oligosaccharide portion of the glycopeptides.

4. Notes

1. Typically 25–500 pmol of the proteolyic digest is injected for the HPLC/ESI-MS analyses, and often the sample is split before entering the mass spectrometer. With a typical split of 1:20, only 5% of the sample is consumed by ESI-MS, and the remaining 95% can be collected for later analyses, such as sequential exoglycosidase digestion, MS/MS, or Edman sequencing.

2. Typical ions which are monitored for oligosaccharides include m/z = 163 (Hexose$^+$ or Hex$^+$), 204 (*N*-acetylhexosamine$^+$ or HexNAc$^+$), 292 (*N*-acetylneuraminic or sialic acid$^+$ or NeuAc$^+$), 325 ([Hex-Hex]$^+$), 366 ([Hex-HexNAc]$^+$), and 657 ([NeuAc-Hex-HexNAc]$^+$). By monitoring for many ions, some sequence information about individual glycopeptides can be obtained in the SIM data, such as the presence of high mannose or complex type structures.

3. It is best to start with approx 400–1000 pmol, depending on the number of digestion steps required, since at least seven different enzymes are required for complete sequencing of a complex-type *N*-linked glycopeptide. In cases where the amount of sample is very limited, just two or three reactions can be performed, yielding enough information to determine the overall size and number of branches of the oligosaccharides. Alternatively, the enzymes can be combined in reactions (the enzymes are all active in 50 m*M* sodium acetate, pH 5.0) *(29)*, although the results of such experiments can be complicated, since some of the enzyme preparations contain protease activity, so that many different glycopeptide species, which differ in the peptide sequence as well as in the oligosaccharide moieties present, can be generated.

4. It is recommended that the enzymes always be tested first on a glycopeptide(s) containing oligosaccharide(s) of known structure (sequence and linkage). This is an important precaution since our experience has been to encounter problems owing to the supply of enzyme preparations with incorrect or additional substrate specificities when compared with the manufacturer's specifications; a particular example has been the *N*-acetylhexosaminidases from *Streptococcus pneumoniae* and chicken liver.

5. The critical part of these analyses is use of the proper final enzyme concentration for the reactions. It is best to start with very small amounts of the enzymes (equal to or less than the manufacturer's recommendation), and then to add more in the next reaction if incomplete digestion has occurred.

6. If the molecular mass of the glycopeptides is 2000 Da or less, MS/MS may be used as an effective alternative. For example, sites of *O*-linked glycosylation have

been determined by MS/MS for glycopeptides with molecular masses between mass 800 and 2000 Da, obtained from yeast-expressed platelet-derived growth factor *(31)* and bovine fetuin *(21)*.

7. The use of neuramindise, followed by the endoglycosidase, *O*-glycosidase, involves solubilization in 50 µL $0.1M$ Na_2HPO_4 (pH 7.2) as buffer, final enzyme concentrations of 0.2 U and 0.01 U/mL, respectively, and incubation times of 17 and 24 h, at 37°C respectively *(18)*. Neuraminidase is added alone first, then the *O*-glycosidase is added to the same tube.

Acknowledgments

We thank Christopher J. Fielding and Patrick Schindler for providing the LCAT/Apo D protein samples. We also thank Frank Masiarz, Zhonghua Yu, and Corey Schwartz for performing the Edman degradation analyses. We gratefully acknowledge David Maltby for technical assistance, and Patrick Schindler and Andreas Ding for the initial development of the SIM electrospray conditions in our laboratory. This work was supported by NIH grant RR 01614, NSF grant DIR 8700766, and NIEHS Superfund Grant ES04705 (to A. L. Burlingame).

References

1. Kobata, A. (1984) The carbohydrates of glycoproteins, in *Biology of Carbohydrates,* vol. 2 (Ginsburg, V. and Robbins, P. W., eds.), Wiley, New York, pp. 87–162.

2. Rosse, W. F. (1993) The glycolipid anchor of membrane surface proteins. *Semin. Hematol.* **30,** 219–231.

3. Miller, M. W. and Hanover, J. A. (1994) Functional nuclear pores reconstituted with β 1-4 galactose-modified *O*-linked *N*-acetylglucosamine glycoproteins. *J. Biol. Chem.* **269,** 9289–9297.

4. Roquemore, E. P., Dell, A., Morris, H. R., Panico, M., Reason, A. J., Savoy, L. A., Wistow, G. J., Zigler, J. S., Earles, B. J., and Hart, G. W. (1992) Vertebrate lens alphacrystallins are modified by *O*-linked *N*-acetylglucosamine. *J. Biol. Chem.* **267,** 555–563.

5. Harris, R. J., Leonard, C. K., Guzzetta, A. W., and Spellman, M. W. (1991) Tissue plasminogen activator has an *O*-linked fucose attached to threonine-61 in the epidermal growth factor domain. *Biochemistry* **30,** 2311–2314.

6. Harris, R. J., Van Halbeek, H., Glushka, J., Basa, L., Ling, V. T., Smith, K. J., and Spellman, M. W. (1993) Identification and structural analysis of the tetrasaccharide NeuAcα(2→6)Galβ(1→4)GlcNAcβ(1→3)Fucα1-*O*-linked to serine 61 of human factor IX. *Biochemistry* **32,** 6539–6547.

7. Settineri, C. A. and Burlingame, A. L. (1995) Mass spectrometry of carbohydrates and glycoconjugates, in *Carbohydrate Analysis: High Performance Liquid Chromatography and Capillary Electrophoresis* (El Rassi, Z., ed.), Elsevier, Amsterdam, pp. 447–514.

8. Dell, A. (1987) FAB-mass spectrometry of carbohydrates. *Adv. Carbohydr. Chem. Biochem.* **45,** 19–72.

9. Townsend, R. R. and Hardy, M. R. (1991) Analysis of glycoprotein oligosaccharides using high-pH anion exchange chromatography. *Glycobiology* 1, 139–147.
10. Fenn, J. B., Mann, M., Meng, C. K., Wong, S. F., and Whitehouse, C. M. (1989) Electrospray ionization for mass spectrometry of large biomolecules. *Science* 246, 64–71.
11. Smith, R. D., Loo, J. A., Edmonds, C. G., Barinaga, C. J., and Udseth, H. R. (1990) New developments in biochemical mass spectrometry: electrospray ionization. *Anal. Chem.* 62, 882–899.
12. Ling, V., Guzzetta, A. W., Canova-Davis, E., Stults, J. T., Hancock, W. S., Covey, T., and Shushan, B. (1991) Characterization of the tryptic map of recombinant DNA derived tissue plasminogen activator by high-performance liquid chromatography-electrospray ionization mass spectrometry. *Anal. Chem.* 63, 2909–2915.
13. Huddleston, M. J., Bean, M. F., and Carr, S.A. (1993) Collisional fragmentation of glycopeptides by electrospray ionization LC/MS and LC/MS/MS: methods for selective detection of glycopeptides in protein digests. *Anal. Chem.* 65, 877–884.
14. Schindler, P. A., Settineri, C. A., Collet, X., Fielding, C. J., and Burlingame, A. L. (1995) Site-specific determination of the *N*- and *O*-linked carbohydrates on human plasma lecithin:cholesterol acyl transferase by HPLC/electrospray mass spectrometry and glycosidase digestion. *Protein Sci.* 4, 791–803.
15. Rush, R. S., Derby, P. L., Smith, D. M., Merry, C., Rogers, G., Rohde, M. F., and Katta, V. (1995) Microheterogeneity of erythropoietin carbohydrate structure. *Anal. Chem.* 67, 1442–1452.
16. Ashton, D. S., Beddell, C. R., Cooper, D. J., Craig, S. J., Lines, A. C., Oliver, R. W., and Smith, M. A. (1995) Mass spectrometry of the humanized monoclonal antibody CAMPATH 1H. *Anal. Chem.* 67, 835–842.
17. Conboy, J. J. and Henion, J. D. (1992) The determination of glycopeptides by liquid chromatography/mass spectrometry with collision-induced dissociation. *J. Am. Soc. Mass Spectrom.* 3, 804–814.
18. Carr, S. A., Huddleston, M. J., and Bean, M. F. (1993) Selective identification and differentiation of *N*- and *O*-linked oligosaccharides in glycoproteins by liquid chromatography-mass spectrometry. *Protein Sci.* 2, 183–196.
19. Hunter, A. P. and Games, D. E. (1995) Evaluation of glycosylation site heterogeneity and selective identification of glycopeptides in proteolytic digests of bovine α_1-acid glycoprotein by mass spectrometry. *Rapid Commun. Mass Spectrom.* 9, 42–56.
20. Settineri, C. A. and Burlingame, A. L. (1994) Strategies for the characterization of carbohydrates from glycoproteins by mass spectrometry, in *Techniques in Protein Chemistry V* (Crabb, J. W., ed.), Academic, San Diego, CA, pp. 97–104.
21. Medzihradszky, K. F., Townsend, R. R., and Burlingame, A. L. (1996) Structural elucidation of *O*-linked glycopeptides by high energy collision-induced dissociation. *J. Am. Soc. Mass Spectrom.* 7, 319–328.
22. Kornfeld, R. and Kornfeld, S. (1976) Comparative aspects of glycoprotein structure. *Ann. Rev. Biochem.* 45, 217–237.

23. Albers, J. J., Cabana, V. G., and Dee Barden Stahl, Y. (1976) Purification and characterization of human plasma lecithin:cholesterol acyltransferase. *Biochemistry* **15**, 1084–1087.
24. Utermann, G., Menzel, H. J., Adler, G., Dieker, P., and Weber, W. (1980) Substitution in vitro of lecithin-cholesterol acyltransferase. Analysis of changes in plasma lipoproteins. *Eur. J. Biochem.* **107**, 225–241.
25. Settineri, C. A. and Burlingame, A. L. (1993) Sequencing of glycopeptides using sequential digestion followed by HPLC/electrospray mass spectrometry, in *Proceedings of the 41st ASMS Conference on Mass Spectrometry and Allied Topics;* ASMS, San Francisco, CA, pp. 94a–94b.
26. Uchida, Y., Tsukada, Y., and Sugimori, T. (1979) Enzymatic properties of neuraminidases from *Arthrobacter ureafaciens*. *J. Biochem.* **86**, 1573–1585.
27. Distler, J. J. and Jourdian, G. W. (1973) The purification and properties of beta-galactosidase from bovine testes. *J. Biol. Chem.* **248**, 6772–6780.
28. Kobata, A. (1979) Use of endo- and exoglycosidases for structural studies of glycoconjugates. *Anal. Biochem.* **100**, 1–14.
29. Sutton, C., O'Neill, J., and Cottrell, J. (1994) Site-specific characterization of glycoprotein carbohydrates by exoglycosidase digestion and laser desorption mass spectrometry. *Anal. Biochem.* **218**, 34–46.
30. Umemoto, J., Bhavanandan, V. P., and Davidson, E. A. (1977) Purification and properties of an endo-α-*N*-acetyl-D-galactosaminidase from *Diplococcus pneumoniae*. *J. Biol. Chem.* **252**, 8609–8614.
31. Settineri, C. A., Medzihradszky, K. F., Masiarz, F. R., Burlingame, A. L., Chu, C., and George-Nascimento, C. (1990) Characterization of *O*-glycosylation sites in recombinant B-chain of platelet-derived growth factor expressed in yeast using liquid secondary ion mass spectrometry, tandem mass spectrometry and Edman sequence analysis. *Biomed. Environ. Mass Spectrom.* **19**, 665–676.

20

Analysis of Complex Protein and Glycoprotein Mixtures by Electrospray Ionization Mass Spectrometry with Maximum Entropy Processing

Brian N. Green, Therese Hutton, and Serge N. Vinogradov

1. Introduction

The electrospray ionization (ESI) spectrum produced from a mixture of proteins and glycoproteins is complicated because each protein in the mixture produces a series of multiply charged ions on a mass-to-charge ratio (m/z) scale. On this scale, the positions of the multiply charged positive ions from a single protein of molecular weight (M_r) are given by: $m/z = (M_r + nH)/n$, where H is the mass of the proton and n is a series of consecutive integers. Thus, each ion in a series differs by one charge from its immediate neighbors. Since a typical protein produces approx 10 multiply charged ions in a series, a 10-component mixture may contain approx 100 peaks. Without some form of deconvolution, whereby all the peaks in the series from each protein are condensed into a single peak on a true mass scale, interpretation is extremely laborious and practically impossible. Various attempts have been made to achieve this aim (1–3) with some success on simple mixtures. However, none of them have been shown to cope effectively with complex mixtures composed of proteins having a wide range of molecular weights and particularly with mixtures where the multiply charged series overlap on the m/z scale.

Over the last 3 years, a method based on maximum entropy (MaxEnt) (4–7) analysis has been developed in a collaboration between John Skilling (MaxEnt Solutions Ltd., Cambridge, UK) and Micromass Ltd. (Altrincham, Cheshire, UK). MaxEnt automatically disentangles the m/z spectrum produced by the mass spectrometer and presents the data from each protein in the mixture as a single peak on a true mol-wt scale, considerably simplifying the spectrum and, thus, facilitating interpretation. Generally, the molecular weights determined

From: *Methods in Molecular Biology, Vol. 61: Protein and Peptide Analysis by Mass Spectrometry*
Edited by: J. R. Chapman Humana Press Inc., Totowa, NJ

by electrospray ionization mass spectrometry (ESI-MS) combined with MaxEnt are within ±0.01% of the values calculated from the sequences. This chapter demonstrates how ESI-MS-MaxEnt is applied to the analysis of complex protein and glycoprotein mixtures exemplified by data from annelid hemoglobins (Hbs). Most of the procedures, however, can be applied to the analysis of any mixture of proteins and glycoproteins.

Annelid Hbs have a characteristic hexagonal bilayer electron microscopic appearance, a high mass (~3.5 MDa) and consist of about 144 functional globin chains (16–18 kDa), which account for about 75% of the total mass, and some 36–40 linker chains (24–32 kDa) necessary for the formation of the native structure. Some of the different globin and linker chains are monomeric, but others form disulfide-bonded dimers at approx 32 kDa (leech, *Macrobdella decora [8]*), trimers at 50–53 kDa (earthworm, *Lumbricus terrestris [9]* and marine worm, *Tylorrhynchus heterochaetus [10]*) or tetramers at approx 66 kDa *(Eudistylia chlorocruorin)*. In addition, some globin and linker chains are glycosylated *(9)*, depending on the species. The mixture of proteins and glycoproteins produced by denaturing earthworm Hb contains about 15 components with molecular weights between 16 and 53 kDa. By using ESI-MS to analyze the denatured native annelid Hb, the masses of all the noncovalently bound proteins present may be determined. Furthermore, by analysis after reduction, analysis after reduction and derivatization, and also analysis after deglycosylation, the masses of all the chains that make up the entire molecule may be established to within ±0.01%. Information is also produced on the mass of the heme group, on the number of cysteines in each chain, and on the chains that are associated with particular multimers.

1.1. Strategy

1. Analyze the native Hb to determine the molecular weights of the heme and all the chains and subunits that exist noncovalently bound together in the original molecule. Note that the native Hb is automatically denatured in the solvent normally used for electrospray analysis.

2. Analyze the native or dehemed Hb following mild reduction. As the reduction proceeds, the appearance of new peaks in the MaxEnt spectrum, concomitant with the disappearance of peaks of higher mass, shows which chains are associated with disulfide-bonded multimers and indicates the masses of the chains. Note that these data give a close, but not necessarily precise determination of these masses, because the reduction of intrachain disulfide bonds may be incomplete. In conjunction with the results from step 3, however, the main purpose of this procedure is to determine the number of cysteines (Cys) in each chain.

3. Analyze the fully reduced and derivatized Hb. In this procedure, free Cys are derivatized to form either carbamidomethylcysteine (Cam) or carboxymethyl-cysteine (Cmc), thereby adding 57.052 Da (+ $CH_2CONH_2 - H$)/Cys or 58.037 Da

(+ CH_2CO_2H – H)/Cys, respectively, to the masses of the chains. By comparing the masses of the chains from this analysis with those from the previous analysis, the number of Cys in each chain may be determined. The accurate masses of the chains in the reduced state may then be determined by subtracting 57.052 Da/Cys (Cam Hb) or 58.037 Da/Cys (Cmc Hb) from the masses of the derivatized chains.

4. Analyze either the deglycosylated Hb or the reduced, derivatized deglycosylated Hb if it is suspected that glycoproteins are present. Because the glycan is generally heterogeneous, glycoproteins are often revealed in the MaxEnt spectrum of the native Hb by the presence of two or more peaks, which differ by the mass of a monosaccharide residue, usually hexose (162.1 Da). The appearance of a new peak in the deglycosylated spectrum concomitant with the disappearance or reduced intensity of the putative glycoprotein peaks allows the mass of the protein to be derived. Furthermore, the mass differences between the glycoproteins and the deglycosylated protein lead to the masses of the glycans, from which the composition of the glycans, in terms of the number and type of monosaccharide residues, often can be inferred.

2. Materials

1. 1% HCOOH (stored stoppered at room temperature).
2. High-performance liquid chromatography (HPLC)-grade H_2O for dilutions (stored at room temperature).
3. HPLC-grade acetonitrile (CH_3CN).
4. 50% Aqueous CH_3CN plus 0.2% formic acid stock solution for dilutions and for washing Rheodyne loops and syringes. Items 1–4 are conveniently stored in 50–100 mL glass bottles with glass stoppers.
5. HPLC pump, capable of providing a low (5 μL/min), pulse-free flow of solvent (50% aqueous CH_3CN) through a Rheodyne injection loop and into the ESI source of a quadrupole mass spectrometer. Alternatively, an infusion pump (Type 22, Harvard Apparatus Inc., South Natick, MA) with a suitable gas-tight Hamilton syringe may be used.
6. MaxEnt software.
7. Stock solutions of 500 pmol/μL horse heart myoglobin (M-1882) and bovine trypsinogen (T-1143) from Sigma (St Louis, MO) in H_2O stored frozen below 0°C in 1.5-mL microcentrifuge tubes. After thawing the stock solution, working solutions for calibration of the *m/z* scale of the mass spectrometer are made by dilution to concentrations of 10 pmol/μL with 50% aqueous CH_3CN, containing 0.2% formic acid. These solutions should also be stored frozen when not in use.
8. Freshly prepared 100 m*M* DTT solution in H_2O.
9. Hb stock solutions. Dissolve approx 0.5 mg of the Hb in a volume of H_2O, which gives a concentration of 5 μg/μL.
10. 6*M* Guanidine hydrochloride.
11. 0.2*M* Tris-HCl, pH 8.5, buffer.
12. 10 m*M* EDTA.
13. 15 m*M* Iodoacetic acid or iodoacetamide.

14. Recombinant *N*-glycosidase F (cloned from *Flavobacterium meningosepticum* and expressed in *Escherichia coli*, EC 3.2.218; 3.5.152).
15. *Diplococcus pneumoniae* *O*-glycosidase (Boehringer Mannheim, Indianapolis, IN).
16. 0.25*M* Sodium phosphate buffer (pH 7.0), containing 1 m*M* EDTA.

3. Methods and Discussion

3.1. Mass Spectrometry (see Note 1)

3.1.1. Sample Solutions

All the analyses may be made from working solutions of the samples at a concentration of 0.5 μg/μL in 50% aqueous acetonitrile containing 0.2% formic acid.

1. Take the sample, which is generally initially dissolved in water at a concentration of 5 μg/μL and dilute aliquots appropriately to make the working solutions for mass spectrometric analysis.
2. Add 0.2–0.5% formic acid to solubilize any samples that tend to be only partially soluble in water at neutral pH.

3.1.2. Mass Spectrometer Scanning Parameters

1. Set the cone voltage ramp to increase from 30 V at *m/z* 600 to 90 V at *m/z* 2500 (*see* Note 2).
2. Acquire data by scanning over the *m/z* range 550–2500 in 10 s (*see* Note 3).
3. Sum the data (Multi-Channel Analyzer [MCA] mode of data collection) over 5–10 min.

3.1.3. Mass Spectrometer Resolution (see Note 4)

1. Set the ion energy to the lowest value that is still consistent with achieving adequate sensitivity. In practice, the ion energy is adjusted to a value that reduces the intensity of a peak (e.g., the *m/z* 1305 peak from myoglobin) to 80–90% of the intensity recorded with an ion energy setting that gives maximum sensitivity (e.g., 2–3 V).
2. Adjust the resolution settings to give a peak width at half height of 0.7–0.8 Da measured on the *m/z* 1305 peak from myoglobin.

3.1.4. Mass Scale Calibration

1. Make a separate introduction of horse heart myoglobin (*see* Note 5) (Sigma M-1882), M_r = 16,951.5 Da *(11)*, at a concentration of 10 p*M*/μL in 50% aqueous acetonitrile containing 0.2% formic acid.
2. Acquire data under the same conditions as for the samples (Section 3.1.2.).
3. Consult the mass spectrometer manufacturer's instruction manual for a detailed description of calibration. Calibration from *m/z* 600 to at least *m/z* 2120 is normally achieved.

3.1.5. Processing the Data Using MaxEnt

1. Baseline subtract the raw data.
2. Process the subtracted data (*see* Note 6) with MaxEnt using a "peak-width at half-height" parameter measured from intense singlet peak(s) around the middle of the multiply charged distribution. Initially, carry out a survey MaxEnt analysis with an output mass range of 2–70 kDa and an output resolution of 5 Da/channel in order to establish the approximate masses of significant (>~5%) components.
3. Halt the processing once it becomes apparent where significant components lie on the mol-wt scale (*see* Note 7), usually after 4 or 5 iterations (approx 5 min).
4. Process a definitive MaxEnt analysis to convergence using 1 Da/channel resolution over an output mass range (or a number of ranges) chosen to include the components found in the survey (*see* Note 8).
5. Centroid the data to determine the masses accurately, and measure the areas under the peaks to obtain relative intensity data (*see* Note 9).

3.2. Preparation and Analysis of Mildly Reduced Native or Dehemed Hb

Treatment of the Hb with dithiothreitol (DTT) should reduce any disulfide bonds present. The objective is to determine the masses of those chains that naturally exist as disulfide-bonded multimers and, in conjunction with the data from Section 3.3., determine the number of Cys in each chain.

1. Prepare a fresh solution of 100 mM DTT ($M_r = 154.3$) in water.
2. Add an appropriate volume of the 100 mM DTT solution to a 5 µg/µL water solution of Hb to give a concentration of 10 mM. If acid was used to assist solubility, raise the pH to 7.0–8.5 by adding ammonium bicarbonate solution before adding the DTT.
3. Incubate at room temperature.
4. After suitable time intervals (e.g., 1, 5, 10, and 30 min), remove 10-µL aliquots from the DTT-treated Hb solution, and dilute with 20 µL of 1% aqueous HCOOH, 20 µL H$_2$O, and 50 µL of CH$_3$CN to make working solutions for ESI analysis.
5. Analyze by ESI-MS as soon as possible after preparing the working solutions, using the mass spectrometer conditions given in Sections 3.1.2.–3.1.3.
6. Process the recorded data by MaxEnt as described in Section 3.1.5.

3.3. Preparation and Analysis of Reduced and Derivatized Hb

Complete reduction (with DTT), followed by Cam or Cmc of the native or dehemed Hb *(12)*, allows the number of Cys present in each chain to be determined and, therefore, the masses of the reduced chains to be precisely measured. Analysis of the reduced Hb as outlined in Section 3.2. is required in order to determine the number of Cys.

1. Prepare a fresh solution of 100 mM DTT in water.
2. Add an appropriate volume of the 100 mM DTT solution to a 5 µg/µL water solution of Hb to give a 10-mM concentration in the Hb solution.

3. Incubate at 50°C for 2 h.
4. Either carboxymethylate the reduced Hb using 15 mM iodoacetic acid or carbamidomethylate it using 15 mM iodoacetamide for 15 min. This reaction is performed in 0.2M Tris-HCl (pH 8.5) buffer containing 6M guanidine hydrochloride and 10 mM EDTA.
5. Remove excess reagents from the derivatized protein mixture, before analysis by ESI-MS, using one of the two methods (a or b) that follows:
   a. Dialyze the derivatized Hb preparation overnight against distilled water at 4°C *(13)* and lyophilize the resulting solution containing the proteins.
   b. Place the derivatized Hb preparation on top of an appropriate cut-off membrane filter (e.g., 10 kDa cut-off microconcentrator, Amicon, Beverly, MA) and centrifuge almost to dryness. Wash the retained protein by placing 300 µL of 1% aqueous HCOOH in the top of the filter and centrifuging again. Repeat the washing procedure and finally dissolve the retained proteins in 0.5% HCOOH.
6. Dilute the purified product appropriately (0.5 µg/µL in 50% aqueous CH_3CN plus 0.2% HCOOH) for ESI analysis.
7. Analyze by ESI-MS using the mass spectrometer conditions given in Sections 3.1.2.–3.1.3.
8. Process the recorded data by MaxEnt as outlined in Section 3.1.5.

3.4. Deglycosylation

Carry out deglycosylation using recombinant N-glycosidase F (cloned from *F. meningosepticum* and expressed in *E. coli*, EC 3.2.218; 3.5.152) or *D. pneumoniae* O-glycosidase (Boehringer Mannheim) in 0.25M sodium phosphate buffer (pH 7.0), containing 1 mM EDTA, overnight at 37°C *(14)*.

3.5. Results and Interpretation (see Note 10)

3.5.1. Tylorrhynchus heterochaetus Hb

Figure 1A illustrates the complexity of the original *m/z* data from native *Tylorrhynchus* Hb and Fig. 1B the simplification resulting from the MaxEnt processing. In essence, Fig. 1B shows a monomer (chain I), three probable linker degradation products (L3–5), a trimer (T), consisting of three already known and sequenced globin chains (IIA, IIB, and IIC), and a linker dimer (D2 = 2L2).

The MaxEnt processed spectrum after mild reduction of the native Hb with 2 mM DTT for 30 min is shown in Fig. 1C. When this spectrum is compared with Fig. 1B, it can be seen that three new peaks have appeared at approx 16.7 kDa, corresponding to chains IIA, IIB, and IIC, as well as the two already known and sequenced linkers (L1 and L2) at 26.3 and 28.2 kDa. Concomitantly, the intensity of the dimer has decreased significantly, and the trimer has almost disappeared. In addition, a new peak is seen at approx 33.5 kDa, corresponding in mass to the trimer minus chain IIA.

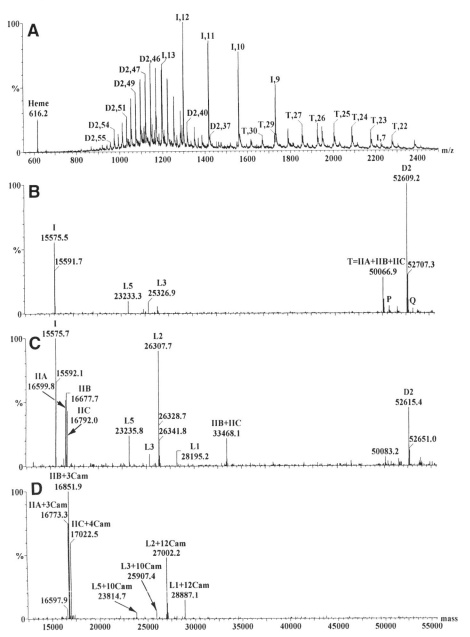

Fig. 1. (A) The original ESI-MS data from native *Tylorrhynchus* Hb and **(B)** the same data after MaxEnt analysis. Peaks P and Q are T and D2 plus heme, respectively (*see* Note 10). The single, small peak, located between L5 and L3 at a mass of 24,835.4, represents component L4. **(C)** The MaxEnt spectrum from the native Hb after reduction with 2 m*M* DTT for 30 min and **(D)** after reduction and carbamidomethylation of the dehemed Hb.

Table 1
Derivation of the Molecular Weights (*see* Note 11)
of the Chains in *Tylorrhynchus* Hb

Chain	Measured mass, Da		No. of Cys		Corrected mass, Da[d]
	Red/Cam[a]	Reduced[b]	ESI[c]	Sequence	
IIA	16,773.1	16,599.6	3	3	16,601.9
IIB	16,851.6	16,677.8	3	3	16,680.4
IIC	17,022.2	16,791.7	4	4	16,794.0
L5	23,814.3	23,234.4	10	10	23,243.8
L3	25,906.5	25,325.8	10	10	25,336.0
L2	27,001.8	26,306.8	12	12	26,317.2
L1	28,886.8	28,192.8	12	12	28,202.2

[a]From reduced, carbamidomethylated data. Mean of two determinations.
[b]From mildly reduced data. Single determination after 10-min DTT treatment.
[c]Difference between Red/Cam and reduced masses divided by 57 and rounded to the nearest whole number.
[d]Red/Cam mass corrected for carbamidomethylation (57.052 Da/Cys). Values represent measured masses with Cys residues reduced.

Figure 1D, from the fully reduced and carbamidomethylated Hb, shows peaks corresponding to all the chains, except the monomer, increased in mass by 57.052 Da times the number of Cys in each chain. It is not understood why there is no peak that corresponds to the underivatized or to the derivatized monomer, nor is it understood why there is no peak in the spectrum of the native Hb (Fig. 1B) corresponding to the dimer of L1.

Table 1 summarizes the way in which the experimentally determined masses from the reduced and carbamidomethylated Hb (Fig. 1D) and from the mildly reduced Hb (Fig. 1C) are used first to determine the number of Cys in the various chains and then to derive the accurate masses of the reduced chains (corrected mass). The separate steps in the method are as follows:

1. Calculate the number of Cys in each chain by dividing the difference between the reduced/Cam masses and reduced masses by 57 (*see* Note 12) and rounding to the nearest whole number (58 would be used for Cmc-derivatized data).
2. Once the number of Cys has been established, derive the masses of the reduced chains by subtracting 57.052 Da/Cys (58.037 Da/Cys for Cmc data) from the experimentally determined reduced/derivatized masses.

Table 2 summarizes, in the "Native Hb mass" and "Reduced mass" columns respectively, the mol-wt data obtained from the native Hb (e.g., Fig. 1B) and those derived, as described earlier, in Table 1. The masses calculated from the sequences are also listed and, except for chain I, show excellent agreement

Table 2
Molecular Weights (see Note 11)
of the Chains and Subunits in Tylorrhynchus Hb

Chain/subunit		Native Hb mass, Da[a,b]	Chain	Reduced mass, Da[a,c]	Sequence mass, Da[d]	Correlation[e]
Monomer	I	15,575.4 ± 1.0	I	ND	15704.9 (15)	—
Linker	L5	23,233.8 ± 2.0	L5	23,243.8 ± 2.0	23,242.8[f]	L5 − 10H = 23,233.7 ± 2.0
	L4	24,835.4 ± 3.0	L4	ND	24,843.8[f]	—
	L3	25,326.9 ± 2.0	L3	25,336.0 ± 2.0	25,335.4[f]	L3 − 10H = 25,325.9 ± 2.0
Dimer	D2	52,609.4 ± 4.0	L2	26,317.2 ± 2.0	26,316.2 (16)	2L2 − 24H = 52,610.2 ± 4.0
	D1	ND	L1	28,202.2 ± 2.0	28,200.5 (16)	—
Trimer	T	50,068.4 ± 4.0	IIA	16,601.9 ± 1.0	16,602.0 (17)	IIA + IIB + IIC − 10H = 50,066.2 ± 2.0
			IIB	16,680.4 ± 1.0	16,680.0 (18)	
			IIC	16,794.0 ± 1.0	16,794.9 (19)	

[a]ND, not determined.
[b]Mean of three determinations on native Hb ± estimated error.
[c]Corrected mass from Table 1 ± estimated error.
[d]Calculated from the sequence with Cys reduced.
[e]Calculated from reduced chain masses with allowances for 2 hydrogens (2 × 1.00794)/disulfide bond.
[f]L3, L4, and L5 are possibly degradation products: L1 (29-253), L1 (34-253), and L2 (30-236), respectively.

(within 0.006%) with the masses derived from the ESI-MS data. The discrepancy between the measured and sequence masses for chain I indicates that the sequence is in error. The correlation column shows how the masses of the proteins in the native Hb can be derived from the masses of the reduced chains by making assumptions about the number of disulfide bonds in the native Hb. For example, the reduced mass of L5 (23,243.8 Da) minus 10 hydrogens for 5 disulfide bonds gives 23,233.7 Da, in close agreement with the mass determined from the native Hb (23,233.8 Da). This suggests that 8 or 10 of the L5 cysteines in the native Hb exist as disulfide bonds, since, in either case, the two masses would still be within experimental error of one another. The situation is similar for L3, except that 3–5 disulfide bonds would fit the experimental errors. In the case of the dimer (D2), 12 disulfide bonds would give the best agreement between the two methods, but 10 and 11 would also agree within the experimental errors.

With the trimer, there is excellent agreement between the sum of the masses of the reduced chains minus 10 hydrogens for 5 disulfide bonds (probably two interchain and three intrachain) and the mass determined directly from the native Hb. Furthermore, in the mildly reduced data, there was no significant mass increase from the earliest appearance of the chains (after 10 min) to when the experiment was discontinued after 60 min. In fact, three determinations (after 10, 30, and 60 min) gave 16,599.9 ± 0.3, 16,677.8 ± 0.1, and 16,791.9 ± 0.2 Da (± SD) for chains IIA, IIB, and IIC, respectively, suggesting that, after reducing the disulfide bonds holding the trimer together, the chains each remain with one intrachain disulfide bond more resistant to reduction than the two interchain bonds.

3.5.2. Trimer of Lumbricus terrestris

Figure 2A shows the MaxEnt-derived spectrum from the trimer of *Lumbricus* isolated by HPLC. Two series of peaks (T1–T4 and S1–S4) are apparent, each series being clearly connected by mass differences that are close to the mass of a hexose residue (162.1 Da), suggesting that the trimer is glycosylated. Figure 2B shows the MaxEnt spectrum from the trimer after reduction by 5 mM DTT for 30 min at 40°C. It shows two peaks, corresponding to chains **b** and **c**, and a group of peaks (**a**1–**a**4) that exhibits a similar pattern to that of the trimer. These latter peaks correspond to glycosylated forms of chain **a**, confirming that only chain **a** is glycosylated. Figure 2C shows the MaxEnt spectrum from the dehemed, reduced, and carbamidomethylated trimer. In essence, it shows chains **b** and **c** increased in mass by three carbamidomethyl groups and chains **a**1–**a**4 increased by four carbamidomethyl groups.

Figure 2D is the spectrum from the dehemed, reduced, carbamidomethylated, and deglycosylated trimer. Even though deglycosylation is incomplete, the

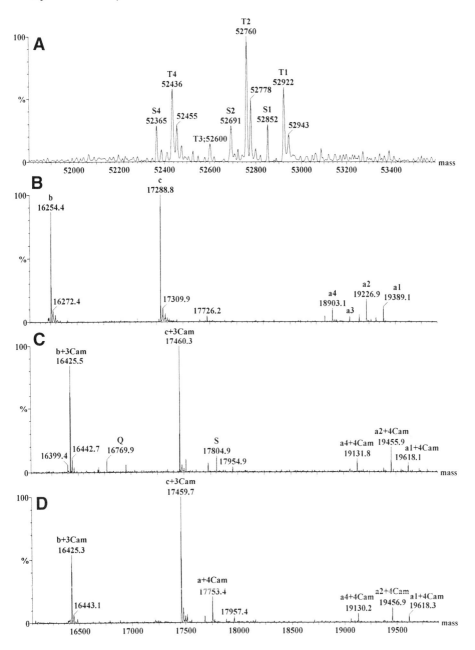

Fig. 2. MaxEnt spectra from the isolated trimer of *Lumbricus* Hb. **(A)** The native trimer; **(B)** after reduction of the trimer by 5 m*M* DTT for 30 min; **(C)** from dehemed, reduced, and carbamidomethylated trimer; and **(D)** following partial deglycosylation of C. In C, peaks Q and S are adducts (*see* Note 10) (+345 Da) of (b + 3Cam) and (c + 3Cam), respectively. Their origin has not been established.

Table 3
Masses (see Note 11) of the Trimer of Lumbricus
and Its Component Chains

Chain/ subunit	Measured mass, Da[a]	Corrected mass, Da[b]	Sequence mass, Da[c]	Assignment
b	16,425.4 ± 1.0	16,254.2	16,254.3 (20)	
c	17,460.1 ± 1.0	17,288.9	17,288.8 (20)	
a	17,753.2 ± 1.0	17,524.0	17,524.9 (20)	
a4	19,131.0 ± 1.0	18,902.8		a4 − a = 1378.8
				$HexNAc_2Hex_6$ = 1379.2
a3	19,294.0 ± 1.0	19,065.8		a3 − a = 1541.8
				$HexNAc_2Hex_7$ = 1541.4
a2	19,455.6 ± 1.0	19,227.4		a2 − a = 1703.4
				$HexNAc_2Hex_8$ = 1703.5
a1	19,618.6 ± 1.0	19,390.4		a1 − a = 1866.4
				$HexNAc_2Hex_9$ = 1865.7
T4	52,435.4 ± 4.0			b + c + a4 − $10H^d$ = 52,435.
T3	52,598.5 ± 4.0			b + c + a3 − $10H^d$ = 52,598.
T2	52,760.0 ± 4.0			b + c + a2 − $10H^d$ = 52,760.
T1	52,922.6 ± 4.0			b + c + a1 − $10H^d$ = 52,923.

[a]Mean of three determinations ± estimated error. Masses of chains b and c (+3Cam) and chain a1–a4 (+4Cam) from dehemed, reduced, and carbamidomethylated sample. Chain a additional deglycosylated. Masses T1–T4 are from isolated trimer.

[b]Corrected for carbamidomethylation (57.052 Da/Cam). Values represent measured reduced chain masses. Chain a also corrected for conversion of Asn to Asp on deglycosylation.

[c]Calculated with Cys reduced.

[d]To allow for the formation of five disulfide bonds.

appearance of a peak corresponding to chain **a** plus four carbamidomethyl groups allows the mass of chain a to be derived. The masses of the glycans can also be derived by difference. For example, the mass of the glycan on **a2** is 19,227.4 − 17,524.0 = 1703.4 Da, which suggests the composition $HexNAc_2Hex_8$ (see Note 13) (calculated residue mass = 1703.5 Da). The results from these analyses are summarized in Table 3.

Referring to Table 3, there is good agreement (within experimental error) between the measured reduced molecular weights ("Corrected mass") of the three chains and their masses calculated from the sequences ("Sequence mass"). Data in the last column show the masses of the glycan residues, derived by subtracting the mass of the deglycosylated chain **a** from the masses of the four glycosylated forms of chain **a** (**a**1–**a**4). These mass differences suggest the four glycan compositions that are listed together with their calculated masses. Finally, the last column also lists the molecular weights of the four trimers calculated from the measured, reduced masses of their constituent chains. These data agree well with the molecular weights determined directly from the isolated trimer.

3.6. Conclusions

ESI-MS-MaxEnt is a powerful combination when applied to the analysis of complex protein mixtures. It is unrivaled in the precision with which molecular weights can be determined (0.01%) and in the mass resolution (1000–2000) it provides. We have shown that, by directly analyzing denatured *Tylorrhynchus* Hb, it provides a wealth of information on the molecular weights of the subunits and chains present in relatively short experiment times. Moreover, by applying well-developed chemical techniques to the native or dehemed hemoglobin, substantial additional information is provided. However, in the case of *Lumbricus* Hb, it was necessary to isolate the trimer by HPLC in order to establish its composition. Nevertheless, direct analysis of the mixtures of proteins produced by the use of various chemical procedures on the isolated trimer gave an unambiguous solution. Although data from annelid hemoglobins have been used in this chapter to illustrate the power of the technique, ESI-MS-MaxEnt is equally applicable to many other problems in biochemistry.

4. Notes

1. These details apply essentially to all electrospray instruments, although some of the terms apply specifically to the Micromass Quattro and Micromass Platform quadrupole mass spectrometers. The data presented in Section 3.5. were obtained using a VG Quattro II.
2. Application of the "cone voltage ramp" is recommended to optimize the sensitivity over the m/z scale. On Micromass instruments, the cone voltage (sampling orifice to skimmer potential difference) for optimum sensitivity increases with m/z. The cone voltage ramp varies the cone voltage linearly with m/z.
3. A minimum m/z scan range of 550–2500 is recommended. This is to ensure that components are not missed.
4. It is always advisable to employ the highest mass resolution consistent with achieving adequate sensitivity, even at the cost of taking a longer time to make the acquisition. Two instrument parameters affect the resolution on quadrupole instruments, namely the ion energy at which the ions travel through the quadrupole analyzer and the RF/DC ratio (radio-frequency potential to direct-current potential ratio). This latter parameter, on Micromass instruments, uses arbitrary values (related to the RF/DC ratio), called LM Res and HM Res, and generally set to about 16.0/15.5.
5. Bovine trypsinogen (Sigma T-1143, $M_r = 23,980.9$ Da) is beneficial as a calibrant when analyzing isolated components from Hbs, which appear at high m/z (above m/z 1800), e.g., the trimer of *lumbricus* Hb. A general-purpose calibrant for pro-

tein analysis that covers the m/z range from 600–3000 is a mixture of myoglobin (5 pM/µL) and trypsinogen (10 pM/µL).

6. Process only that part of the m/z spectrum that contains significant multiply charged data and particularly exclude the heme peak and its adducts if present.

7. At this stage, artifacts, particularly at half and twice the mass of the major peaks, may still be apparent at low (~5–10%) levels, but generally become insignificant on well-resolved data as the MaxEnt procedure approaches convergence.

8. Note that all the significant peaks revealed by the survey must be encompassed by the output mass ranges used for the definitive MaxEnt; otherwise artifacts may be produced.

9. For a more detailed description of processing by MaxEnt, consult the mass spectrometer manufacturer's instruction manual.

10. A number of peaks occur in the electrospray spectra of Hbs, which here are termed adducts. These entities are usually held to be noncovalently bound, although some are chemical modifications. A major reason for describing them is to ensure that they are not interpreted as component proteins. Most of them occur at well-defined mass increments above the mass of the protein with which they are associated. A good indicator for an adduct is that, in a given spectrum, the same mass difference between the masses of the major protein peaks and minor peaks tends to recur. A feature of electrospray spectra is that there are generally more minor peaks on the high mass side of the major peaks than on the low mass side. This is an area where interpretation can become ill defined.

   Some adducts are:

   a. Alkali metal adducts: These mainly represent the addition of Na and K. In the MaxEnt spectrum, these peaks are found 22 and 38 u above the mass of the protein, and not 23 and 39 u higher as the atomic weights of Na and K would suggest. This is because a proton is removed whenever a singly charged adduct ion is added. Thorough desalting of the sample generally reduces alkali metal adducts to a low level.

   b. Heme adducts (M_r usually 616.50): These are common at a low level in the spectra obtained from native Hbs, but may not be associated with all the proteins present. Heme adduction accounts for peaks P and Q in Fig. 1B from native *Tylorrhynchus*. Heme adducts always appear on both α- and β-chains at about the 5% level in the spectra obtained from diluted whole human blood. In the case of human Hb, the mass difference tends to be slightly lower (1–2 u) than the M_r of the heme (616.50), suggesting that the adducts may be covalently bound.

   c. Miscellaneous adducts (mainly chemical modifications): Oxidation or hydroxylation (+16.00 u), acetylation (+42.04 u), formylation (+28.01 u), and cysteinylation (+119.14 u) are some examples, although not necessarily observed with Hb. It has been suggested that the +119 u peak associated with the β-chain when "old" human blood is analyzed by ESI-MS is owing to cysteinylation. This has not been observed with annelid Hbs.

Some adducts at +16–18 u appear when the Hb is treated with DTT. These grow in intensity with respect to the protein itself on prolonged exposure to DTT. Note that generally, an increase of +16–18 u will not be resolved from +22 u (Na), so there is often a peak (or peaks) of rather indeterminate mass (depending on the relative intensities of its components) around +16–22 u, which may be the result of oxidation or hydroxylation.

d. "+98" Adducts: These were observed with *Tylorrhynchus* Hb (Fig 1B, trimer and dimer), and are commonly observed with many commercial preparations of proteins from natural sources. They are not covalently bound and are believed to result from the addition of either phosphoric or sulfuric acid *(21)*.

11. Molecular weights are average values based on the following atomic weights of the elements: C = 12.011, H = 1.00794, N = 14.00674, O = 15.9994, and S = 32.066 *(22)*.

12. Using 57 tends to give values that are slightly high (0–2%), probably because many of the intrachain disulfide bonds are still intact in the mildly reduced data. Nevertheless, for all practical purposes, the result is sufficiently precise to give an unambiguous value for the number of Cys residues.

13. ESI-MS using a quadrupole instrument cannot distinguish between the various monosaccharide isoforms present in a glycan. For example, glucose, galactose, and mannose all have the same residue mass (162.14 Da), and in this chapter are referred to as hexoses, since only MS has been used to establish their composition. *N*-acetylhexoseamine (HexNAc, 203.195 Da) is similarly defined.

References

1. Mann, M., Meng, C. K., and Fenn, J. B. (1989) Interpreting mass spectra of multiply charged ions. *Anal. Chem.* **61,** 1702–1708.

2. Cottrell, J. C., Green, B. N., and Jarvis, S. A. (1991) Applications of an Improved Technique for the Transformation of the Electrospray Spectra of Proteins on to a True Molecular Weight Scale. *Proceedings of the 39th Conference on Mass Spectrometry and Allied Topics*, Nashville, TN, pp. 234,235.

3. Hagen, J. J. and Monnig, C. A. (1994) Method for estimating molecular mass from electrospray spectra. *Anal. Chem.* **66,** 1877–1883.

4. Ferrige, A. G., Seddon, M. J., and Jarvis, S. A. (1991) Maximum entropy deconvolution in electrospray mass spectrometry. *Rapid Commun. Mass Spectrom.* **5,** 374–379.

5. Ferrige, A. G., Seddon, M. J., Green, B. N., Jarvis, S. A., and Skilling, J. (1992) Disentangling electrospray spectra with maximum entropy. *Rapid Commun. Mass Spectrom.* **6,** 707–711.

6. Evershed, R. P., Robertson, D. H. L., Beynon, R. J., and Green, B. N. (1993) Application of electrospray ionisation mass spectrometry with maximum-entropy analysis to allelic "fingerprinting" of major urinary peptides. *Rapid Commun. Mass Spectrom.* **7,** 882–885.

7. Shackleton, C. H. L. and Witkowska, H. E. (1991) Mass spectrometry in the characterization of variant hemoglobins, in *Mass Spectrometry: Clinical and Biomedical Applications,* vol. 2 (Desiderio, D. M., ed.), Plenum, New York, pp. 135–199.

8. Weber, R. E., Malte, H., Braswell, E. H., Oliver R. W. A., Green, B. N., Sharma, P. K., Kuchumov, A., and Vinogradov, S. N. (1995) Mass spectrometric composition, molecular mass and oxygen binding of *Macrobdella decora* hemoglobin and its tetramer and monomer subunits. *J. Mol. Biol.* **251**, 703–720.

9. Martin, P. D., Kuchumov, A. R., Green, B. N., Oliver, R. W. A., Braswell, E. H., Wall, J. S., and Vinogradov, S. N. (1996) Mass spectrometric composition and molecular mass of *Lumbricus terrestris* hemoglobin: a refined model of its quaternary structure. *J. Mol. Biol.* **255**, 154–169.

10. Green, B. N., Suzuki, T., Gotoh, T., Kuchumov, A. R., and Vinogradov, S. N. (1995) Electrospray ionization mass spectrometric determination of the complete polypeptide chain composition of *Tylorrhynchus heterochaetus* hemoglobin. *J. Biol. Chem.* **270**, 18,209–18,211.

11. Zaia, J., Annan, R. S., and Biemann, K. (1992) The correct molecular weight of myoglobin, a common calibrant for mass spectrometry. *Rapid Commun. Mass Spectrom.* **6**, 32–36.

12. Crestfield, A. M., Moore, S., and Stein, W. H. (1963) The preparation and enzymatic hydrolysis of reduced and *S*-carboxymethylated proteins. *J. Biol. Chem.* **239**, 622–627.

13. Teale, F. W. J. (1959) Cleavage of the haem-protein link by acid methylethylketone. *Biochim. Biophys. Acta.* **35**, 43–45.

14. Tarentino, A. L., Gomez, C. M., and Plummer, T. H., Jr. (1985) Deglycosylation of asparagine linked glycans by peptide: *N*-glycosidase F. *Biochemistry* **24**, 4665–4671.

15. Suzuki, T., Takagi, T., and Gotoh, T. (1982) Amino acid sequence of the smallest polypeptide chain containing heme of the extracellular hemoglobin from the polychaete *Tylorrhynchus heterochaetus*. *Biochim. Biophys. Acta.* **708**, 253–259.

16. Suzuki, T., Takagi, T., and Gotoh, T. (1990) Primary structure of two linker chains of the extracellular hemoglobin from the polychaete *Tylorrhynchus heterochaetus*. *J. Biol. Chem.* **265**, 12,168–12,177.

17. Suzuki, T. and Gotoh, T. (1986) The complete amino acid sequence of giant multisubunit hemoglobin from the polychaete *Tylorrhynchus heterochaetus*. *J. Biol. Chem.* **261**, 9257–9267.

18. Suzuki, T., Yasunaga, H., Furukohri, T., Nakamura, K. and Gotoh, T. (1985) Amino acid sequence of polypeptide chain IIB of extracellular hemoglobin from the polychaete *Tylorrhynchus heterochaetus*. *J. Biol. Chem.* **260**, 11,481–11,487.

19. Suzuki, T., Furukohri, T., and Gotoh, T. (1985) Subunit structure of extracellular hemoglobin from the polychaete *Tylorrhynchus heterochaetus* and amino acid sequence of the constituent polypeptide chain (IIC). *J. Biol. Chem.* **260**, 3145–3154.

20. Fushitani, K., Matsuura, M. S. A., and Riggs, A. F. (1988) The amino acid sequences of chains a, b and c that form the trimer subunit of the extracellular hemoglobin from *Lumbricus terrestris*. *J. Biol. Chem.* **263**, 6502–6517.

21. Chowdhury, S. K., Katta, V., Beavis, R. C., and Chait, B. T. (1990) Origin and removal of adducts (molecular mass = 98 U) attached to peptide and protein ions in electrospray mass spectra. *J. Am. Soc. Mass Spectrom.* **1**, 382–388.

22. IUPAC Commission on Atomic Weights and Isotopic Abundances. (1993) Atomic weights of the elements 1991. *J. Phys. Chem. Ref. Data* **22(6)**, 1571–1584.

21

The Use of Databases in Searching the Literature of Biological Mass Spectrometry

Ronald W. A. Oliver and Michael P. Carrier

1. Introduction

Now that mass spectrometric techniques can be applied to a much wider range of molecules, the problem of searching the literature in order to find out whether or not the mass spectrum of a particular analyte has been reported has increased significantly. For example, many reports of mass spectral data now may be found in journals with which the mass spectrometrist is almost certainly unfamiliar. In this connection, it might be thought that the ready availability of a number of computerized bibliographic databases would mean that this problem can easily be solved by using the most appropriate database. The purpose of the present chapter is to show, by means of a detailed consideration of the reporting of a key biochemical analyte, human hemoglobin, that this is not the case, since a number of key references are overlooked. The reasons for this finding, which confirm previous Salford studies on the efficiency of literature databases for small molecules *(1,2)*, are first that none of the databases cover 100% of the scientific literature and, second, no system is 100% efficient in its abstracting of experimental measurements published in scientific papers. The reason for the choice of human hemoglobin to illustrate this data searching problem is that one of us had conducted joint research with the late Professor M. Barber on the measurement of the mass spectra of intact globins using fast-atom bombardment-mass spectrometry techniques *(3)*, and this research interest is still maintained by the first author *(4)*.

It has been suggested that if all of the world's scientific literature, except for the Proceedings of the Royal Society, were destroyed, then succeeding generations could retrieve most, if not all, of the current scientific knowledge from these proceedings. An analogous claim has been made for the key importance

From: *Methods in Molecular Biology, Vol. 61: Protein and Peptide Analysis by Mass Spectrometry*
Edited by: J. R. Chapman Humana Press Inc., Totowa, NJ

of publications concerned with the study of hemoglobin as a record of most aspects of molecular biology and biochemistry, and we would now suggest that the history of the measurement of hemoglobin provides a similar record of the development of modern mass spectrometry. An inspection of the Appendix lends credence to the latter claim since it shows how all of the suitable ionization techniques were immediately applied to a study of this key biochemical analyte as soon as they were available.

2. Materials

The following widely used bibliographic databases have been evaluated, namely:

1. BIDS: using the ISI file on Bath Information Data Service (BIDS) based at the University of Bath, accessed via a PC connected to the UK JANET Network.
2. Chemical Abstracts: using the CA File on STN International, accessed via a PC and a modem.
3. Medline: using KNOWLEDGE INDEX on CompuServe, accessed via a PC and modem.

A 132 reference list (*see* Appendix) compiled by the authors over the last 7 years using both manual literature searches and computerized searches of these three bibliographic databases, was used as the definitive standard. Published conference abstracts were excluded from this list.

3. Method

The search terms employed were as follows:

1. "Human," *and*
2. "Hb or Hemoglobin?" (the question mark covers the possibility of hemoglobins), *and*
3. "Mass spectrum?" (the question mark covers all possible endings), *and not*
4. "Xenobiotic."

The validity of every reference retrieved from the three databases employed was assessed manually by comparing the computer outputs with the definitive bibliography forming the Appendix. To each reference in the latter, a checklist abbreviation was then added (*see* Key to Appendix) from which the total number of correct references yielded by the various databases was calculated.

BIDS is the smallest and most recent of the databases. The maximum number of retrieved references that could be expected from a search using BIDS is 128 rather than the 132 references listed in the Appendix, since the BIDS database does not contain references prior to 1981. In fact, only 34 correct references were found using this database. Although this kind of performance might be acceptable for an entry into a subject, e.g., at the undergraduate level, it would hardly be suitable as, for example, the basis of a research exercise. In the latter case, it is most likely that use would be made of one of the two major bibliographic databases assessed here, namely, *Chemical Abstracts* (CA) and Medline. Thus, the finding that these bases both retrieved only about 50% of the references listed in the Appendix (CA, 68 correct references, and Medline, 65 correct references) gives cause for concern. Even the use of a combination of CA and Medline resulted in a total of only 91 correct references, i.e., about a 70% success rate.

None of the databases yielded references to the majority of the abstracted proceedings of research conferences subsequently published in book form, viz., Appendix refs. *44, 50, 51, 59, 62, 63, 92, 93, 112, 113.* The other understandably missed references were publications that were not abstracted, such as the often valuable booklets produced by instrument manufacturers: Appendix ref. *124*, or articles in "free" journals, Appendix ref. *68*. All but 2 of the 15 references to articles published in foreign languages, namely refs. *24* and *40*, were successfully retrieved. The remainder of the missed references may be ascribed to the incompleteness of the abstracting processes used by the compilers of the databases. As usual with computer searching of bibliographic databases, it was found that all of the databases used in this study also yielded a number of incorrect references. The actual numbers of incorrect references (and corresponding percentages, expressed relative to the number correctly retrieved) are: BIDS 26 (76%), CA 42 (62%), and Medline 47 (72%).

In view of these findings, which are consistent with the findings of previous scientific literature searching assessments performed at Salford *(1,2)*, we recommend that mass spectrometrists studying molecules of biological interest should not rely on one of the databases alone, but should search both of the major databases used here. It should also be recognized that even when this has been done, it is likely that a considerable number of other references to spectral data will exist. Our experience indicates that many of these missing references may be found by a careful study of the references in the abstracted articles. Some idea of the number and range of journals in the field of biology, which now publish mass spectral data, may be obtained from the present study where, of the 132 references to human hemoglobin listed in the Appendix, 119 were published in a total of 60 journals whose titles are listed in Table 1.

4. Conclusion

In conclusion, the present findings provide evidence for the statement that the availability of bibliographic databases, which can be searched by computer, does not completely solve the problem of finding whether or not mass spectral

Table 1
List of Journals

Acta Haematol.	*Int. J. Mass Spectrom. Ion Proc.*
Am. J. Hematol.	*Iyo Masu Kenkyukai Koenshu*
Anal. Biochem.	*J. Am. Chem. Soc.*
Anal. Chem.	*J. Am. Soc. Mass Spectrom.*
Anal. Lett.	*J. Biol. Chem.*
Analusis	*J. Chem. Soc. Chem. Commun.*
Ann. Hematol.	*J. Chromatogr.*
Art. Cells Blood Subst. Immob. Biotechnol.	*J. Exp. Med.*
Biochemistry	*J. Mol. Biol.*
Biochem. Biophys. Res. Commun.	*J. Org. Chem.*
Biochim. Biophys. Acta	*J. Toxicol. Environ. Health.*
Biol. Chem. Hoppe Seyler	*Kawasaki Med. J.*
(Biol.) Biomed. (Environ.) Mass Spectrom.	*Kuromatogurafi*
Biomed. Biochem. Acta	*Lab. Equip. Digest*
Br. J. Haematol.	*Mass Spectrom. Rev.*
Chemtracts: Org. Chem.	*Mol. Biol.*
Chim. Ind. (Milan)	*N. Engl. J. Med.*
Chim. Oggi	*Org. Mass Spectrom.*
Chromatographia	*Pept. Res.*
Clin. Biochem.	*Proc. Natl. Acad. Sci. USA*
Clin. Chem.	*Protein Sci.*
Clin. Chem. Enzym. Commun.	*Rapid Commun. Mass Spectrom.*
Clin. Chim. Acta	*Rinsho Byori*
Eur. J. Biochem.	*Shengwu Huaxue Zazhi*
Eur. Mass Spectrom.	*Spectros. Int. J.*
Ital. Chim. Clin.	*Springer Proc. Phys.*
Haematologica	*Tanpakushitsu Kakusan Koso*
Hemoglobin	*Toxicol. Appl. Pharmacol.*
Hoppe-Seyler's Z Physiol. Chem.	*Trends Anal. Chem.*
Human Genet.	*Yeast*

data for a particular analyte have been published. If, however, the user recognizes the existence of the problem and follows the procedure outlined in this chapter, then the shortcomings of the commercially available bibliographic databases may be accommodated successfully.

Acknowledgment

We thank J. Silbering of the Department of Drug Abuse at the Karolinska Institute, Stockholm for his assistance and interest in this study.

References

1. Brooks, C. T. (1980) *Computer Production and Analysis of a Fully Indexed Bibliographic Database on Oestrogen Methodologies.* PhD Thesis, University of Salford, UK.
2. Blunden-Ellis, J. C. P. (1984) *Bibliographic Analysis of the Literature of the HPLC Methods for Separating PTH-Amino Acids.* MSc Thesis, University of Salford, UK.
3. Barber, M., Bell, D., Morris, M., Tetler, L. W., Woods, D., Gordon, D. B., Garner, G. V., Airey, C. J., Croft, L. R., and Oliver, R. W. A. (1988) Mass Spectral Analysis of Aberrant Haemoglobins. 11th International MS conference, Bordeaux, France, September.
4. Green, B. N., Quaife, R., Hassounah, F. H., and Oliver R. W. A. (1996) On the combined use of mass spectrometric and genetic analytical techniques to identify hemoglobin variants. *Hemoglobin,* in press.

Appendix

A definitive bibliography on the application of mass spectrometry to the study of human hemoglobins and a "checklist" of retrieval performance using three bibliographic databases.*

1. Bookchin, R. M. and Gallop, P. M. (1968) Structure of hemoglobin A_{1c}—nature of the *N*-terminal β-chain blocking group. *Biochem. Biophys. Res. Commun.* **32,** 86–93. XXX
2. Morris, H. R. and Williams, D. H. (1972) The identification of a mutant peptide of an abnormal haemoglobin by mass spectrometry. *J. Chem. Soc. Chem. Commun.* 114–116. XXX
3. Wada, Y., Fugita, T., Hayashi, A., Matsuo, T., Katakuse, I., and Matsuda, H. (1980) I. Structural analysis of human hemoglobin variants by field desorption mass spectrometry. *Iyo Masu Kenkyukai Koenshu* **5,** 135–140. (Japanese) XCX
4. Matsuo, T., Matsuda, H., Katakuse, I., Wada, Y., Fujita, T., and Hayashi, A. (1981) Field desorption mass spectra of tryptic peptides of human hemoglobin chains. *Biomed. Mass Spectrom.* **8,** 25–30. BCX
5. Wada, Y., Hayashi, A., Fujita, T., Matsuo, T., Katakuse, I., and Matsuda, M. (1981) Structural analysis of human hemoglobin variants with field desorption mass spectrometry. *Biochim. Biophys. Acta.* **667,** 233–241. BCM
6. Aschauer, H., Schafer, W., Sanguansermsri, T., and Braunitzer, G. (1981) Human embryonic haemoglobins. Ac-Ser-Leu-Thr- is the N-terminal sequence of the zeta-chains. *Hoppe Seyler's Z. Physiol. Chem.* **362,** 1657–1659. XCM
7. Matsuo, T., Katakuse, I., Matsuda, H., Wada, Y., Fujita, T., and Hayashi, A. (1981) II. Field desorption mass spectra of peptide mixture of human globin digested by thermolysin and staphyloccal protease. *Iyo Masu Kenkyukai Koenshu* **6,** 107–110. (Japanese) XCX

*Key to symbols used: B, references found in BIDS; C, reference found in *Chemical Abstracts*; M, reference found in Medline; X, reference not retrieved using search terms given in text.

8. Wada, Y., Hayashi, A., Matsuo, T., Sakurai, T., Matsuda, H., and Higuchi, T. (1982) Structural analysis of human hemoglobin variants by mass spectrometry. III. Comparison of FD and FAB mass spectra and identification of homozygosity for T-gamma globin by mass spectrometry. *Iyo Masu Kenkyukai Koenshu* **7**, 123–126. (Japanese) XCX

9. Wada, Y., Hayashi, A., Matsuo, T., Sakurai, T., Matsuda, H., and Katakuse, I. (1983) Structural analysis of human hemoglobin variants by mass spectrometry. IV. Characterisation of two new hemoglobin variants by FDMS and molecular SIMS. *Iyo Masu Kenkyukai Koenshu* **8**, 209–212. (Japanese) XCX

10. Wada, Y., Hayashi, A., Fujita, T., Matsuo, T., Katakuse, I., and Matsuda, H. (1983) Structural analysis of human hemoglobin variants by mass spectrometry. *Int. J. Mass Spectrom. Ion Phys.* **48**, 209–212. BCX

11. Wada, Y., Hayashi, A., Masanori, F., Katakuse, I., Ichihara, T., Nakabushi, H., Matsuo, T., Sakurai, T., and Matsuda, H. (1983) Characterisation of a new fetal haemoglobin variant, Hb F Izumi $^{A\gamma}$ (6) Glu → Gly, by molecular secondary ion mass spectrometry. *Biochim. Biophys. Acta.* **749**, 244–248. XCM

12. Katakuse, I., Ichihara, T., Nakabushi, H., Matsuo, T., Matsuda, H., Wada, Y., and Hayashi, A. (1984) Secondary ion mass spectra of tryptic peptides of human hemoglobin chains. *Biomed. Mass Spectrom.* **11**, 386–391. BXM

13. Pucci, P., Carestia, C., Fioretti, G., Mastrobuoni, A. M., and Pagano, L. (1985) Protein finger printing by FAB. Characterisation of normal and variant human haemoglobins. *Biochem. Biophys. Res. Commun.* **130**, 84–90. XCM

14. Wada, Y., Hayashi, A., Matsuo, T., Matsuda, H., and Katakuse, I. (1985) Structural analysis of human hemoglobin variants by mass spectrometry. V. Characterisation of a foetal hemoglobin variant in the extract from dried blood on filter paper. *Iyo Masu Kenkyukai Koenshu* **10**, 87–90. (Japanese) XCX

15. Fujita, S., Ohta, Y., Saito, S., Kobayashi, Y., Naritomi, Y., Kawaguchi, K., Imamura, T., Wada, Y., and Hayashi, A. (1985) Hemoglobin A_2 Honai $\alpha_2\delta_2(90)$ Glu → Val: a new delta chain variant. *Hemoglobin* **9(6)**, 597–607. XXX

16. Wada, Y., Hayashi, A., Katakuse, I., Matsuo, T., and Matsuda, H. (1985) Application of glycinamidation to the peptide mapping using secondary ion mass spectrometry. *Biomed. Mass Spectrom.* **12**, 122–126. XXX

17. Rahbar, S., Louis, J., Lee, T., and Asmerom, Y. (1985) Hemoglobin North Chicago, $\beta(36)$ Pro → Ser. A new high affinity hemoglobin. *Hemoglobin* **9**, 559–576. XXM

18. Boissel, J.-P., Kasper, T. J., Shah, S. C., Malone, J. I., and Bunn, H. F. (1986) Amino-terminal processing of proteins. Hemoglobin South Florida. *Proc. Natl. Acad. Sci. USA* **82**, 8448–8452. XXM

19. Wada, Y., Fujita, T., Kidoguchi, K., and Hayashi, A. (1986) Foetal haemoglobin variants in 80,000 Japanese neonates: high prevalence of Hb F Yamaguchi $^{A\gamma}$T(80) Asp → Asn. *Hum. Genet.* **72**, 196–202. XXX

20. Rahbar, S., Lee, T. D., Baker, J. A., Rabinowitz, L. T., Asmerom, Y., Legesse, K., and Ranney, H. M. (1986) Reverse phase HPLC and SIMS. A strategy for identification of ten human hemoglobin variants. *Hemoglobin* **10**, 379–400. BCM

21. Blouquit, Y., Rhoda, M. D., Delanoe-Garin, J., Rosa, R., Prome, J. C., Poyart, C., Puzo, G., Bernassaus, J. M., and Rosa, J. (1986) Glycerated hemoglobin $\alpha_2\beta_2(82)N$-ε-glyceryllysine: a new post-translational modification occurring in erythrocyte bisphosphoglyceromutase deficiency. *J. Biol. Chem.* **261,** 6758–6764. XCX

22. Matsuo, T., Sakurai, T., Katakuse, I., Matsuda, H., Wada, Y., and Hayashi, A. (1986) Amino acid sequencing of peptide mixtures: structural analysis of human hemoglobin variants (Digit Printing Method). *Springer Proc. Phys.* **9,** 113–117. XCX

23. Castagnola, M., Landolfi, R., Rossetti, D., DeAngelis, F., and Ceccarelli, S. (1986) Determination of abnormal hemoglobins by the combined use of reversed phase high performance liquid chromatography and fast atom bombardment mass spectrometry. *Anal. Lett.* **19,** 1793–1807. XCX

24. Wada, Y. (1986) Structural analysis of variant protein by mass spectrometry. *Iyo Masu Kenkyukai Koenshu* **11,** 55–60. (Japanese) XXX

25. Hidaka, K. and Iuchi, I. (1986) Hemoglobin J-Norfolk found in the Kobe District. *Kawasaki Med. J.* **12,** 97–99. (Japanese) XCX

26. Hayashi, A., Wada, Y., Matsuo, T., Katakuse, L., and Matsuda, H. (1987) Neonatal screening and mass spectrometric analysis of hemoglobin variants in Japan. *Acta. Haematol.* **78,** 114–118. XXM

27. Wada, Y., Ikkala, E., Imai, K., Matsuo, T., Matsuda, H., Lehtinen, M., Hayashi, A., and Lehmann, H. (1987) Structure and function of a new hemoglobin variant Hb. Meilahti $\alpha_2\beta_2(36)$Pro → Thr, characterised by mass spectrometry. *Acta. Haematol.* **78,** 109–113. XCM

28. Prome, D., Prome, J. C., Blouquit, Y., Lacombe, C., Rosa, J., and Robinson, J. D. (1987) FAB mapping of proteins: detection of mutation sites in abnormal human hemoglobins. *Spectros. Int. J.* **5,** 157–170. XCX

29. Wada, Y. and Hayashi, A. (1987) Structural analysis of hemoglobin by mass spectrometry: a review. *Tanpakushitsu Kakusan Koso* **32,** 697–703. (Japanese) XXM

30. Blouquit, Y., Rhoda, M. D., Delanoe-Garin, J., Rosa, R., Prome, J. C., Poyart, C., Puzo, G., Bernassaus, J. M., and Rosa, J. (1987) Glycerated hemoglobin $\alpha_2\beta_2(82)$ N-ε-glyceryllysine: a new post-translational modification occuring in erythrocyte bisphosphoglyceromutase deficiency. *Biomed. Biochim. Acta.* **46,** S202–S206. XXX

31. Castognola, M., Dobasz, M., Landolfi, R., Pascali, V. L., deAngelis, F., Vettore, L., and Perona, G. (1988) Determination of neutral haemoglobin variants by immobilized pH gradient, reversed-phase high-performance liquid chromatography and fast-atom bombardment mass spectrometry—the case of Hb Torino $\alpha(43)$Phe → Val. *Biol. Chem. Hoppe Seyler.* **369,** 241–246. XXM

32. DeBiasi, R., Spiteri, D., Caldora, M., Iodice, R., Pucci, P., Malorni, A., Ferranti, P., and Marino, G. (1988) Identification by fast atom bombardment mass spectrometry of Hb Indianapolis $\beta(112)$Cys → Arg in a family from Naples, Italy. *Hemoglobin* **12,** 323–336. XXM

33. Keitt, A. S. and Jones, R. T. (1988) The variant fetal hemoglobin F Texas I is abnormally acetylated. *Am. J. Hematol.* **28,** 47–52. XCM

34. Prome, D., Prome, J. C., Pratbernou, F., Blouquit, Y., Galacteros, F., Lacombe, C., Rosa, J., and Robinson, J. D. (1988) Identification of some abnormal haemoglobins by fast atom bombardment mass spectrometry and fast atom bombardment tandem mass spectrometry. *Biomed. Environ. Mass Spectrom.* **16,** 41–44. XCM

35. Wada, Y., Hayashi, A., Matsuo, T., and Sakurai, T. (1988) Analysis of protein variants by high performance mass spectrometer. *Iyo Masu Kenkyukai Koenshu* **13,** 187–190. (Japanese). XCX

36. Rahbar, S., Rosen, R., Nozari, G., Lee, T. D., Asmeron, Y., and Wallace, R. B. (1988) Hemoglobin. Pasadena. *Am. J. Hematol.* **27,** 204–208. XCM

37. Pucci, P., Ferranti, P., Marino, G., and Malorni, A. (1989) Characterisation of abnormal human haemoglobins by fast atom bombardment mass spectrometry. *Biomed. Environ. Mass Spectrom.* **18,** 20–26. XCM

38. Molchanova, T. P., Mirgorodskaya, O. A., Abaturov, L. V., Podtelezhnikov, A. V., Yu, Tokarev, N., and Grachev, S. A. (1989) Location of amino acid substitutions in human hemoglobin. Mass spectrometric rapid analysis of tryptic peptides. *Mol. Biol. (USSR)* **23,** 225–239. (Russian) XCM

39. Wada, Y., Hayashi, A., Oka, Y., Matsuo, T., Sakurai, T., Matsuda, H., and Katakuse, I. (1989) Mass spectrometric characterisation of a haemoglobin variant, haemoglobin Riyadh. *Int. J. Mass Spectrom. Ion Proc.* **91,** 79–84. XCX

40. Liu, S., Ren, B., He, W., Wen, H., and Weng, Q. (1989) New methods for determining amino acid sequences: FAB-mass spectroscopy. *Shengwu Huaxue Zazhi* **5,** 97–101. (Chinese) XXX

41. Wada, Y., Fujita, T., Hayashi, A., Sakurai, T., and Matsuo, T. (1989) Structural analysis of protein variants by mass spectrometry. Characterisation of haemoglobin Providence using a grand-scale mass spectrometer. *Biomed. Environ. Mass Spectrom.* **18,** 563–565. XXM

42. Wada, Y., Matsuo, T., and Sakurai, T. (1989) Structure elucidation of hemoglobin variants and other proteins by digit-printing method. *Mass Spectrometry Rev.* **8,** 379–434. XXX

43. Stachowiak, K. and Dyckes, D. F. (1989) Peptide mapping using thermospray LC/MS detection: rapid identification of hemoglobin variants. *Pept. Res.* **2,** 267–274. XXM

44. Pucci, D., Marino, G., Ferranti, P., and Malorni, A. (1989) Identification by FAB-MS of hemoglobin Indianapolis in a family from Naples. *Adv. Mass Spectrom.* **11,** 1428,1429. XXX

45. Malorni, A., Pucci, P., Ferranti, P., and Marino, G. (1989) Mass spectrometric analysis of human hemoglobin variants. *Chim. Oggi* **7,** 57–60. XCX

46. Pucci, P., Ferranti, P., Malorni, A., and Marino, G. (1989) Spettrometria di massa FAB nello studio delle emoglobinopatie. *G. Ital. Chim. Clin.* **14,** 115–121. (Italian) XCX

47. Foldi, J., Horanyi, M., Szelenyi, J. G., Hollan, S. R., Aseeva, E. A., Lutsenko, I. N., Spivak, V. A., Toth, O., and Rozynov, B. V. (1989) Hemoglobin Siriraj found in the Hungarian population. *Hemoglobin* **13**, 177–180. XXX

48. Ferranti, P. (1989) Caratterizzazione di varianti genetiche dell'emoglobina umana mediante spettrometria di massa FAB. *Chim. Ind.* **71**, Pt 1–2 (ParteI) 89,90 (Milan) and Pt 7–8 (Parte II), 221–231. (Italian) BXX

49. Lacombe, C., Prome, D., Blonquit, Y., Bardakdjian, J., Arous, N., Mrad, A., Prome, J.-C., and Rosa, J. (1990) New results of hemoglobin variants structure determination by FAB-mass spectrometry. *Hemoglobin* **14**, 529–548. BXM

50. Green, B. N., Oliver, R. W. A., Falick, A. M., Shackleton, C. H. L., Roitman, F., and Witkowska, H. E. (1990) Electrospray MS, LSIMS for the rapid detection and characterisation of variant haemoglobins, in *Biological Mass Spectrometry* (Burlingame, A. L. and McCloskey, J. A., eds.), Elsevier, Amsterdam, pp. 129–146. XXX

51. Hillenkamp, F., Karas, M., Ingendoh A., and Stahl, B. (1990) Matrix assisted UV-laser desorption/ionization. A new approach to mass spectrometry of large biomolecules, in *Biological Mass Spectrometry* (Burlingame, A. L. and McCloskey, J. A., eds.), Elsevier, Amsterdam, pp. 49–60. XXX

52. Pucci, P., Ferranti, P., Malorni, A., and Marino, G. (1990) FAB-MS analysis of haemoglobin variants: use of V-8 protease in the identification of HbM Hyde Park and Hb San Jose. *Biomed. Environ. Mass Spectrom.* **19**, 568–572. XCM

53. Falick, A. M., Shackleton, C. H. L., Green, B. N., and Witkowska, H. E. (1990) Tandem mass spectrometry in the clinical analysis of variant hemoglobins. *Rapid Commun. Mass Spectrom.* **4**, 396–400. XXM

54. Williamson, D., Nutkins, J., Rosthoj, S., Brennan, S. O., Williams, D. H., and Carrell, R. W. (1990) Characterisation of Hb Aalborg, a new unstable hemoglobin variant, by FAB mass spectrometry. *Hemoglobin* **14**, 137–145. XCM

55. Hill, R. E. (1990) The widening horizons of bioanalytical mass spectrometry. *Clin. Chim. Acta.* **194**, 1–17. XXM

56. Cappiello, A., Palma, P., Papayannopoulos, A., and Biemann, K. (1990) Efficient introduction of HPLC fractions into a high performance tandem mass spectrometer. *Chromatographia* **30**, 477–483. BCX

57. Frigeri, F., Pandolfi, G., Camera, A., Rotoli, B., Ferranti, P., Malorni, A., and Pucci, P. (1990) Hemoglobin G San-Jose: identification by mass spectrometry. *Clin. Chem. Enzym. Comm.* **3**, 289–294. XXX

58. Shackleton, C. H. L., Falick, A. M., Green, B. N., and Witkowska, H. E. (1991) Electrospray MS in the clinical diagnosis of variant haemoglobins. *J. Chromatogr.* **562**, 175–190. XCM

59. Falick, A. M., Witkowska, H. E., Labin, B. H., Nagel, R. L., and Shackleton, C. H. L. (1991) Identification of variant haemoglobins by tandem mass spectrometry, in *Techniques in Protein Chemistry II* (Villafranca, J. J., ed.), Academic, San Diego, CA, pp. 557–565. XXX

60. Oliver, R. W. A. and Green, B. N. (1991) On the application of electrospray-mass spectrometry to the characterisation of abnormal or variant haemoglobins. *Trends Anal. Chem.* **10**, 85–91. XXX

61. Ferranti, P., Malorni, A., Pucci, P., Fanali, S., Nardi, A., and Ossicini, L. (1991) Capillary zone electrophoresis and mass spectrometry for the characterization of genetic variants of human hemoglobin. *Anal. Biochem.* **194,** 1–8. XCM

62. Matsuda, H., Matsuo, T., Katakuse, I., and Wada, Y. (1991) Investigation of amino acid mutations by high resolution mass spectrometry, in *Mass Spectrometry of Peptides* (Desiderio, D. M., ed.), CRC, Boca Raton, FL, pp. 221–256. XXX

63. Lee, T. D. and Rahbar, S. (1991) The mass spectral analysis of hemoglobin variants, in *Mass Spectrometry of Peptides* (Desiderio, D. M., ed.), CRC, Boca Raton, FL, pp. 257–274. XXX

64. Prome, D., Blouquit, Y., Ponthus, C., Prome, J. C., and Rosa, J. (1991) Structure of the human adult hemoglobin minor fraction Al_b by electrospray and S.I.M.S. Pyruvic acid as amino-terminal blocking group. *J. Biol. Chem.* **266,** 13,050–13,054. BCM

65. Covey, T. R., Huang, E. C., and Henion, J. D. (1991) Structural characterization of protein tryptic peptides via liquid chromatography/MS and collision-induced dissociation of their doubly charged molecular ions. *Anal. Chem.* **63,** 1193–1200. BCM

66. Jensen, O. N., Hojrup, P., and Roepstorff, P. (1991) Plasma desorption mass spectrometry as a tool in characterization of abnormal proteins. Application to variant human hemoglobins. *Anal. Biochem.* **199,** 175–183. XCM

67. Johansson, I. M., Huang, E. H., Henion, J. D., and Zweigenbaum, J. (1991) Capillary electrophoresis atmospheric pressure ionization mass spectrometry for the characterizatfion of peptides. *J. Chromatogr.* **554,** 311–327. BCM

68. Oliver, R. W. A. (1991) LC and MS in the diagnosis of haemoglobin disorders. *Lab Equip. Dig.* **29,** 9–11. XXX

69. Marsh, G., Masino, G., Pucci, P., Ferranti, P., Malorni, A., Kaeda, J., Marsh, J., and Luzzatto, L. (1991) A third instance of the high oxygen affinity variant, Hb Heathrow [β103(G5)Phe → Leu]: identification of the mutation by mass spectrometry and by DNA analysis. *Hemoglobin* **15,** 43–51. XXM

70. Petrilli, P., Sepe, C., and Pucci, P. (1991) A new procedure for peptide alignment in protein sequence determination using FAB mass spectral data. *Biol. Mass Spectrom.* **20,** 115–120. XCX

71. Jensen, O. N. and Roepstorff, P. (1991) Application of reversed phase high performance liquid chromatography and plasma desorption mass spectrometry for the characterisation of a hemoglobin variant. *Hemoglobin* **15,** 497–507. XXM

72. Jensen, O. N., Roepstorff, P., Rozynov, B., Horanyi, M., Szelenyi, J., Hollan, S. R., Aseeva, E. A., and Spivak, V. A. (1991) Plasma desorption mass spectrometry of haemoglobin tryptic peptides for the characterisation of a Hungarian α-chain variant. *Biol. Mass Spectrom.* **20,** 579–584. BXM

73. Imai, K., Fushitani, K., Miyazaki, C. J., Ishimori, K., Kitagawa, T., Wada, Y., Morimoto, H., Morishima, I., Shih, D. T-b., and Tame, J. (1991) Site-directed mutagenesis in haemoglobin. *J. Mol. Biol.* **218,** 769–778. XXX

74. Witkowska, H. E., Lubin, B. H., Beuzard, Y., Baruchel, S., Esseltine, D. W., Vishinsky, E. P., Kleman, K. M., Bardakdjian-Michau, J., Pinkoski, L., Cahn, S., Roitman, E., Green, B. N., Falick, A. M., and Shackleton, C. H. L. (1991) Sickle cell disease in a patient with sickle cell trait and compound heterozygosity for hemoglobin S and hemoglobin Quebec-Chori. *N. Engl. J. Med.* **325,** 1150–1154. XXX

75. Prome, J. C. (1991) Characterisation of post-translational modifications of proteins by mass spectrometry: some selected problems. *Analusis* **19,** 79–84. BCX

76. Manning, L. R., Morgan, S., Beavis, R. C., Chait, B. T., Manning, J. R., Hess, J. R., Cross, M., Currell, D. L., Marini, M. A., and Winslow, R. B. (1991) Preparation, properties and plasma retention of human hemoglobin derivatives; comparison of uncrosslinked carboxymethylated hemoglobin with crosslinked tetrameric hemoglobin. *Proc. Natl. Acad. Sci. USA* **88,** 3329–3333. BCM

77. Goldberg, D. E., Slater, A. F. G., Beavis, R., Chait, B., Cerami, A., and Henderson, G. B. (1991) Hemoglobin degradation in the human malaria pathogen Plasmodium falciparum; a catabolic pathway initiated by a specific aspartic protease. *J. Exp. Med.* **173,** 961–969. BCM

78. Lubin, B. H., Witkowska, H. E., and Kleman, K. (1991) Laboratory diagnosis of hemoglobinopathies. *Clin. Biochem.* **24,** 363–374. XXM

79. Chowdhury, S. K., Katta, V., and Chait, B. T. (1991) Electrospray ionization mass spectrometric analysis of proteins, in *Methods and Mechanisms for Producing Ions from Large Molecules* (Standing, K. G. and Ens, W., eds.), Plenum, New York (NATO Advanced Science Institute, Series B, vol. 269), pp. 201–210. XCX

80. Ishimori, K., Imai, K., Miyazaki, G., Kitagawa, T., Wada, Y., Morimoto, H., and Morishima, I. (1992) Site-directed mutagenesis in hemoglobin. *Biochemistry* **31,** 3256–3264. XXX

81. Brennan, O., Shaw, J., Allen, J., and George, P. M. (1992) Beta 141 Leu is not deleted in the unstable haemoglobin Atlanta-Coventry but is replaced by a novel amino acid of mass 129 daltons. *Br. J. Haematol.* **81,** 99–103. XCM

82. Wada, Y. (1992) Mass spectrometry in the integrated strategy for the structural analysis of protein variants. *Biol. Mass Spectrom.* **21,** 617–624. XXX

83. De Caterina, M., Esposito, P., Grimaldi, E., Di Mario, G., Scopacasa, F., Ferranti, P., Parlapiano, A., Malorni, A., Pucci, P., and Marino, G. (1992) Characterization of hemoglobin Lepore variants by advanced mass spectrometric procedures. *Clin. Chem.* **38,** 1444–1448. XXM

84. Ferrige, A. G., Seddon, M. J., Green, B. N., Jarvis, S. A., and Skilling, J. (1992) Disentangling electrospray spectra with maximum entropy. *Rapid Commun. Mass Spectrom.* **6,** 707–711. XXX

85. Wada, Y., Matsuo, T., Papayannopoulos, I. A., Costello, C. E., and Biemann, K. (1992) Fast atom bombardment and tandem mass spectrometry for the characterisation of hemoglobin variants including a new variant. *Int. J. Mass Spectrom. Ion Processes* **122,** 219–229. BCX

86. De Angioletti, M., Maglione, G., Ferranti, P., De Bonis, C., Lacerra, G., Scarallo, A., Pagano, L., Fioretti, G., Cutolo, R., Malorni, A., Pucci, P., and Carestia, C. (1992) Hemoglobin City of Hope in Italy. Association of the gene with haplotype IX. *Hemoglobin* **16,** 27–34. XCM

87. Wada, Y., Tamura, J., Musselman, B. D., Kassel, D. B., Sakurai, T., and Matsuo, T. (1992) Electrospray ionization mass spectra of hemoglobin and transferrin by a magnetic sector mass spectrometer. *Rapid Commun. Mass Spectrom.* **6,** 9–13. BCM

88. Rotoli, B., Camera, A., Fontana, R., Frigeri, F., Pandolfi, G., Vecchione, R., Poggi, V., Longo, G., Carestia, C., De Angiolestti, M., Lacerra, G., Pucci, P., Marino, G., Ferranti, P., Malorni, A., Romano, R., and Formisano, S. (1992) Hb-Hyde Park. A de novo mutation identified by mass spectrometry and DNA analysis. *Haematologica* **77,** 110–118. XXM

89. Vassaur, C., Blouquit, Y., Kister, J., Prome, D., Kavanaugh, J. S., Rogers, P. H., Guillemin, C., Arnone, A., Galacteros, F., Poyart, C., Rosa, J., and Wajcman, H. (1992) Hemoglobin Thionville. *J. Biol. Chem.* **267,** 12,682–12,691. XXX

90. Coghlan, D., Jones, G., Denton, K. A., Wilson, M. T., Chan, B., Harris, R., Woodrow, J. R., and Ogden, J. E. (1992) Structural and functional characterisation of recombinant human, haemoglobin A expressed in *Saccharomyces cerevisiae*. *Eur. J. Biochem.* **207,** 931–936. XXX

91. Malorni, A., Pucci, P., Ferranti, P., and Marino, G. (1992) Characterisation of human hemoglobin variants by mass spectrometry, in *Mass Spectrometry in the Biological Sciences: A Tutorial* (Gross, M. L., ed.), Kluwer, Dordrecht, Germany, pp. 325–332. CBX

92. Suwanrumpha, S., McClean, M. A., Fink, S. W., Wilder, C., Stachowiak, K., Dyckes, D. F., and Freas, R. B. (1992) Determination of biomolecules by using liquid chromatography and thermospray mass spectrometry, in *Mass Spectrometry in the Biological Sciences: A Tutorial* (Gross, M. L., ed.), Kluwer, Dordrecht, Germany, pp. 281–301. XXX

93. Roepstorff, P. (1992) Plasma desorption mass spectrometry; principles and applications to protein studies, in *Mass Spectrometry in the Biological Sciences: A Tutorial* (Gross, M. L., ed.), Kluwer, Dordrecht, Germany, pp. 213–227. XXX

94. Seta, K., Hail, M., Mylchreest, I., and Okuyama, T. (1992) Structural analysis of hemoglobin variant by microbore column HPLC/ESI/TSQMS. *Kuromatogurafi* **13,** 349,350. (Japanese) XCX

95. Mosca, A., Paleari, R., Rubino, F. M., Zecca, L., De Bellis, G., Debernardi, S., Baudo, F., Cappellini, D., and Fiorelli, G. (1993) Hemoglobin Abrusso identified by mass spectrometry and DNA analysis. *Hemoglobin* **17,** 261–268. XCM

96. Witkowska, H. E., Bitsch, F., and Shackleton, C. H. L. (1993) Expediting variant hemoglobin characterization by combined HPLC/electrospray mass spectrometry. *Hemoglobin* **17,** 227–242. XXM

97. Light-Wahl, K. J., Loo, J. A., Edmonds, C. G., Smith, R. D., Witkowska, H. E., Shackleton, C. H. L., and Wu, C. S. C. (1993) Collisionally activated dissocia-

tion and tandem mass-spectrometry of intact hemoglobin beta-chain variant proteins with electrospray ionization. *Biol. Mass Spectrom.* **22**, 112–120. BCM

98. Shen, T.-J., Ho, N. T., Simplaceanu, V., Zou, M., Green, B. N., Tam, M. F., and Ho, C. (1993) Production of unmodified human adult hemoglobin in *Escherichia coli. Proc. Natl. Acad. Sci.* **90**, 8108–8112. BCM

99. Lane, P. A., Witkowska, H. E., Falick, A. M., Houston, M. L., and McKinna, J. D. (1993) Hemoglobin D Ibadan-beta thalassemia: detection by neonatal screening and confirmation by electrospray ionization mass spectrometry. *Am. J. Hematol.* **44**, 158–161. XXM

100. Brennan, S. O., Shaw, J. G., George, P. M., and Huisman, T. H. J. (1993) Post-translational modification of beta 141 Leu associated with the β(75)Leu → Pro mutation in hemoglobin Atlanta. *Hemoglobin* **17**, 1–7. XCM

101. Ferranti, P., Parlapiano, A., Malorni, A., Pucci, P., Marino, G., Cossu, G., Manca, L., and Masala, B. (1993) Hemoglobin Oziero: a new alpha chain variant (α(71) Ala → Val) characterization using FAB- and electrospray-mass spectrometric techniques. *Biochim. Biophys. Acta.* **1162**, 203–208. BCM

102. De Llano, J. J. M., Jones, W., Schneider, K., Chait, B. T., Benjamin, L. J., and Weksler, B. (1993) Biochemical and functional properties of recombinant human sickle hemoglobin expressed in yeast. *J. Biol. Chem.* **268**, 27,004–27,011. XCX

103. Wilson, J. B., Brennan, S. O., Allen, J., Shaw, J. G., Gu, L.-H., and Huisman, T. H. J. (1993) The M gamma chain of human fetal hemoglobin is an A gamma chain with an in vitro modification of gamma 141 leucine to hydroxyleucine. *J. Chromatogr. Biomed. Appl.* **617**, 37–42. XCM

104. De Llano, J. J. M., Schneewind, O., Stetler, G., and Manning, J. M. (1993) Recombinant human sickle hemoglobin expressed in yeast. *Proc. Natl. Acad. Sci. USA* **90**, 918–922. BCM

105. Li, Y.-T., Hsieh, Y.-L., Henion, J. T., and Ganem, B. (1993) Studies on heme binding in myoglobin, hemoglobin and cytochrome C by ion spray mass spectrometry. *J. Am. Soc. Mass Spectrom.* **4**, 631-637. XXX

106. Loo, J. A., Ogorzalek-Loo, R. R., and Andrews, P.C. (1993) Primary to quaternary protein structure determination with electrospray ionization and magnetic sector mass spectrometry. *Org. Mass Spectrom.* **28**, 1640–1649. XXX

107. Ganem, B. and Henion, J. D. (1993) Detecting non-covalent complexes of biological macromolecules: new applications of ion-spray mass spectrometry. *Chemtracts Org. Chem.* **6**, 1–22. XXX

108. Wajcman, H., Kalmes, G., Groff, P., Prome, D., Riou, J., and Galacteros, F. (1993) Hemoglobin Melusine; (α(114)Pro → Ser). A new neutral hemoglobin variant. *Hemoglobin* **17**, 397–405. XXX

109. Wajcman, H., Kister, J., Prome, D., Galacteros, F., and Gilsanz, F. (1993) Hb Villaverde (β(89)Ser → Thr): the structural modification of an intrasubunit contact is responsible for a high oxygen affinity. *Biochim. Biophys. Acta* **1225**, 89–94. BCM

110. Springer, D. L., Bull, R. J., Goheen, S. C., Sylvester, D. M., and Edmonds, C. G. (1993) Electrospray ionisation mass spectrometric characterization of acrylamide adducts to hemoglobin. *J. Toxicol. Environ. Health* **40**, 161–176. XCM

111. Bergmark, E., Calleman, C. J., He, F., and Costa, L. G. (1993) Determination of hemoglobin adducts in humans occupationally exposed to acrylamide. *Toxicol. Appl. Pharmacol.* **120,** 45–54. XXM

112. Shackleton, C. H. L. and Witkowska, H. E. (1994) Mass spectrometry in the characterisation of variant hemoglobins, in *Mass Spectrometry: Clinical and Biomedical Applications,* vol. 2 (Desiderio, D. M., ed.), Plenum, New York, pp. 135–199. XXX

113. Wada, Y. and Matsuo, T. (1994) Structure determination of aberrant proteins, in *Biological Mass Spectrometry, Present and Future* (Matsuo, T., Caprioli, R. M., Gross, M. L., and Seyama, Y., eds.), Wiley, Chichester, UK, pp. 369–399. XXX

114. Light-Wahl, K. L., Schwartz, B. L., and Smith, R. D. (1994) Observations on the non covalent quaternary associations of proteins by electrospray ionisation mass spectrometry. *J. Am. Chem. Soc.* **116,** 5271–5278. BCX

115. De Llano, J. J. and Manning, J. M. (1994) Properties of a recombinant hemoglobin double mutant. *Protein Sci.* **3,** 1206–1212. BCM

116. Yanase, H., Cahill, S., De Llano, J. J., Manning, L. R., Schneider, K., Chait, B. T., Vandegriff, K. D., Winslow, R. M., and Manning, J. M. (1994) Properties of a recombinant hemoglobin with aspartic acid 99 (beta) substituted by lysine. *Protein Sci.* **3,** 1213–1223. BCM

117. Wajcman, H., Kister, J., M'Rad, A., Soummer, A. M., and Galacteros, F. (1994) Hemoglobin Cemenelum: $\alpha(92)$Arg → Trp) a hemoglobin variant of the alpha 1/beta 2 interface that displays a moderate increase in oxygen affinity. *Ann. Hematol.* **68,** 73–76. XXX

118. Bakhtiar, R., Wu, Q., Hofstadler, S. A., and Smith, R. D. (1994) Charge state specific facile gas-phase cleavage of Asp 75-Met 76 peptide bond in the alpha chain of human apohemoglobin probed by electrospray ionization mass spectrometry. *Biol. Mass Spectrom.* **23,** 707–710. BCM

119. Ferranti, P., Malorni, A., and Pucci, P. (1994) Structural characterisisation of hemoglobin variants using capillary electrophoresis and fast atom bombardment mass spectrometry. *Methods Enzymol.* **231,** 45–65. XXM

120. Ishimori, K., Hashimoto, M., Imai, K., Fushitani, K., Miyazaki, G., Morimoto, H., Wada, Y., and Morishima, I. (1994) Site directed mutagenesis in hemoglobin. *Biochemistry* **33,** 2546–2553. XXM

121. Konishi, Y. and Feng, R. (1994) Conformational stability of heme proteins in vacuo. *Biochemistry* **33,** 9706–9711. BCM

122. Bonaventura, C., Bonaventura, J., Stevens, R., and Millington, D. (1994) Acrylamide in polyacrylamide gels can modify proteins during electrophoresis. *Anal. Biochem.* **222,** 44–48. BXX

123. Rao, M. J., Schneider, K., Chait, B. T., Chao, T. L., Keller, H., Anderson, S., Manjula, B. N., Kumar, R., and Acharya, A. S. (1994) Recombinant hemoglobin A produced in transgenic swine: structural equivalence with human hemoglobin A. *Art. Cells Blood Subs. Immob. BioTech.* **22,** 695–770. BCM

124. Woolfit, A. R. and Bott, P. A. (1994) The analysis of native protein complexes by electrospray ionisation on an Autospec. *Fisons Instruments Application Notes* no. 36. XXX

125. Kluger, R. and Song, Y. (1994) Changing a protein into a general acylating reagent. *J. Org. Chem.* **59,** 733–736. BXX

126. Ferranti, P., Barone, F., Pucci, P., Malorni, A., Marino, G., Pilo, G., Manca, L., and Masala, B. (1994) HbF-Sassari-a novel G-gamma variant with Thr at (75), characterised by MS techniques. *Hemoglobin* **18,** 307–315. BCM

127. Gilbert, S. C., Van Urk, H., Greenfield, A. J., McAvoy, M. J., Denton, K. A., Coghlan, D., Jones, B. D., and Mead, D. J. (1994) Increase in copy number of an integrated vector during continuous culture of Hansenula polymorpha expressing functional human hemoglobin. *Yeast* **10,** 1569–1580. BCM

128. Kim, H.-W., Shen, T.-J., Sun, D. P., Ho, N. T., Madrid, M., and Ho, C. (1995) A novel low oxygen affinity recombinant hemoglobin (α(96) Val → Trp). *J. Mol. Biol.* **248,** 867–882. XXX

129. Shimizu, A. and Nakanishi, T. (1995) Applications of mass spectrometry for clinical laboratory test. *Rinsho Byori* **43,** 8–18 (Japanese). XXM

130. Yanase, H., Manning, L. R., Vandegriff, K., Winslow, R. M., and Manning, J. M. (1995) A recombinant human hemoglobin with asparagine-102(β) substituted by alanine has a limiting low oxygen affinity, reduced marginally by chloride. *Protein Sci.* **4,** 21–28. BCM

131. Hofstadler, S. A., Swanek, F. D., Gale, D. C., Ewing, A. G., and Smith, R. D. (1995) Capillary electrophoresis-electrospray ionization Fourier transform ion cyclotron resonance mass spectrometry. *Anal. Chem.* **67,** 1477–1480. XCM

132. Prome, D., Prome, J. C., Gale, D. C., and Rosa, J. (1995) Characterization of new amino-terminal blocking groups in the normal human adult hemoglobin. *Eur. Mass. Spectrom.* **1,** 195–201. XCM

Appendices

You never know when a good appendix may prove useful—"I happened to have a copy of the affidavit with me in the appendix to . . . Zollner's 'Transcendental Physics.'" (Joseph Cook, *Occident* lect. v).

Appendix I: Glossary of Abbreviations and Acronyms

The following terms and their contracted forms, together with those to be found in Appendices III–VI, are used in this volume specifically or are in general use.

AA	Amino acid.
ACN	Acetonitrile.
ACTH	Adrenocorticotrophic hormone.
AP	Alkaline phosphatase.
APCI	Atmospheric pressure chemical ionization (also in APCI-MS).
API	Atmospheric pressure ionization.
Apo-D	Apolipoprotein D.
B	Magnetic sector—also used in instrument configurations (e.g., BEEB and BEqQ) and in linked-scanning modes (e.g., B/E).
BaP	Benzo[a]pyrene.
BE	β-Endorphin.
BSA	Bovine serum albumin.
Cam	Carbamidomethylcysteine.
CD	Circular dichroism.
cDNA	Complementary DNA.
CF-FAB	Continuous flow-fast atom bombardment.
CHAPS	3-[(3-cholamidopropyl)dimethylammonio]-1-propanesulphonate.
CI	Chemical ionization.
CID	Collision-induced dissociation.
CMC	Critical micelle concentration.
Cmc	Carboxymethylcysteine.
ConA	Concanavalin A.
ΔCS	Capillary-skimmer voltage difference in ESI.
CTAB	Cetyltrimethylammonium bromide.
CZE/MS	Capillary electrophoresis-mass spectrometry.
Da	Dalton (unit of molecular mass).
DC	Direct current (also as d.c.).
DCI	Desorption chemical ionization.
DHB	2,5-Dihydroxybenzoic acid (gentisic acid).
DLI	Direct liquid introduction.

Dm	Distamycin.
DMHAA	Dimethylhexylamineacetyl.
DMOAA	Dimethyloctylamineacetyl.
E	Electrostatic sector—also used in instrument configurations (e.g., EBE and EBEqQ) and in linked-scanning modes (e.g., B/E).
e^-	Electron.
e^-_{th}	Electron with thermal energy.
EDC	1-(3-Dimethylaminopropyl)-3-ethylcarbodiimide.
EDTA	Ethylenediaminetetraacetic acid.
EI	Electron ionization (also in EI-MS).
ESI	Electrospray ionization (also in ESI-MS).
eV	Electron volts.
FAB	Fast-atom bombardment (also in FAB-MS).
FD	Field desorption.
FMOC	9-Fluorenylmethoxycarbonyl chloride.
FSC	Fused-silica capillary.
FSH	Follicle-stimulating hormone.
FTICR	Fourier transform ion cyclotron resonance (see also FTMS).
FTMS	Fourier transform mass spectrometry (see also FTICR).
FWHM	Full width at half-maximum intensity.
GC	Gas chromatography.
GC/MS	Gas chromatography-mass spectrometry (also in GC/EI-MS, etc.).
G6PDH	Glucose-6-phosphate dehydrogenase.
Hb	Hemoglobin.
HPLC	High-performance liquid chromatography (also in HPLC/MS, HPLC/UV, etc.).
HSA	Human serum albumin.
IAA	Iodoacetic acid.
IP	Ionization potential.
IR	Infrared.
IR-MALDI-MS	Infrared matrix-assisted laser desorption/ionization.
IS	Ion-spray (also in IS-MS).
ITMS	Ion trap mass spectrometry.
K_d	Dissociation constant.
kDa	Kilodalton (10^3 Da).
keV	Kiloelectron volts.
LC	Liquid chromatography.
LCAT	Lecithin:cholesterol acyltransferase.
LC/MS	Liquid chromatography-mass spectrometry (also in LC/APCI-MS, etc.).
LDAO	Lauryl dimethylamine oxide.
LDMS	Laser desorption mass spectrometry.
LH	Luteinizing hormone.
LSIMS	Liquid secondary-ion mass spectrometry.
$M^{+\bullet}$	Molecular ion-radical—positively charged.

$(M + H)^+$	Protonated molecular ion—positively charged (also written as MH^+).
$(M + NH_4)^+$	Ammoniated molecular ion—positively charged.
$(M + Na)^+$	Sodiated molecular ion—positively charged.
$(M - H)^+$	Molecular ion after loss of hydrogen radical—positively charged.
$M^{-\bullet}$	Molecular ion-radical—negatively charged.
$(M + Cl)^-$	Chloride addition to molecular ion—negatively charged.
$(M - H)^-$	Molecular ion after loss of hydrogen radical—negatively charged.
$[M - n]$	Loss of n mass units from the molecular ion (e.g., $[M - 17]$).
M_r	Relative molecular mass (*see also* mol wt and RMM).
MALDI	Matrix-assisted laser desorption/ionization (also in MALDI-TOFMS).
MCA	Multichannel analyzer.
MCP	Microchannel plate.
MDa	Megadalton (10^6 Da).
ME	Methionine enkephalin.
MES	4-Morpholinoethanesulphonic acid.
MIKES	Mass-analyzed ion kinetic energy scan.
MSn	Component mass spectrometer of an MSMS instrument (e.g., MS1, MS2).
MSMS	Mass spectrometry-mass spectrometry or tandem mass spectrometry.
$(MS)^2$	As MSMS.
$(MS)^n$	A multiple-stage MSMS experiment.
mol wt	Molecular weight (*see also* M_r).
m/z	Mass-to-charge ratio.
m-NBA	*m*-Nitrobenzyl alcohol.
NH_4OAc	Ammonium acetate.
4-NMPA	*N*-Methyl-4-pyridiniumacetyl (also 3-NMPA).
NMR	Nuclear magnetic resonance.
NP40	Nonidet P40.
oa	Orthogonal acceleration.
ODS	Octadecylsilyl.
OPCP	Opioid peptide-containing protein.
OVA	Ovalbumin.
PA	Proton affinity.
PAD	Postacceleration detector.
PAGE	Polyacrylamide gel electrophoresis (also in 2D-PAGE).
PDMS	Plasma desorption mass spectrometry (also in PD).
PEEK	Polyetherether ketone.
PEG	Polyethylene glycol.
PFK	Perfluorokerosene.
PFTBA	Perfluorotri-*n*-butylamine.
PITC	Phenylisothiocyanate.
POD	Horse radish peroxidase.

POMC	Pro-opiomelanocortin.
ppm	Parts per million.
PSD	Postsource decay (also as in PSD-MALDI).
PVDF	Polyvinylidene difluoride.
PyrA	Pyridiniumacetyl.
Q	Quadrupole mass filter—also used in instrument configurations (e.g., QqQ and BEqQ) and in linked-scanning modes (e.g., B^2/Q).
q	Quadrupole mass filter used as a collision cell (*see* preceding definition for an example).
RE	Recombination energy.
REMPI	Resonance-enhanced multiphoton ionization.
RF	Radio frequency (also as r.f.).
rhuEPO	Recombinant human erythropoietin.
RIA	Radioimmunoassay.
RMM	Relative molecular mass (alternatively rmm and *see* M_r)
RP	Resolving power.
RPLC	Reversed-phase liquid chromatography (also as RP-HPLC).
SCF	Stem cell factor.
SDS	Sodium dodecyl sulphate.
SFC/MS	Supercritical fluid chromatography-mass spectrometry.
SIM	Selected-ion monitoring.
SIMS	Secondary-ion mass spectrometry.
S/N	Signal-to-noise ratio.
SP	Substance P.
SRM	Selected-reaction monitoring.
TEAF	Triethylamine/formic acid.
TFA	Trifluoroacetic acid.
THTA	Tetrahydrothiopheniumacetyl.
TIC	Total-ion current.
TLC	Thin-layer chromatography.
TMA	Trimethylamine
TMAA	Trimethylammoniumacetyl.
TMPB	Trimethylpyryliumbutyl.
TOF	Time-of-flight (instrument, also in TOFMS).
TPCK	1-Tosylamide-2-phenylethyl chloromethyl ketone.
TPPE	Triphenylphosphoniumethyl.
TQ	Triple quadrupole (instrument, also in TSQ and TQMS).
Tris	Tris(hydroxymethyl)aminomethane.
TS	Thermospray (ionization).
u	Unified atomic mass unit (*see also* Da).
UV	Ultraviolet.
UV-MALDI-MS	Ultraviolet matrix-assisted laser desorption/ionization.
WT	Wild type.

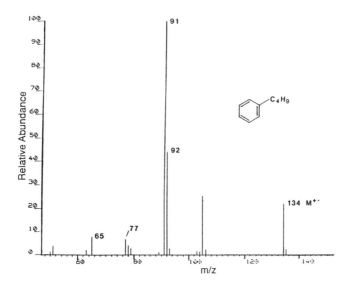

Fig. 1. Electron ionization spectrum of *n*-butylbenzene (mol wt 134).

Appendix II: Mass and Abundance Values

1. Molecular Weight

In the spectrum of *n*-butylbenzene (M_r = 134), reproduced in Fig. 1, the most intense peak in the molecular ion region (i.e., the highest mass region of the spectrum) is taken to be the molecular ion. The nominal mass-to-charge (*m/z*) value of this ion (134) corresponds, for a singly charged ion (cf. Section 2.), to the nominal molecular weight, or relative molecular mass (M_r) of the sample. The higher mass peak in the molecular ion region represents the presence of heavier isotopes (*see* Section 3.).

The nominal molecular weight is simply calculated on an atomic weight scale where C = 12, H = 1, O = 16, N = 14, and so on, and is a perfectly satisfactory parameter for use in the description of lower mol-wt analytes. As the molecular weight increases, however, the weight calculated on the basis of accurate monoisotopic atomic weights (e.g., ^{12}C = 12.0000, ^{1}H = 1.00783, *see* Table 1) increasingly diverges from the simple nominal molecular weight. This difference is particularly evident where large numbers of hydrogen atoms, which provide a relatively large positive fractional contribution, are present. For example, a saturated hydrocarbon with the formula $C_{42}H_{86}$ has a nominal molecular weight of 590 but an accurate molecular weight of 590.6730 (i.e., actually closer to 591 than 590 as a nominal value). In this case, provided that an instrument with sufficient accuracy of mass assignment was available, a

Table 1 (Appendix II)
Mass and Abundance Values

Element	Isotopically averaged mass[a]	Isotope	Isotopic mass	Isotopic abundance, %
H	1.00794	^{1}H	1.007835[b]	99.985
		^{2}H	2.01410	0.015
C	12.011	^{12}C	12.00000	98.90
		^{13}C	13.003355[b]	1.10
N	14.00674	^{14}N	14.00307	99.634
		^{15}N	15.00011	0.366
O	15.9994	^{16}O	15.99491	99.762
		^{17}O	16.99913	0.038
		^{18}O	17.99916	0.200
F	18.99840	^{19}F	18.99840	100
Na	22.98977	^{23}Na	22.98977	100
Si	28.0855	^{28}Si	27.97693	92.23
		^{29}Si	28.97649	4.67
		^{30}Si	29.97377	3.10
P	30.97376	^{31}P	30.97376	100
S	32.066	^{32}S	31.97207	95.02
		^{33}S	32.97146	0.75
		^{34}S	33.96787	4.21
		^{36}S	35.96708	0.02
Cl	35.4527	^{35}Cl	34.96885	75.77
		^{37}Cl	36.96590	24.23
K	39.0983	^{39}K	38.96371	93.258
		^{40}K	39.96400	0.012
		^{41}K	40.96183	6.730

[a]Equivalent to atomic weight.
[b]Not rounded off where result is equivocal.

recorded mass of 590.7 would be more satisfactory than the use of a nominal mass such as 590.

At still higher molecular weight, further changes occur in the molecular ion region and, indeed, in all of the high mass regions of the spectrum. These changes have two major causes. First, the effect of an increasing relative abundance of isotopic molecules (principally the effect of ^{13}C) at higher mass and second, the increasing difficulty in resolving adjacent (nominal) mass peaks at higher mass. In addition to these effects, the divergence of accurate and nominal mass, described in the preceding paragraph, continues to increase. As an example of the effect of isotopes, Fig. 2 shows the molecular ion region that

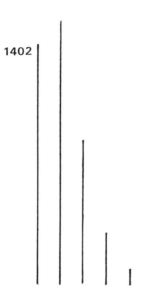

1402

Fig. 2. Molecular ion region of a saturated hydrocarbon, $C_{100}H_{202}$, nominal mol wt 1402. The accurate mol wt is 1403.581 and the isotopically averaged mol wt is 1404.704. Mass increases from left to right.

might be expected in the spectrum of a saturated hydrocarbon with the composition $C_{100}H_{202}$. In Fig. 2 it can be seen that the peak at nominal mass 1403, representing the presence of a single ^{13}C atom, is now more abundant than the lowest mass peak, which corresponds to the presence of the lightest isotopes of carbon and hydrogen at m/z 1402. This trend continues much further with an increase in molecular weight as can be seen in the high-mass data of Fig. 10 in Chapter 8.

The data in Fig. 10 of Chapter 8 (traces A and B) have been plotted as they would be input to the data system and not in the "stick" format used in data system presentation (e.g., Figs. 1 and 2 of this Appendix). Thus, it is evident that these high-mass data are such that a resolution of 15,000 (i.e., very far from routine operation) is required, even to separate the nominal mass components as shown in trace A. Usually, such data would be recorded at low resolution, resulting in a continuous profile which represents all of the isotopic distribution, as shown in trace B. The m/z value which represents the position of this overall profile can be measured, however, in the same way as, for example, any of the separate isotope peaks shown in trace A at high resolution.

The position of any overall profile or individual peak is most representatively measured as its centroid (center of gravity), but a peak-top value, particularly a value measured after curve fitting round the apex, is a close approximation. In an example, such as trace B from Fig. 10 in Chapter 8, where instrumental resolution is insufficient to resolve adjacent nominal mass peaks, the measurement of a mass for the overall isotopic profile corresponds to the determination of an isotopically averaged molecular weight (i.e., one calculated using C = 12.011, H = 1.00794, etc; *see* Table 1). In this case, the isotopically averaged molecular weight is 15,867.2 Da, whereas calculation of the monoisotopic molecular weight (i.e., using C = 12.00000, H = 1.00783, etc.) gives 15,857.2 Da. The peak that corresponds to this monoisotopic molecular weight is, in fact, so small that it is barely visible. An empirical relationship, average isotopic weight = monoisotopic weight × 1.000655 ± 0.15 *(1)*, can be usefully employed.

The isotopically averaged molecular weight is a characteristic measure, in just the same way as the monoisotopic molecular weight, and is almost universally used as a means of identifying higher mol-wt materials (e.g., proteins). For peptides with a molecular weight to about 2000 Da, and for other materials of a similar molecular size, however, a monoisotopic molecular weight (preferably with a single figure after the decimal point) is ideally used. A figure after the decimal point can distinguish, for example, between lipids (high fractional mass owing to many hydrogens) and carbohydrates (low fractional mass owing to a low number of hydrogens) with the same nominal molecular weights. In practice, the ability to measure a monoisotopic molecular weight of this magnitude is dependent on the use of an instrument with sufficient mass resolution, such as a double-focusing magnetic sector instrument (*see* Chapter 2, Section 3.). If resolution of the isotopic components cannot be achieved, then an isotopically averaged value is also used at these intermediate molecular weights.

Finally, it should be recognized that most ionization techniques for higher mol-wt molecules (e.g., FAB, MALDI, and ES) will invariably lead to the recording of $(M + H)^+$ ions in the positive-ion mode and $(M - H)^-$ ions in the negative-ion mode (or multicharged versions of these) rather than a true molecular ion, $M^{+\bullet}$. Ideally, such measurements should be reported after conversion to the corresponding mass of the neutral molecule by subtraction or addition of the mass of hydrogen from or to the measured value.

2. Mass-to-Charge Ratio

As is evident from the discussion of mass analyzers in Chapter 2 (Section 3.), the quantity measured by these analyzers is a mass-to-charge (m/z) ratio, rather than the actual mass, for each ion. In many cases, the ions recorded in a mass spectrum will, in fact, be singly charged. Under these circumstances, the m/z ratio and the mass are identical so that, for example, the m/z ratio at the molecular ion of an analyte is numerically equal to its molecular weight or relative molecular mass (M_r), measured in Dalton.

As is also evident from Section 2.5. in Chapter 2, however, some ionization techniques, especially electrospray (ES) or ion-spray (IS), produce multiply charged ions whose *m/z* values are not directly equal to the ion mass. This tendency is also evident, although to a much lesser extent, with other techniques such as MALDI and FAB. In all these cases, the number of charges carried generally increases with an increase in analyte molecular weight. The situation with ES, or IS, is more complex since (in the positive-ion mode) a continuous series of multiprotonated molecular ion species, which often does not include a singly charged ion, is recorded, each species bearing one more charge than the adjacent species (e.g., *see* Chapter 10, Fig. 1). An equivalent situation exists in the negative-ion mode.

The molecular weight of an analyte may readily be determined from ES or IS spectra from the knowledge that adjacent peaks in the series differ by only one charge. If two adjacent peaks from the series are selected, then, for an ion of measured *m/z* ratio, m_i (which originates from an analyte with molecular weight M_r and which carries an unknown number of charges, i, supplied by proton attachment), m_i and M_r are related by Eq. (1), where M_h is the mass of a proton. The *m/z* ratio, m_{i+1}, for an ion that carries one additional proton attached to the same analyte molecule is then given by Eq. (2), so that i may be determined from the two *m/z* values using Eq. (3), which is derived from Eq. (1) and (2).

$$m_i = (M_r + iM_h)/i \tag{1}$$

$$m_{i+1} = [M_r + (i+1)M_h]/(i+1) \tag{2}$$

$$i = (m_{i+1} - M_h)/(m_i - m_{i+1}), \text{ or } i \approx (m_{i+1} - 1)/(m_i - m_{i+1}) \tag{3}$$

For charge-carrying adducts other than protons (e.g., Na^+ addition in the positive-ion mode or proton abstraction in the negative-ion mode), M_h may be replaced by M_a where M_a is the adduct ion mass. Once the charge state, i, has been calculated using Eq. (3), each peak in the series of multiply charged molecular ions can be used to provide a separate estimate of the molecular weight, M_r, and these estimates averaged.

So-called deconvolution algorithms offer another method of processing electrospray data. In deconvolution, contributions from a charge-state series are summed into a single peak (without adduct ion) which is plotted on a true mass scale rather than on an *m/z* scale. The use of higher mass resolution offers an alternative method of assigning charge state so long as any loss of instrumental sensitivity at higher resolution is not excessive. Thus, with resolution sufficient to resolve individual isotopic peaks, the number of these in an *m/z* unit unequivocally defines the number of charges, without the need to identify other peaks in the same series.

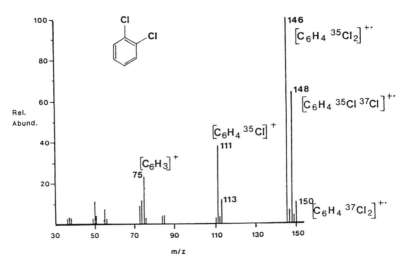

Fig. 3. Electron ionization spectrum of 1,2-dichlorobenzene (mol wt based on ^{35}Cl is 146).

3. Isotopic Abundance Distribution

Isotopic abundances corresponding to a single element with two isotopes can be calculated from the expansion of $(A + B)^N$ where A is the natural abundance of the lighter isotope, B is the natural abundance of the heavier isotope, and N is the number of atoms of the element concerned. For example, for chlorine, where $A = .76$ for ^{35}Cl and $B = .24$ for ^{37}Cl (*see* Table 1), the relative abundances for an ion that contains two chlorine atoms ($N = 2$) are derived from the expansion of $(.76 + .24)^2 = [.76^2 + 2\,(.76 \times .24) + .24^2]$ which equates to fractional values of .58, .36, and .06. These relative intensities are seen at m/z 146, 148, and 150 in Fig. 3.

With an element that has three isotopes, e.g., oxygen (although the abundance of ^{17}O is low) or sulfur (discounting the very low-abundance ^{36}S isotope) (*see* Table 1), the abundances for different combinations of the three isotopes, where A, B, and C are the respective abundances of the lightest to heaviest isotopes, are now calculated from the expansion of $(A + B + C)^N$, where N is again the number of atoms of the element concerned. For example, the result of this expansion when $N = 2$ is $(A^2 + 2AB + B^2 + 2AC + 2BC + C^2)$ which, for oxygen, equates to fractional values of (.995, 8×10^{-4}, 2×10^{-7}, 4×10^{-3}, 2×10^{-6}, and 4×10^{-6}).

A simple consideration of the nominal masses of the oxygen isotopes (viz. $M_A = 16$, $M_B = 17$, and $M_C = 18$) shows that the respective masses for O_2 in the preceding expansion are, 32, 33, 34, 34, 35, and 36. Thus, in practice (i.e., at nominal mass resolution), the resulting relative abundance values are

.995, 8×10^{-4}, 4×10^{-3}, 2×10^{-6}, and 4×10^{-6} at m/z 32–36. In fact, the only appreciable contribution in this distribution, apart from that at m/z 32, is approx 0.4% at m/z 34 (i.e., about double the ^{18}O contribution from a single oxygen atom and almost entirely originating from $^{16}O^{18}O$).

When more than one element is present, the relative abundances are now given by the expansion of the product of a number of appropriate expressions for two elements, each with two isotopes. For example, the appropriate expression is $(A_x + B_x)^N \times (A_y + B_y)^M$ where A_x and B_x are the natural abundances of the lighter and heavier isotopes for element x, A_y and B_y are the equivalent values for element y, and N and M are the respective numbers of atoms of elements x and y that are present. Evaluation of such expressions becomes increasingly complex but, fortunately, a knowledge of exact relative abundances is not needed in most mass spectrometric analyses. It is sufficient for peptide and protein analysis to understand in a qualitative manner how the isotopic distribution changes as the total number of atoms contained in a molecule is increased.

The most obvious effect of an increase in molecular weight with carbon-containing materials is owing to ^{13}C (a relatively abundant isotope) and is seen in Fig. 2 and then in Chapter 8, Fig. 10. In this progression, the lowest mass peak (corresponding to ^{12}C, ^1H, ^{16}O, etc.) becomes less and less significant in abundance until, as in Fig. 10 from Chapter 8, it is barely detectable. A simple explanation of this change is the increasing probability that ^{13}C will be present in any molecule as the overall number of carbon atoms in the molecule increases.

Reference

1. Papayannopoulos, I. A. (1995) The interpretation of collision-induced dissociation tandem mass spectra of peptides. *Mass Spectrom. Rev.* **14,** 49–73.

Appendix III: Mass Values and Abbreviated Forms for Amino Acid and Monosaccharide Residues

The molecular weight of a protein or peptide (M_p), of known amino acid composition, can be calculated by adding the masses of each individual amino acid residue (M_{ai}), using data from Table 2, to the masses of the appropriate N- and C-terminal groups $(M_N M_C)$ as in Eq. (4):

$$M_p = \sum n_i M_{ai} + M_N + M_C \tag{4}$$

where n_i represents the number of occurrences of residue i. For a straightforward protein or peptide with unmodified carboxy- and amino-termini, then M_N = the mass of a hydrogen atom (monoisotopic = isotopic average = 1.008)

Table 2 (Appendix III)
Amino Acid Residues

Amino acid	Three (one)- letter code	Monoisotopic mass	Average mass
Glycine	Gly (G)	57.021	57.052
Alanine	Ala (A)	71.037	71.079
Serine	Ser (S)	87.032	87.078
Proline	Pro (P)	97.053	97.117
Valine	Vat (V)	99.068	99.133
Threonine	Thr (T)	101.048	101.105
Cysteine	Cys (C)	103.009	103.145
Isoleucine	Ile (I)	113.084	113.160
Leucine	Leu (L)	113.084	113.160
Asparagine	Asn (N)	114.043	114.104
Aspartic acid	Asp (D)	115.027	115.089
Glutamine	Gln (Q)	128.059	128.131
Lysine	Lys (K)	128.095	128.174
Glutamic acid	Glu (E)	129.043	129.116
Methionine	Met (M)	131.040	131.199
Histidine	His (H)	137.059	137.142
Phenylalanine	Phe (F)	147.068	147.177
Arginine	Arg (R)	156.101	156.188
Tyrosine	Tyr (Y)	163.063	163.176
Tryptophan	Trp (W)	186.079	186.213
Homoserine lactone	—	83.037	83.090
Homoserine	—	101.048	101.105
Pyroglutamic acid	—	111.032	111.100
Carbamidomethylcysteine	—	160.031	160.197
Carboxymethylcysteine	—	161.147	161.181
Pyridylethylcysteine	—	208.067	208.284

and M_C = the mass of a hydroxyl group (monoisotopic = 17.003, isotopic average = 17.007). Note that the masses of the amino acid residues (these are detailed in Chapter 12 as well as in Table 2) are equivalent to those of the corresponding amino acid less the mass of a water molecule.

Just as for amino acids (Table 2), the molecular weight of an oligosaccharide (M_{os}) of known monosaccharide composition may be calculated from Eq. (5) using the residue masses (M_{si}) given in Table 3 and adding the mass of a water molecule (M_{H2O}; monoisotopic = 18.011, isotopic average = 18.015) to represent the terminating groups:

$$M_{os} = \sum n_i M_{si} + M_{H_2O} \tag{5}$$

Table 3 (Appendix III)
Monosaccharide Residues

Monosaccharide name	Abbreviation	Monoisotopic mass	Average mass
Deoxypentose	DeoxyPen	116.117	116.047
Pentose	Pen	132.116	132.042
Deoxyhexose[a]	DeoxyHex	146.143	146.058
Hexose[b]	Hex	162.142	162.053
Hexosamine	HexN	161.158	161.069
N-Acetylhexosamine[c]	HexNAc	203.195	203.079
Hexuronic acid	HexA	176.126	176.032
N-Acetylneuraminic acid	NeuAc	291.258	291.095
N-Glycolylneuraminic acid	NeuGc	307.257	307.090

[a]For example, Fucose (Fuc), Rhamnose (Rha).
[b]For example, Mannose (Man), Glucose (Glc), Galactose (Gal).
[c]For example, Glucosamine (GlcNAc), Galactosamine (GalNAc).

where n_i represents the number of occurrences of residue i. Again, the masses of sugar residues (these are detailed in Chapters 19–21 as well as in Table 3) are equivalent to those of the corresponding sugar less the mass of a water molecule.

Appendix IV: Peptide Fragmentation

Peptides are generally based on a linear arrangement of repeating units, —NH—CHR—CO—, which differ only in the nature of the side chain, R. Fragmentation of peptide molecular ions (or, more usually, $[M + H]^+$ ions in the positive-ion mode), which has been principally investigated under high-energy collision-induced dissociation conditions (e.g., Chapter 3), can be described, therefore, in terms of a limited number of fragmentation paths, each of which has been assigned a letter-based code (Chapter 3).

Thus, fragmentation types are denoted a_n, b_n, c_n, and d_n when the positive charge is retained on the peptide N-terminal fragment and v_n, w_n, x_n, y_n, and z_n when the charge is retained by the C-terminal fragment. In this convention, the subscript refers to the position in the chain of amino acids where fragmentation occurs whereas the letter denotes the fragmentation type. Equivalent fragmentation, which is induced under conditions other than high-energy collision-induced dissociation (Chapter 3), is also described using this same letter convention.

1. N-Terminal charge retention:

   a. The a, b, and c ions are formed by main-chain fragmentation:

$$\text{H}_2\text{N--CHR}_1\text{--CO--NH--CHR}_2\text{--CO--NH--CHR}_3\text{--CO--NH--CHR}_4\text{--CO--NH--CHR}_5\text{-........-CHR}_n\text{--CO}_2\text{H}$$

(labels: a_4, b_4, c_4, 2H)

   b. The d ions are produced by side-chain fragmentation:

$$\text{H}_2\text{N--CHR}_1\text{--CO--NH--CHR}_2\text{--CO--NH--CHR}_3\text{--CO--NH--CH--CO--NH---CHR}_5\text{-.........}$$

(label: a_4; side chain: CH with R' and R'')

$$\text{H}_2\text{N--CHR}_1\text{--CO--NH--CHR}_2\text{--CO--NH--CHR}_3\text{--CO--NH--CH}$$

(label: d_4; side chain: CH with R' and R'', H)

For example, for the N-terminal sequence QHTV (*see* Table 2 for an explanation of amino acid abbreviations) in the peptide QHTVTTTTK, a_4 = m/z 438 (R_1 = C_3H_6NO; $R_2 = C_4H_5N_2$; $R_3 = C_2H_5O$; $R_4 = C_3H_7$; $R_5 = R_3$), b_4 = m/z 466, and c_4 = m/z 483. A d_4 ion is also seen at m/z 424 owing to the loss of CH_2 (14 u) from the valine (R' = R" = CH_3) β-carbon (1).

2. C-Terminal charge retention:

   a. The x, y, and z ions are formed by main-chain fragmentation:

$$\text{H}_2\text{N--CHR}_n\text{-......-CHR}_5\text{--CO--NH--CHR}_4\text{--CO--NH--CHR}_3\text{--CO--NH--CHR}_2\text{--CO--NH--CHR}_1\text{--CO}_2\text{H}$$

(labels: x_4, y_4, z_4, 2H)

b. The v and w ions are produced by side-chain fragmentation:

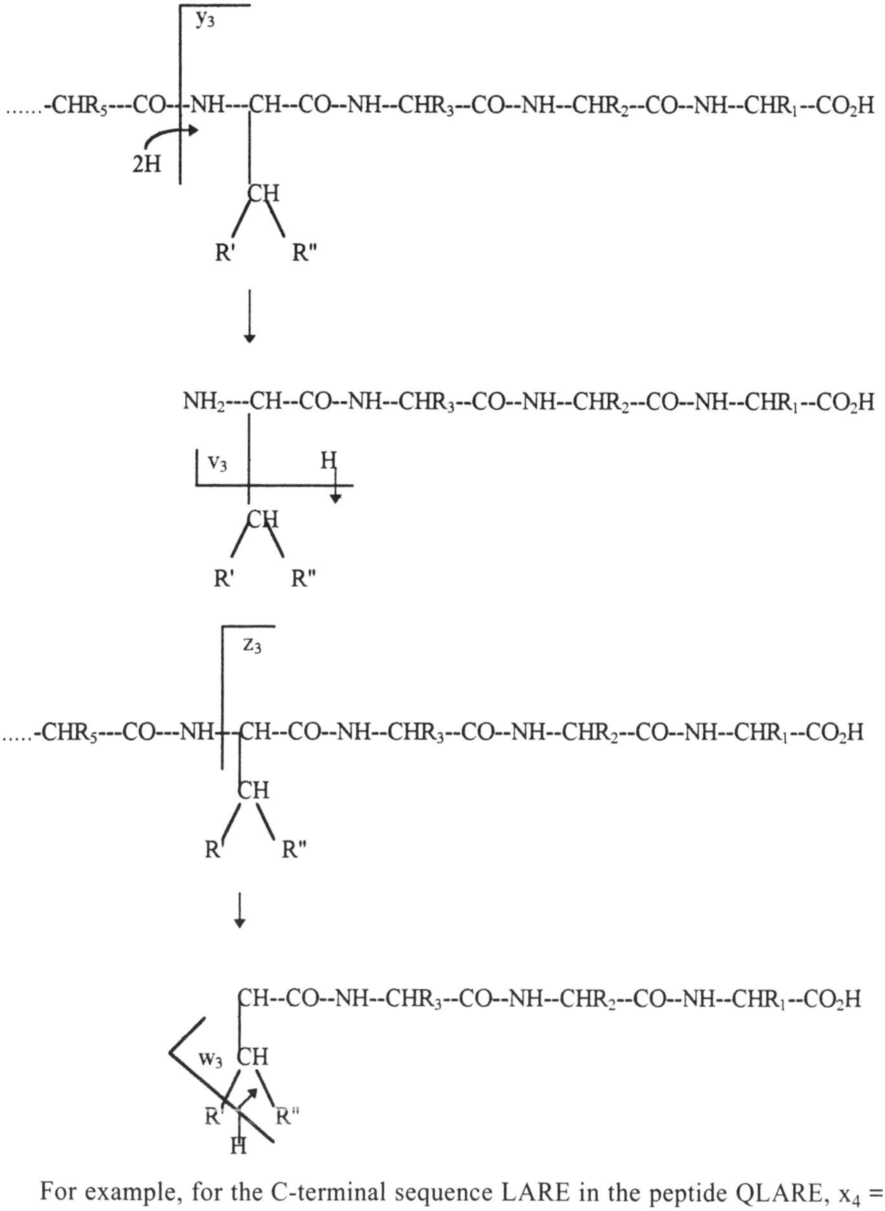

$$y_3$$

$$\ldots\text{-}CHR_5\text{---}CO\text{-}\vert\text{-}NH\text{---}CH\text{--}CO\text{--}NH\text{--}CHR_3\text{--}CO\text{--}NH\text{--}CHR_2\text{--}CO\text{--}NH\text{--}CHR_1\text{--}CO_2H$$

2H

CH

R'   R"

↓

$$NH_2\text{---}CH\text{--}CO\text{--}NH\text{--}CHR_3\text{--}CO\text{--}NH\text{--}CHR_2\text{--}CO\text{--}NH\text{--}CHR_1\text{--}CO_2H$$

$$v_3 \qquad H$$

CH

R'   R"

$$z_3$$

$$\ldots\text{-}CHR_5\text{---}CO\text{---}NH\text{-}\vert\text{-}CH\text{--}CO\text{--}NH\text{--}CHR_3\text{--}CO\text{--}NH\text{--}CHR_2\text{--}CO\text{--}NH\text{--}CHR_1\text{--}CO_2H$$

CH

R'   R"

↓

$$CH\text{--}CO\text{--}NH\text{--}CHR_3\text{--}CO\text{--}NH\text{--}CHR_2\text{--}CO\text{--}NH\text{--}CHR_1\text{--}CO_2H$$

$$w_3 \quad CH$$

R'   R"

H

For example, for the C-terminal sequence LARE in the peptide QLARE, $x_4 =$ m/z 514 ($R_1 = C_3H_5O_2$; $R_2 = C_4H_{10}N_3$; $R_3 = CH_3$; $R_4 = C_4H_9$), y_4 would appear at m/z 488, but is not significant, and $z_4 = m/z$ 471. A v_4 ion at m/z 430, owing to the loss of C_4H_{10} (58 u), from the leucine (R' = C_3H_7; R" = H) side chain, and a w_4 ion at m/z 429 owing to the loss of C_3H_6 (42 u) from the leucine β-carbon side-chain also are observed (1).

3. Nonsequential fragmentation: In addition to the ion-types already described, other ions, which do not result from sequential fragmentation, may be formed (viz. internal acyl ions and internal immonium ions).

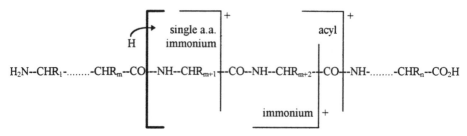

For example, in the peptide GIPTLLLFK, an acyl ion corresponding to the internal sequence PT is seen at m/z 199 ($R_{m+1} = C_3H_5$; $R_{m+2} = C_2H_5O$) whereas a single amino acid (a.a.) immonium ion which corresponds to proline (P) is found at m/z 70 *(2)*.

References

1. Baldwin, M. A., Falick, A. M., Gibson, B. W., Prusiner, S. B., Stahl, N., and Burlingame, A. L. (1990) Tandem mass spectrometry of peptides with N-terminal glutamine. *J. Am. Soc. Mass Spectrom.* **1**, 258–264.
2. Biemann, K. (1990) Sequencing of peptides by tandem mass spectrometry and high-energy collision-induced dissociation, in *Methods in Enzymology*, vol. 193 (McCloskey, J. A., ed.), Academic, New York, pp. 455–479.

Appendix V: Peptides, Proteins, and Proteolytic Agents

A number of peptide and protein materials, which are readily available from suppliers in good purity, are commonly used as mass standards in high-mol-wt mass spectrometry. A selection of the most useful of these materials, covering a range of molecular weights, is listed in Table 4 together with the corresponding mol-wt information.

Table 5 provides a brief summary of specific proteolytic agents. References to Chapters 8 and 14 of this book which offer descriptions of the use of a particular agent are provided where appropriate, otherwise a literature reference is substituted in the table.

References

1. Kanda, F., Yoshida, S., Okumura, T., and Takamatsu, T. (1995) Asparaginyl endopeptidase mapping of proteins with subsequent matrix-assisted laser desorption/ionization mass spectrometry. *Rapid Commun. Mass Spectrom.* **9**, 1095–1100.
2. Vestling, M. M., Kelly, M. A., and Fenselau C. (1994) Optimization by mass spectrometry of a tryptophan-specific protein cleavage reaction. *Rapid Commun. Mass Spectrom.* **8**, 786–790.

Table 4 (Appendix V)
Peptide and Protein Standards
for Mass Spectrometry

Peptide/protein	Isotopically averaged molecular weight
Angiotensin II (human)	1046.2
Substance P (human)	1347.7
Insulin (bovine)	5733.6
Cytochrome c (equine)	12,360.1
RNase A (bovine)	13,682.2
Myoglobin (equine)	16,951.5[a]
Trypsinogen (bovine)	23,980.9

[a]The observed signal corresponds to apomyoglobin.

Table 5 (Appendix V)
Enzymatic and Chemical Proteolytic Agents[a]

Cleavage agent	Cleavage site[b]	References
Asparaginyl endopeptidase Asp-C	Asp...X	(1)
BNPS skatole	Tryp...X	(2)
Carboxypeptidase B	C-terminal residue, especially Arg and Lys	Chapter 14
Chymotrypsin	Phe...X, Tyr...X, Trp...X and other hydrophobic residues	Chapter 14
Cyanogen bromide	Met...X	Chapter 14
Endoproteinase Asp-N	X...Asp	Chapter 8
Endoproteinase Glu-C (*Staphyllococcus aureus* V8 protease)	Glu...X	Chapter 8
Endoproteinase Lys-C	Lys...X (X ≠ Pro)	Chapter 8
Trypsin	Lys...X, Arg...X (X ≠ Pro)	Chapter 8

[a]Details of glycosidases (rather than proteases) used in structural elucidation can be found in Chapters 17–20.

[b]Cleavage sites are indicated by "...". In this convention, the C-terminus is assumed to be to the right, so that "Lys...X" indicates cleavage on the C-terminal side of lysine whereas X...Asp indicates cleavage on the N-terminal side of aspartic acid.

Table 6 (Appendix VI)
Simple and Binary Matrices for MALDI

Matrix material	Applications
2,5-Dihydroxybenzoic acid[a] (gentisic acid) (2,5-DHB) (DHB)	Wide range including proteins *(2)*, lipids *(3,4)*, and oligosaccharides (Ch. 18)
DHB + 5-methoxysalicylic acid (9:1 [w/w])	(Glyco)proteins *(5,6)*, membrane proteins *(7)*
DHB + 1-hydroxyisoquinoline (3:1 [w/w])	Oligosaccharides *(8)*
3,5-Dimethoxy-4-hydroxycinnamic acid[a] (sinapinic acid)	Proteins (Ch. 13), glycoproteins *(9)*, and membrane proteins (Ch. 16)
α-Cyano-4-hydroxycinnamic acid (CHCA)[a]	Peptides *(10)*, lipids *(3)*
3-Hydroxypicolinic acid (3-HPA)[a]	Oligonucleotides *(11)*
3-HPA + picolinic acid (1:8 [*M/M*])	Oligonucleotides *(12)*
6,7-Dihydroxycoumarin (esculetin)	Lipids *(3)*

[a]These matrices are the most commonly used.

Appendix VI: Matrix-Assisted Laser Desorption/Ionization (MALDI)—Matrices and Contaminants

Details of a number of matrix materials for MALDI, including matrices whose use is described within the main body of this book, are given in Table 6. Table 7, based on information originally published by Mock et al. *(1)*, gives some indication of the concentration level at which common contaminants may adversely affect the recording of a useful MALDI spectrum. Such adverse effects include signal suppression, peak broadening (often owing to adduct formation), and erratic signal production.

References

1. Mock, K. K., Sutton, C. W., and Cottrell, J. S. (1992) Sample immobilization protocols for matrix-assisted laser desorption mass spectrometry. *Rapid Commun. Mass Spectrom.* **4,** 233–238.
2. Strupat, K., Karas, M., and Hillenkamp, F. (1991) 2,5-Dihydroxybenzoic acid: a new matrix for laser desorption/ionization mass spectrometry. *Int. J. Mass Spectrom. Ion Proc.* **111,** 89–102.
3. Harvey, D. J. (1995) Matrix-assisted laser desorption/ionization mass spectrometry of sphingo- and glycosphingolipids. *J. Mass Spectrom.* **30,** 1311–1324.
4. Harvey, D. J. (1995) Matrix-assisted laser desorption/ionization mass spectrometry of phospholipids. *J. Mass Spectrom.* **30,** 1333–1346.
5. Karas, M., Bahr, U., Strupat, K., Hillenkamp, F., Tsarbopoulos, A., and Pramanik, B. N. (1995) Matrix dependence of metastable fragmentation of glycoproteins in MALDI TOF mass spectrometry. *Anal. Chem.* **67,** 675–679.

Table 7 (Appendix VI)
Concentration of Sample Contaminants Tolerated
in MALDI Operation

Contaminant	Maximum recommended concentration[a]
Urea	$0.5M$
Guanidine-HCl	$0.5M$
Glycerol	1%
Alkali metal salts	$0.1M$
Tris buffer	$0.05M$
NH_4HCO_3	$0.05M$
Phosphate buffer	$0.01M$
Detergents other than SDS	0.1%
SDS	0.01%

[a]The contaminant materials in this table are arranged in order of increasing detrimental effect. In general, all reasonable efforts should be made to reduce excessive contamination, although, in practice, any adverse effects will be very much dependent on the choice of matrix and other instrumental variables.

6. Vestal, M. L., Juhasz, P., and Martin, S. A. (1995) Delayed extraction matrix-assisted laser desorption time-of-flight mass spectrometry. *Rapid Commun. Mass Spectrom.* **9,** 1044–1050.
7. Rosinke, B., Strupat, K., Hillenkamp, F., Rosenbusch, J., Dencher, N., Krüger, U., and Galla, H.-J. (1995) Matrix-assisted laser desorption/ionization mass spectrometry (MALDI-MS) of membrane proteins and non-covalent complexes. *J. Mass Spectrom.* **30,** 1462–1468.
8. Mohr, M. D., Börnsen, K. O., and Widmer, H. M. (1995) Matrix-assisted laser desorption/ionization mass spectrometry: improved matrix for oligosaccharides. *Rapid Commun. Mass Spectrom.* **9,** 809–814.
9. Nelson, R. W., Dogruel, D., and Williams P. (1994) Mass determination of human immunoglobulin IgM using matrix-assisted laser desorption/ionization time-of-flight mass spectrometry. *Rapid Commun. Mass Spectrom.* **8,** 627–631.
10. Beavis, R. C., Chaudary, T., and Chait, B. T. (1992) α-Cyano-4-hydroxy cinnamic acid as a matrix for matrix-assisted laser desorption mass spectrometry. *Org. Mass Spectrom.* **27,** 156–158.
11. Wu, K. J., Steding, A., and Becker, C. H. (1993) Matrix-assisted laser desorption time-of-flight mass spectrometry of oligonucleotides using 3-hydroxy-picolinic acid as an ultra-violet sensitive matrix. *Rapid Commun. Mass Spectrom.* **7,** 142–146.
12. Taranenko, N. I., Tang, K., Allman, S. L., Ch'ang, L. Y., and Chen, C. H. (1994) 3-Aminopicolinic acid as a matrix for laser desorption mass spectrometry of biopolymers. *Rapid Commun. Mass Spectrom.* **8,** 1001–1006.

Appendix VII: Current References to the Use of Mass Spectrometry in Protein and Peptide Analysis

This final appendix provides an update to the main text by listing references to current literature in this fast-moving field. These references are mostly taken from the appropriate section of the mass spectrometry current awareness service which is regularly compiled for the *Journal of Mass Spectrometry*. Further references, taken directly from the *Journal of Mass Spectrometry* and from *Rapid Communications in Mass Spectrometry*, have been added to those in the current awareness listings. Most references selected were published in 1995. Coverage extends to the end of 1995 for the above journals but stops a little earlier in the year for other journals. The listed references have been categorized under the general headings proteins or peptides while the protein references have been further divided into "primary structure," "monitoring," "secondary structure," "interaction," and "physical chemistry."

A survey of the 1995 current awareness listings suggests the following principal sources for articles concerned with the use of mass spectrometry in protein and peptide analysis. This list of the "top 10" journals, which carry over 80% of the published articles, is presented in order of decreasing frequency of citation:

1. *Journal of Mass Spectrometry;*
2. *Rapid Communications in Mass Spectrometry;*
3. *Journal of the American Society for Mass Spectrometry;*
4. *Analytical Chemistry;*
5. *Journal of Chromatography* (Parts A/B);
6. *Analytical Biochemistry;*
7. *European Mass Spectrometry;*
8. *Journal of the American Chemical Society;*
9. *Journal of Biological Chemistry;* and
10. *Protein Science.*

Protein Primary Structure

Aaserud, D. J., Little, D. P., O'Connor, P. B., and McLafferty, F. W. (1995) Distinguishing *N*- and *C*-terminus ions for mass spectrometry sequencing of proteins without prior degradation. *Rapid Commun. Mass Spectrom.* **9(10),** 871.

Allmaier, G., Schaffer, C., Messner, P., Rapp, U., and Mayer-Posner, F. J. (1995) Accurate determination of the molecular weight of the major surface layer protein isolated from *Clostridium thermosaccharolyticum* by time-of-flight mass spectrometry. *J. Bacteriol.* **177(5),** 1402.

Ashton, D. S., Beddell, C. R., Cooper, D. J., Craig, S. J., Lines, A. C., Oliver, R. W. A., and Smith, M. A. (1995) Mass spectrometry of the humanized monoclonal antibody CAMPATH 1H. *Anal. Chem.* **67(5),** 835.

Bai, J., Qian, M. G., Liu, Y. H., Liang, X. L., and Lubman, D. M. (1995) Peptide mapping by CNBr degradation on a nitrocellulose membrane with analysis by matrix assisted laser desorption/ionization mass spectrometry. *Anal. Chem.* **67(10),** 1705.

Banks, J. F. (1995) Separation and analysis of proteins by perfusion liquid chromatography and electrospray ionization mass spectrometry. *J. Chromatogr. A.* **691(1–2),** 325.

Bennett, K. L., Hick, L. A., Truscott, R. J. W., Sheil, M. M., and Smith, S. V. (1995) Optimum conditions for electrospray mass spectrometry of a monoclonal antibody (letter). *J. Mass Spectrom.* **30(5),** 769.

Bihoreau, N., Veillon, J. F., Ramon, C., Scohyers, J. M., and Schmitter, J. M. (1995) Characterization of a recombinant antihaemophilia-A factor (factor VIII-II) by matrix-assisted laser desorption/ionization mass spectrometry. *Rapid Commun. Mass Spectrom.* **9(15),** 1584.

Blackledge, J. A. and Alexander, A. J. (1995) Polyethylene membrane as a sample support for direct matrix-assisted laser desorption/ionization mass spectrometric analysis of high mass proteins. *Anal. Chem.* **67(5),** 843.

Bonner, R. and Shushan, B. (1995) The characterization of proteins and peptides by automated methods. *Rapid Commun. Mass Spectrom.* **9(11),** 1067.

Bonner, R. and Shushan, B. (1995) Error-tolerant protein database searching using peptide product-ion spectra. *Rapid Commun. Mass Spectrom.* **9(11),** 1077.

Boots, J. W. P., Van Dongen, W. D., Verheij, H. M., De Haas, G. H., Haverkamp, J., and Slotboom, A. J. (1995) Identification of the active site histidine in *Staphylococcus hyicas* lipase using chemical modification and mass spectrometry. *BBA-Protein Struct. Mol. Enzymol.* **1248(1),** 27.

Borthwick, E. B., Connell, S. J., Tudor, D. W., Robins, D. J., Shneier, A., Abell, C., and Coggins, J. R. (1995) *Escherichia coli* dihydrodipicolinate synthase: characterization of the imine intermediate and the product of bromopyruvate treatment by electrospray mass spectrometry. *Biochem. J.* **305(2),** 521.

Bouchon, B., Jaquinod, M., Klarskov, K., Trottein, F., Klein, M., Van Dorsselaer, A., Bischoff, R., and Roitsch, C. (1994) Analysis of the primary structure and post-translational modifications of the *Schistosoma mansoni* antigen Smp28 by electrospray mass spectrometry. *J. Chromatogr. B-Bio Med. Appl.* **662(2),** 279.

Burlet, O., Yang, C. Y., Guyton, J. R., and Gaskell, S. J. (1995) Tandem mass spectrometric characterization of a specific cysteic acid residue in oxidized human apoprotein B_{100}. *J. Am. Soc. Mass Spectrom.* **6(4),** 242.

Cerpa-Poljak, A., Jenkins, A., and Duncan, M. W. (1995) Recovery of peptides and proteins following matrix-assisted laser desorption/ionization mass spectrometry. *Rapid Commun. Mass Spectrom.* **9(3),** 233.

Chowdhury, S. K., Doleman, M., and Johnston, D. (1995) Fingerprinting proteins coupled with polymers by mass spectrometry: investigation of polyethylene glycol-conjugated superoxide dismutase. *J. Am. Soc. Mass Spectrom.* **6(6),** 478.

Chowdhury, S. K., Eshraghi, J., Wolfe, H., Forde, D., Hlavac, A. G., and Johnston D. (1995) Mass spectrometric identification of amino acid transformations during oxidaion of peptides and proteins: modifications of methionine and tyrosine. *Anal. Chem.* **67(2),** 390.

Chowdhury, S. K., Vavra, K. J., Brake, P. G., Banks, T., Falvo, J., Wahl, R., Eshraghi, J., Gonvea, G., Chait, B. T., and Vestal, C. H. (1995) Examination of recombinant truncated mature human fibroblast collagenase by mass spectrometry: identification of differences with the published sequence and determination of stable isotope incorporation. *Rapid Commun. Mass Spectrom.* **9(7)**, 563.

Clauser, K. R., Hall, S. C., Smith, D. M., Webb, J. W., Andrews, L. E., Tran, H. M., Epstein, L. B., and Burlingame, A. L. (1995) Rapid mass spectrometric peptide sequencing and mass matching for characterization of human melanoma proteins isolated by two-dimensional PAGE. *Proc. Natl. Acad. Sci. USA* **92(11)**, 5072.

Cordwell, S. J., Wilkins, M. R., Cerpa-Poljak, A., Gooley, A. A., Duncan, M., Williams, K. L., and Humphery-Smith, I. (1995) Cross-species identification of proteins separated by two-dimensional gel electrophoresis using matrix-assisted laser desorption/ionisation time-of-flight mass spectrometry and amino acid composition. *Electrophoresis* **16(3)**, 438.

Davison, A. J. and Davison, M. D. (1995) Identification of structural proteins of channel catfish virus by mass spectrometry. *Virology* **206(2)**, 1035.

Dillen, L., Zhang, X. Y., Claeys, M., Liang, F., DePotter, W. P., VanDongen, W., and Esmans, E. L. (1995) Isolation and MS characterization of an oxidized form of vasostatin I, an N-terminal chromogranin A-derived protein, from bovine chromaffin cells. *J. Mass Spectrom.* **30(11)**, 1599.

Fabris, D., Vestling, M. M., Cordero, M. M., Doroshenko, V. M., Cotter, R. J., and Fenselau, C. (1995) Sequencing electroblotted proteins by tandem mass spectrometry. *Rapid Commun. Mass Spectrom.* **9(11)**, 1051.

Farmer, T. B. and Caprioli, R. M. (1995) Mass discrimination in matrix-assisted laser desorption ionization time-of-flight mass spectrometry: a study using cross-linked oligomeric complexes. *J. Mass Spectrom.* **30(9)**, 1245.

Feistner, G. J., Faull, K. F., Barofsky, D. F., and Roepstorff, P. (1995) Mass spectrometric peptide and protein charting. *J. Mass Spectrom.* **30(4)**, 519.

Fisichella, S., Foti, S., Maccarrone, G., and Saletti R. (1995) Tryptic peptide mapping of sequence 299–585 of human serum albumin by high-performance liquid chromatography and fast atom bombardment mass spectrometry. *J. Chromatogr. A.* **693(1)**, 33.

Gimon-Kinsel, M. E., Kinsel, G. R., Edmondson, R. D., and Russell, D. H. (1995) Photodissociation of high molecular weight peptides and proteins in a two-stage linear time-of-flight mass spectrometer. *J. Am. Soc. Mass Spectrom.* **6(7)**, 578.

Glocker, M. O., Arbogast, B., and Deinzer, M. L. (1995) Characterization of disulfide linkages and disulfide bond scrambling in recombinant human macrophage colony stimulating factor by fast-atom bombardment mass spectrometry of enzymatic digests. *J. Am. Soc. Mass Spectrom.* **6(8)**, 638.

Green, B. N., Suzuki, T., Gotoh, T., Kuchumov, A. R., and Vinogradov, S. N. (1995) Electrospray ionization mass spectrometric determination of the complete polypeptide chain composition of *Tylorrhynchus hererochaetus* hemoglobin. *J. Biol. Chem.* **270(31)**, 18,209.

Griffin, P. R., MacCoss, M. J., Eng, J. K., Blevins, R. A., Aaronson, J. S., and Yates, J. R. (1995) Direct database searching with MALDI-PSD spectra of peptides. *Rapid Commun. Mass Spectrom.* **9(15)**, 1546.

Guan, Z. Q., Campbell, V. L., and Laude, D. A. (1995) Charge state assignment from Schiff-base adducts in low resolution electrospray mass spectra of protein mixtures and dissociation products. *J. Mass Spectrom.* **30(1)**, 119.

Haebel, S., Jensen, C., Andersen, S. O., and Roepstorff, P. (1995) Isoforms of a cuticular protein from larvae of the meal beetle, *Tenebrio molitor*, studied by mass spectrometry in combination with Edman degradation and two-dimensional polyacrylamide gel elestrophoresis. *Protein Sci.* **4(3)**, 394.

He, S. M., Pan, S. H., Wu, K. L., Amster, I. J., and Orlando, R. (1995) Analysis of normal human fetal eye lens crystallins by high-performance liquid chromatography-mass spectrometry. *J. Mass Spectrom.* **30(3)**, 424.

High, K. A., Methven, B. A., McLaren, J. W., Siu, K. W. M., Wang, J., Klaverkamp, J. F., and Blais, K. S. (1995) Physico-chemical characterization of metal binding proteins using HPLC-ICP-MS, HPLC-MA-AAS and electrospray-MS. *Fresenius J. Anal. Chem.* **351(4–5)**, 393.

Hofstadler, S. A., Swanek, F. D., Gale, D. C., Ewing, A. G., and Smith, R. D. (1995) Capillary electrophoresis/electrospray ionization Fourier transform ion cyclotron resonance mass spectrometry for direct analysis of cellular proteins (letter). *Anal. Chem.* **67(8)**, 1477.

Hunter, A. P. and Games, D. (1995) Evaluation of glycosylation site heterogeneity and selective identification of glycopeptides in proteolytic digests of α_1-acid glycoprotein by mass spectrometry. *Rapid Commun. Mass Spectrom.* **9(1)**, 42.

Jespersen, S., Hojrup, P., Andersen, S. O., and Roepstorff, P. (1994) The primary structure of an endocuticular protein from two locust species, *Locusta migratoria* and *Schistocerca gregaria*, determined by a combination of mass spectrometry and automatic Edman degradation. *Comp. Biochem. Physiol. Pt. B* **109(1)**, 125.

Kanda, F., Yoshida, S., Okumura, T., and Takamatsu, T. (1995) Asparaginyl endopeptidase mapping of proteins with subsequent matrix-assisted laser desorption/ionization mass spectrometry. *Rapid Commun. Mass Spectrom.* **9(12)**, 1095.

Karas, M., Bahr, U., Strupat, K., Hillenkamp, F., Tsarbopoulos, A., and Pramanik, B. N. (1995) Matrix dependence of metastable fragmentation of glycoproteins in MALDI TOF mass spectrometry (letter). *Anal. Chem.* **67(3)**, 675.

Kilby, G. W., Truscott, R. J. W., Aquilina, J. A., Sheil, M. M., Riley, M. L., and Harding, J. J. (1995) Electrospray ionisation mass spectrometry of lens crystallins: verification and detection of errors in protein sequences of bovine γ-crystallins. *Eur. Mass Spectrom.* **1(2)**, 203.

King, T. B., Colby, S. M., and Reilly, J. P. (1995) High resolution MALDI-TOF mass spectra of three proteins obtained using space-velocity correlation focusing (letter). *Int. J. Mass Spectrom. Ion Proc.* **145(1–2)**, L1.

Klabunde, T., Stahl, B., Suerbaum, H., Hahner, S., Karas, M., Hillenkamp, F., Krebs, B., and Witzel, H. (1994) The amino acid sequence of the red kidney bean Fe(III)-

Zn(II) purple acid phosphatase: determination of the amino acid sequence by a combination of matrix-assisted laser desorption/ionization mass spectrometry and automated Edman sequencing. *Eur. J. Biochem.* **226(2),** 369.

Knierman, M. D., Coligan, J. E., and Parker, K. C. (1994) Peptide fingerprints after partial acid hydrolysis: analysis by matrix-assisted laser desorption/ionization mass spectrometry. *Rapid Commun. Mass Spectrom.* **8(12),** 1007.

Krebs, J. F., Siuzdak, G., Dyson, H. J., Stewart, J. D., and Benkovic, S. J. (1995) Detection of a catalytic antibody species acylated at the active site by electrospray mass spectrometry. *Biochemistry* **34(3),** 720.

Krell, T., Pitt, A. R., and Coggins, J. R. (1995) The use of electrospray mass spectrometry to identify an essential arginine residue in type II dehydroquinases. *FEBS Lett.* **360(1),** 93.

Laidler, P., Cowan, D. A., Hider, R. C., Keane, A., and Kicman, A. T. (1995) Tryptic mapping of human chorionic gonadotropin by matrix-assisted laser desorption/ionization mass spectrometry. *Rapid Commun. Mass Spectrom.* **9(11),** 1021.

Leonil, J., Molle, D., Gaucheron, F., Arpino, P., Guenot, P., and Maubois, J. L. (1995) Analysis of major bovine milk proteins by on-line high-performance liquid chromatography and electrospray ionization mass spectrometry. *Lait* **75(3),** 193.

Li, Y. M., Brostedt, P., Hjerten, S., Nyberg, F., and Silberring, J. (1995) Capillary liquid chromatography-fast atom bombardment mass spectrometry using a high-resolving cation exchanger, based on a continuous chromatographic matrix: application to studies on neuropeptide peptidases. *J. Chromatogr. B-Biomed Appl.* **664(2),** 426.

Mann, F. and Wilm, M. (1995) Electrospray mass spectrometry for protein characterization. *Trends Biochem. Sci.* **20(6),** 219.

McIlhinney, R. A. J. and Harvey, D. J. (1995) Determination of N-terminal myristoylation of proteins using a combined gas chromatographic/mass spectrometric assay of derived myristoylglycine: electron impact-induced fragmentation of acylglycine derivatives. *J. Mass Spectrom.* **30(6),** 900.

Ming, D., Markley, J. L., and Hellekant, G. (1995) Quantification of cysteinyl sulfhydryl residues in peptides and proteins by ESI-MS or MALDI-MS. *Biotechniques* **18(5),** 808.

Mirgorodskaya, E. P., Mirgorodskaya, O. A., Dobretsov, S. V., Shevchenko, A. A., Dodonov, A. F., Kozlovskiy, V. I., and Raznikov, V. V. (1995) Electrospray mass spectrometric study of melittin trypsinolysis by a kinetic approach. *Anal. Chem.* **67(17),** 2864.

Murphy, C. M. and Fenselau, C. (1995) Recognition of the carboxy terminal peptide in cyanogen bromide digests of proteins. *Anal. Chem.* **67(9),** 1644.

Nabuchi, Y., Kuboniwa, H., Takasu, H., Asoh, Y., and Ushio, H. (1995) Peptide mapping of recombinant human parathyroid hormone by enzymatic digestion and subsequent fast-atom bombardment mass spectrometry. *Rapid Commun. Mass Spectrom.* **9(4),** 257.

Nairn, J., Krell, T., Coggins, J. R., Pitt, A. R., Fothergill-Gilmore, L. A., Walter, R., and Price, N. C. (1995) The use of mass spectrometry to examine the formation and hydrolysis of the phosphorylated form of phosphoglycerate mutase. *FEBS Lett.* **359(2–3)**, 192.

Nakanishi, T., Shimizu, A., Okamoto, N., Ingendoh, A., and Kanai, M. (1995) Analysis of serum protein precipitated with antiserum by matrix-assisted laser desorption ionization/time-of-flight and electrospray ionization mass spectrometry as a clinical laboratory test. *J. Am. Soc. Mass Spectrom.* **6(9)**, 854.

Nelson, R. W., Dogruel, D., Krone, J. R., and Williams, P. (1995) Peptide characterization using bioreactive mass spectrometer probe tips. *Rapid Commun. Mass Spectrom.* **9(14)**, 1380.

Nelson, R. W., Dogruel, D., and Williams, P. (1995) Detection of human IgM at *m/z* ~1*M* Da (letter). *Rapid Commun. Mass Spectrom.* **9(7)**, 625.

O'Connor, P. B., Speir, J. P., Senko, M. W., Little, D. P., and McLafferty, F. W. (1995) Tandem mass spectrometry of carbonic anhydrase (29 kDa). *J. Mass Spectrom.* **30(1)**, 88.

Packman, L. C., Mewies, M., and Scrutton, N. S. (1995) The flavinylation reaction of trimethylamine dehydrogenase: analysis by directed mutagenesis and electrospray mass spectrometry. *J. Biol. Chem.* **270(22)**, 13,186.

Pedersen, J., Andersen, J., Roepstorff, P., Filimonova, M., and Biedermann, K. (1995) Characterizaion of Serratia marcescens nuclease isoforms by electrospray mass spectrometry. *Biochemistry-Engl. Tr.* **60(3)**, 341.

Pedersen, J., Andersen, G., Roepstorff, P., Filimonova, M. N., and Bidermann, K. (1995) *Serratia marcescens* nucleases. I. Comparison of natural and recombinant nucleases by electrospray mass spectrometry. *Bioorg. Khim.* **21(5)**, 330 (Russian, English Abstract).

Pedersen, J., Filimonova, M., Roepstorff, P., and Biedermann, K. (1995) Isoforms of *Serratia marcescens* nuclease produced by native and recombinant strains. Comparative characterization by plasma desorption mass spectrometry. *Biochemistry-Engl. Tr.* **60(3)**, 331.

Pedersen, J., Filimonova, M. N., Roepstorff, P., and Bidermann, K. (1995) *Serratia marcescens* nucleases. II. Analysis of primary structures by peptide mapping in combination with desorption mass spectrometry. *Bioorg. Khim.* **21(5)**, 336.

Perera, I. K., Perkins, J., and Kantartzoglou, S. (1995) Spin coated samples for high resolution matrix-assisted laser desorption/ionization time-of-flight mass spectrometry of large proteins. *Rapid Commun. Mass Spectrom.* **9(2)**, 180.

Petillot, Y., Forest, E., Meyer, J., and Moulis, J. M. (1995) Observation of holoprotein molecular ions of several ferredoxins by electrospray-ionization-mass spectrometry. *Anal. Biochem.* **228(1)**, 56.

Petillot, Y., Golinelli, M. P., Forest, E., and Meyer, J. (1995) Electrospray-ionization mass spectrometry of molecular variants of a [2Fe-2S] ferredoxin. *Biochem Biophys Res Commun* **210(3)**, 686.

Potts, W., Van Horn, R., Anderson, K., Blake, T., Garver, E., Joseph, G., Dreyer, G., Shu, A., Heys, R., and Fong, K. L. (1995) Characterization of the metabolites of the peptidomimetic human immunodeficiency virus type I protease inhibitor SK-and-F 107461 in rats using liquid chromatography/mass spectrometry. *Drug Metab. Disposition* **23(8)**, 799.

Prome, D., Prome, J. C., Blouquit, Y., and Rosa, J. (1995) Characterization of new amino-terminal blocking groups in the normal human adult hemoglobin Hb A_{1b}. *Eur. Mass Spectrom.* **1(2)**, 195.

Qian, M. G. and Lubman, D. M. (1995) Analysis of tryptic digests using microbore HPLC with an ion trap storage reflectron time-of-flight detector. *Anal. Chem.* **67(17)**, 2870.

Resing, K. A., Johnson, R. S., and Walsh, K. A. (1995) Mass spectrometric analysis of 21 phosphorylation sites in the internal repeat of rat profilaggrin precursor of an intermediate filament associated protein. *Biochemistry-USA* **34(29)**, 9477.

Resing, K. A., Mansour, S. J., Hermann, A. S., Johnson, R. S., Candia, J. M., Fukasawa, K., Van de Woude, G. F., and Ahn, N. G. (1995) Determination of v-Mos-catalyzed phosphorylation sites and autophosphorylation sites on MAP kinase kinase by ESI/MS. *Biochemistry-USA* **34(8)**, 2610.

Rider, M. H., Puype, M., Van Damme, J., Gevaert, K., De Boeck, S., D'Alayer, J., Rasmussen, H. H., Celis, J. E., and Vandekerckhove, J. (1995) An agarose-based gel-concentration system for microsequence and mass spectrometric character-ization of proteins previously purified in polyacrylamide gels starting a low pico-mole levels. *Eur. J. Biochem.* **230(1)**, 258.

Rosinke, B., Strupat, K., Hillenkamp, F., Rosenbusch, J., Dencher, N., Kruger, U., and Galla, H. J. (1995) Matrix-assisted laser desorption/ionization mass spectrometry (MALDI-MS) of membrane proteins and non-covalent complexes. *J. Mass Spectrom.* **30(10)**, 1462.

Rouimi, P., Debrauwer, L., and Tulliez, J. (1995) Electrospray ionization-mass spec-trometry as a tool for characterization of glutathione S-transferase isozymes. *Anal. Biochem.* **229(2)**, 304.

Rudiger, A., Rudiger, M., Weber, K., and Schomburg, D. (1995) Characterization of post-translational modifications of brain tubulin by matrix-assisted laser desorp-tion/ionization mass spectrometry: direct one-step analysis of a limited subtilisin digest. *Anal. Biochem.* **224(2)**, 532.

Seielstad, D. A., Carlson, K. E., Katzenellenbogen, J. A., Kushner, P. J., and Greene, G. L. (1995) Molecular characterization by mass spectrometry of the human estrogen receptor ligand-binding domain expressed in *Escherichia coli*. *Mol. Endocrinol.* **9(6)**, 647.

Speir, J. P., Senko, M. W., Little, D. P., Loo, J. A., and McLafferty, F. W. (1995) High-resolution tandem mass spectra of 37–67 kDa proteins. *J. Mass Spectrom.* **30(1)**, 39.

Svoboda, M., Meister, W., and Vetter, W. (1995) A method for counting disulfide bridges in small proteins by reduction with mercaptoethanol and electrospray mass spectrometry. *J. Mass Spectrom.* **30(11)**, 1562.

Thiede, B., Wittmann-Leibold, B., Bienert, M., and Krause, E. (1994) MALDI-MS for C-terminal sequence determination of peptides and proteins degraded by carboxy-peptidase Y and P. *FEBS Lett.* **357(1)**, 65.

Thulin, C. D. and Walsh, K. A. (1995) Identification of the amino terminus of human filaggrin using differential LC/MS techniques: implications for profilaggrin pro-cessing. *Biochemistry-USA* **34(27)**, 8687.

Wendt, H., Durr, E., Thomas, R. M., Przybylski, M., and Bosshard, H. R. (1995) Characterization of leucine zipper complexes by electrospray ionization mass spectrometry. *Protein Sci.* **4(8)**, 1563.

Westman, A., Huth-Fehre, T., Demirev, P., and Sundqvist, B. U. R. (1995) Sample morphology effects in matrix-assisted laser desorption ionization mass spectrometry of proteins. *J. Mass Spectrom.* **30(1)**, 206.

Wilcox, M. D., Schey, K. L., Busman, M., and Hildebrandt, J. D. (1995) Determination of the complete covalent structure of the γ_2 subunit of bovine brain G proteins by mass spectrometry. *Biochem. Biophys. Res. Commun.* **212(2)**, 367.

Woods, A. S., Huang, A. Y. C., Cotter, R. J., Pasternack, G. R., Pardoll, D. M., and Jaffee, E. M. (1995) Simplified high-sensitivity sequencing of a major histocompatibility complex class I-associated immunoreactive peptide using matrix-assisted laser desorption/ionization mass spectrometry. *Anal. Biochem.* **226(1)**, 15.

Wu, Q. Y., Van Orden, S., Cheng, X. H., Bakhtiar, R., and Smith, R. D. (1995) Characterization of cytochrome *c* variants with high-resolution FTICR mass spectrometry: correlation of fragmentation and structure. *Anal. Chem.* **67(14)**, 2498.

Yeh, H. I., Hsieh, C. H., Wang, L. Y., Tsai, S. P., Hsu, H. Y., and Tam, M. F. (1995) Mass spectrometric analysis of rat liver cytosolic glutathione S-transferases: modifications are limited to *N*-terminal processing. *Biochem. J.* **308(1)**, 69.

Young, N. M., Watson, D. C., and Thibault, P. (1995) Mass spectrometric analysis of genetic and post-translational heterogeneity in the lectins jacalin and *Maclura pomifera* agglutinin. *Glycoconjugate J.* **12(2)**, 135.

Young, N. M., Watson, D. C., Yaguchi, M., Adar, R., Arango, R., Rodriguez-Arango, E., Sharon, N., Blay, P. K. S., and Thibault, P. (1995) *C*-terminal post-translational proteolysis of plant lectins and their recombinant forms expressed in *Escherichia coli*: characterization of "ragged ends" by mass spectrometry. *J. Biol. Chem.* **270(6)**, 2563.

Youngquist, R. S., Fuentes, G. R., Lacey, M. P., and Keough, T. (1995) Generation and screening of combinatorial peptide libraries designed for rapid sequencing by mass spectrometry. *J. Am. Chem. Soc.* **117(14)**, 3900.

Zhong, F. and Zhao, S. (1995) A study of factors influencing the formation of singly charged oligomers and multiply charged monomers by matrix-assisted laser desorption/ionization mass spectrometry. *Rapid Commun. Mass Spectrom.* **9(7)**, 570.

Zhu, Y. F., Lee, K. L., Tang, K., Allman, S. L., Taranenko, N. I., and Chen, C. H. (1995) Revisit of MALDI for small proteins. *Rapid Commun. Mass Spectrom.* **9(13)**, 1315.

Protein Monitoring

Borromeo, V., Berrini, A., Secchi, C., Brambilla, G. F., and Cantafora, A. (1995) Matrix-assisted laser desorption mass spectrometry for the detection of recombinant bovine growth hormone in sustained-release form. *J. Chromatogr. B-Biomed. Appl.* **669(2)**, 366.

Clerc, F. F., Monegier, B., Faucher, D., Cuine, F., Pourcet, C., Holt, J. C., Tang, S. Y., Van Dorsselaer, A., Becquart, J., and Vuilhorgne, M. (1994) Primary structure control of recombinant proteins using high-performance liquid chromatography-mass spectrometry and microsequencing. *J. Chromatogr. B-Bio. Med. Appl.* **662(2)**, 245.

Gaucheron, F., Molle, D., Leonil, J., and Maubois, J. (1995) Selective determination of phosphopeptide β-CN(1-25) in a β-casein digest by adding iron: characterization by liquid chromatography with on-line electrospray-ionization mass spectrometric detection. *J. Chromatogr. B-Bio. Med. Appl.* **664(1)**, 193.

Goto, Y., Shima, Y., Morimoto, S., Shoyama, Y., Murakami, H., Kusai, A., and Nojima, K. (1994) Determination of tetrahydrocannabinolic acid-carrier protein conjugate by matrix-assisted laser desorption/ionization mass spectrometry and antibody formation. *Org. Mass Spectrom.* **29(11)**, 668.

Hsieh, F. Y. L., Tong, X., Wachs, T., Ganem, B. F., and Henion, J. (1995) Kinetic monitoring of enzymatic reactions in real time by quantitative high-performance liquid chromatography-mass spectrometry. *Anal. Biochem.* **229(1)**, 20.

Liao, P. C. and Allison, J. (1995) Enhanced detection of peptides in matrix-assisted laser desorption/ionization mass spectrometry through the use of charge-localized derivatives (letter). *J. Mass Spectrom.* **30(3)**, 511.

Marsillo, R., Catinella, S., Seraglia, R., and Traldi, P. (1995) Matrix-assisted laser desorption/ionization mass spectrometry for the rapid evaluation of thermal damage in milk. *Rapid Commun. Mass Spectrom.* **9(6)**, 550.

Mirza, U. A., Chait, B. T., and Lander, H. M. (1995) Monitoring reactions of nitric oxide with peptides and proteins by electrospray ionization-mass spectrometry. *J. Biol. Chem.* **270(29)**, 17,185.

Nakanishi, T., Kishikawa, N., Shimizu, A., Hayashi, A., and Inoue, F. (1995) Assignment of the ions in the electrospray ionization mass spectra of the tryptic digest of the whole sequence of α- and β-chains: a rapid diagnosis for haemoglobinopathy. *J. Mass Spectrom.* **30(12)**, 1663.

Newton, R. P., Evans, A. M., Langridge, J. I., Walton, T. J., Harris, F. M., and Brenton, A. G. (1995) Assay of adenosine 3',5'-cyclic monophosphate-dependent protein kinase activity by quantitative fast atom bombardment mass spectrometry. *Anal. Biochem.* **224(1)**, 32.

Nicola, A. J., Gusev, A. I., Proctor, A., Jackson, E. K., and Hercules, D. M. (1995) Application of the fast-evaporation sample preparation method for improving quantification of angiotensin II by matrix-assisted laser desorption/ionization. *Rapid Commun. Mass Spectrom.* **9(12)**, 1164.

Ogunsola, F. T., Magee, J. T., and Duerden, B. I. (1995) Correlation of pyrolysis mass spectrometry and outer membrane protein profiles of *Clostridium difficile*. *Clin. Infect. Dis.* **20(Suppl. 2)**, S331.

Protein Secondary Structure

Cohen, S. L., Ferre-D'Amare, A. R., Burley, S. K., and Chait, B. T. (1995) Probing the solution structure of the DNA-binding protein Max by a combination of proteolysis and mass spectrometry. *Protein Sci.* **4(6)**, 1088.

Hooke, S. D., Eyles, S. J., Miranker, A., Radford, S. E., Robinson, C. V., Dobson, C. M. (1995) Cooperative elements in protein folding monitored by electrospray ionization mass spectrometry. *J. Am. Chem. Soc.* **117(28)**, 7548.

Jaquinod, M., Halgand, F., Caffrey, M., Saint-Pierre, C., Gagnon, J., Fitch, J., Cusanovich, M., and Forest, E. (1995) Conformational properties of rhodobacter capsulatus cytochrome c_2 wild-type and site-directed mutants using hydrogen/deuterium exchange monitored by electrospray ionization mass spectrometry. *Rapid Commun. Mass Spectrom.* **9(12)**, 1135.

Loo, R. R. O., Dales, N., and Andrews, P. C. (1994) Surfactant effects on protein structure examined by electrospray ionization mass spectrometry. *Protein Sci.* **3(11)**, 1975.

Muir, T. W., Williams, M. J., and Kent, S. B. H. (1995) Detection of synthetic protein isomers and conformers by electrospray mass spectrometry. *Anal. Biochem.* **224(1)**, 100.

Osborn, B. L. and Abramson, F. P. (1995) Stable-isotope dilution studies of an intact protein using high-performance liquid chromatography/chemical reaction interface mass spectrometry. *Anal. Biochem.* **229(2)**, 347.

Packman, L. C. and Berry, A. (1995) A reactive, surface cysteine residue of the class II fructose-1,6-bisphosphate aldolase of *Escherichia coli* revealed by electrospray ionisation mass spectrometry. *Eur. J. Biochem.* **227(1–2)**, 510.

Protein Interaction

Anderegg, R. J. and Wagner, D. S. (1995) Mass spectrometric characterization of a protein–ligand interaction. *J. Am. Chem. Soc.* **117(4)**, 1374.

Bakhtiar, R. and Stearns, R. A. (1995) Studies on non-covalent associations of immunosuppressive drugs with serum albunin using pneumatically assisted electrospray ionization mass spectrometry. *Rapid Commun. Mass Spectrom.* **9(3)**, 240.

Bruce, J. E., Anderson, G. A., Chen, R., Cheng, X., Gale, D. C., Hofstadler, S. A., Schwartz, B. L., and Smith, R. D. (1995) Bio-affinity characterization mass spectrometry. *Rapid Commun. Mass Spectrom.* **9(8)**, 644.

Bruce, J. E., VanOrden, S. L., Anderson, G. A., Hofstadler, S. A., Sherman, M. G., Rockwood, A. L., and Smith, R. D. (1995) Selected ion accumulation of noncovalent complexes in a fourier transform ion cyclotron resonance mass spectrometer. *J. Mass Spectrom.* **30(1)**, 124.

Bruenner, B. A., Jones, A. D., and German, J. B. (1995) Direct characterization of protein adducts of the lipid peroxidation product 4-hydroxy-2-nonenal using electrospray mass spectrometry. *Chem. Res. Toxicol.* **8(4)**, 552.

Cheng, X. H., Chen, R. D., Bruce, J. E., Schwartz, B. L., Anderson, G. A., Hofstadler, S. A., Gale, D. C., Smith, R. D., Gao, J. M., Sigal, G. B., Mammen, M., and Whitesides, G. M. (1995) Using electrospray ionization FTICR mass spectrometry to study competitive binding of inhibitors to carbonic anhydrase. *J. Am. Chem. Soc.* **117(34)**, 8859.

Hamdan, M., Curcuruto, O., and DiModugno, E. (1995) Investigation of complexes between some glycopeptide antibiotics and bacterial cell-wall analogues by electrospray and capillary zone electrophoresis/electrospray mass spectrometry. *Rapid Commun. Mass Spectrom.* **9(10)**, 883.

High, K. A., Blais, J. S., Methven, B. A. J., and McLaren, J. W. (1995) Probing the characteristics of metal-binding proteins using high-performance liquid chromatography/atomic absorption spectroscopy and inductively coupled plasma mass spectrometry. *Analyst* **120(3)**, 629.

Hu, P. and Loo, J. A. (1995) Determining calcium-binding stoichiometry and cooperativity of parvalbumin and calmodulin by mass spectrometry. *J. Mass Spectrom.* **30(8)**, 1076.

Loo, J. A. (1995) Observation of large subunit protein complexes by electrospray ionization mass spectrometry. *J. Mass Spectrom.* **30(1)**, 180.

Moreau, S., Awade, A. C., Molle, D., Le Graet, Y., and Brule, G. (1995) Hen egg white lysozyme–metal ion interactions: investigation by electrospray ionization mass spectrometry. *J. Agr. Food Chem.* **43(4)**, 883.

Schwartz, B. L., Bruce, J. E., Anderson, G. A., Hofstadler, S. A., Rockwood, A. L., Smith, R. D., Chilkoti, A., and Stayton, P. S. (1995) Dissociation of tetrameric ions of noncovalent streptavidin complexes formed by electrospray ionization. *J. Am. Soc. Mass Spectrom.* **6(6)**, 459.

Schwartz, B. L., Gale, D. C., Smith, R. D., Chilkoti, A., and Stayton, P. S. (1995) Investigation of non-covalent ligand binding to the intact streptavidin tetramer by electrospray ionization mass spectrometry. *J. Mass Spectrom.* **30(8)**, 1095.

Tiller, P. R., Mutton, I. M., Lane, S. J., and Bevan, C. D. (1995) Immobilized human serum albumin: liquid chromatography/mass spectrometry as a method of determining drug–protein binding. *Rapid Commun. Mass Spectrom.* **9(4)**, 261.

Weber, R. E., Malte, H., Braswell, E. H., Oliver, R. W. A., Green, B. N., Sharma, P. K., Khuchumov, A., and Vinogradov, S. N. (1995) Mass spectrometric composition, molecular mass and oxygen binding of *Macrobdella decora* hemoglobin and its tetramer and monomer subunits. *J. Mol. Biol.* **251(5)**, 703.

Protein Physical Chemistry

Gross, D. S. and Williams, E. R. (1995) Experimental measurement of coulomb energy and intrinsic dielectric polarizability of a multiply protonated peptide ion using electrospray ionization Fourier-transform mass spectrometry. *J. Am. Chem. Soc.* **117(3)**, 883.

Ishikawa, K., Nishimura, T., Koga, Y., and Niwa, Y. (1994) Role of Coulomb energy in promoting collisionally activated dissociation of multiply charged peptides formed by electrospray ionization. *Rapid Commun. Mass Spectrom.* **8(12)**, 933.

Johnson, R. S., Krylov, D., and Walsh, K. A. (1995) Proton mobility within electrosprayed peptide ions. *J. Mass Spectrom.* **30(2)**, 386.

Johnson, R. S. and Walsh, K. A. (1994) Mass spectrometric measurement of protein-amide hydrogen exchange rates of apo- and holo-myoglobin. *Protein Sci.* **3(12)**, 2411.

Le Blanc, J. C. Y., Wang, J. Y., Guevremont, R., and Siu, K. W. M. (1994) Electrospray mass spectra of protein cations formed in basic solutions. *Org. Mass Spectrom.* **29(11)**, 587.

Loo, R. R. O., Winger, B. E., and Smith, R. D. (1994) Proton transfer reaction studies of multiply charged proteins in a high mass-to-charge ratio quadrupole mass spectrometer. *J. Am. Soc. Mass Spectrom.* **5(12)**, 1064.

Ogorzalek Loo, R. R. and Smith, R. D. (1995) Proton transfer reactions of multiply charged peptide and protein cations and anions. *J. Mass Spectrom.* **30(2)**, 339.

Peptide Structure

Agbas, A., Ahmed, M. S., Millington, W., Cemerikic, B., Desiderio, D. M., Tseng, J. L., and Dass, C. (1995) Dynorphin A(1-8) in human placenta: amino acid sequence determined by tandem mass spectrometry. *Peptides* **16(4)**, 623.

Amankwa, L. N., Harder, K., Jirik, F., and Aebersold, R. (1995) High-sensitivity determination of tyrosine-phosphorylated peptides by on-line enzyme reactor and electrospray ionization mass spectrometry. *Protein Sci.* **4(1)**, 113.

Andren, P. E. and Caprioli, R. M. (1995) In vivo metabolism of substance P in rat striatum utilizing microdialysis/liquid chromatography/micro-electrospray mass spectrometry. *J. Mass Spectrom.* **30(6)**, 817.

Barnhart, E. R., Maggio, V. L., Alexander, L. R., Patterson, D. G., Kesner, L., Gelbaum, L. T., and Ash, R. J. (1995) Fast atom bombardment mass spectrometric detection of dimeric peptides in bacitracin preparations (letter). *J. Mass Spectrom.* **30(8)**, 1201.

Becchi, M., Rebuffat, S., Dugast, J. Y., Hlimi, S., Bodo, B., and Molle, G. (1995) Sequencing of hydrophobic peptides as lithiated adducts using liquid secondary-ion mass spectrometry without tandem mass spectrometry. *Rapid Commun. Mass Spectrom.* **9(1)**, 37.

Castoro, J. A., Wilkins, C. L., Woods, A. S., and Cotter, R. J. (1995) Peptide amino acid sequence analysis using matrix-assisted laser desorption/ionization and Fourier transform mass spectrometry. *J. Mass Spectrom.* **30(1)**, 94.

Chu, Y. H., Kirby, D. P., and Karger, B. L. (1995) Free solution identification of candidate peptides from combinatorial libraries by affinity capillary electrophoresis/mass spectrometry. *J. Am. Chem. Soc.* **117(19)**, 5419.

Cordero, M. M., Cornish, T. J., Cotter, R. J., and Lys, I. A. (1995) Sequencing peptides without scanning the reflectron: post-source decay with a curved-field reflectron time-of-flight mass spectrometer. *Rapid Commun. Mass Spectrom.* **9(13)**, 1356.

D'Agostino, P. A., Hancock, J. R., Provost, L. R., Semchuk, P. D., and Hodges, R. S. (1995) High resolution electrospray mass spectrometry with a magnetic sector instrument: accurate mass measurement and peptide sequencing. *Rapid Commun. Mass Spectrom.* **9(7)**, 597.

Dass, C. and Mahalakshmi, P. (1995) Amino acid sequence determination of phosphoenkephalins using liquid secondary ionization mass spectrometry. *Rapid Commun Mass Spectrom.* **9(12)**, 1148.

Dookeran, N. N. and Harrison, A. G. (1995) Gas-phase H-D exchange reactions of protonated amino acids and peptides with ND_3. *J. Mass Spectrom.* **30(5)**, 666.

Doroshenko, V. M. and Cotter, R. J. (1995) High-performance collision-induced dissociation of peptide ions formed by matrix-assisted laser desorption/ionization in a quadrupole ion trap mass spectrometer. *Anal. Chem.* **67(13),** 2180.

Downard, K. M. and Bieman, K. (1995) Methionine specific sequence ions formed by the dissociation of protonated peptides at high collision energies. *J. Mass Spectrom.* **30(1),** 25.

Goodlett, D. R., Abuaf, P. A., Savage, P. A., Kowalski, K. A., Mukherjee, T. K., Tolan, J. W., Corkum, N., Goldstein, G., and Crowther, J. B. (1995) Peptide chiral purity determination: hydrolysis in deuterated acid, derivatization with Marfey's reagent and analysis using high-performance liquid chromatography-electrospray ionization-mass spectometry. *J. Chromatogr. A.* **707(2),** 233.

Griffiths, W. J., Lindh, I., Bergman, T., and Sjovall, J. (1995) Negative-ion electrospray mass spectra of peptides derivatized with 4-aminonaphthalenesulphonic acid. *Rapid Commun. Mass Spectrom.* **9(8),** 667.

Harada, K., Fujii, K., Mayumi, T., Hibino, Y., Suzuki, M., Ikai, Y., and Oka, H. (1995) A method using LC/MS for determination of absolute configuration of constituent amino acids in peptide: advanced Marfey's method. *Tetrahedron Lett.* **36(9),** 1515.

Hudson, H. R., Pianka, M., Volckman, J. F., and Creaser, C. S. (1994) Fast-atom bombardment mass spectrometry of phosphonopeptides. *Phosphor. Sulfur Silicon* **92(1–4),** 171.

Jang, J. S. and Chang, Y. S. (1994) Characterization of phosphopeptide and triphenylphosphonium derivative by tandem mass spectrometry. *Bull. Kor. Chem. Soc.* **15(12),** 1134.

Kieliszewski, M. J., O'Neill, M., Leykam, J., and Orlando, R. (1995) Tandem mass spectrometry and structural elucidation of glycopeptides from a hydroxyproline-rich plant cell wall glycoprotein indicate that contiguous hydroxyproline residues are the major sites of hydroxyproline *O*-arabinosylation. *J. Biol. Chem.* **270(6),** 2541.

Kolli, V. S. K. and Orlando, R. (1995) Complete sequence confirmation of large peptides by high energy collisional activation of multiply protonated ions. *J. Am. Soc. Mass Spectrom.* **6(4),** 234.

Kosaka, T., Ishikawa, T., and Kinoshita, T. (1995) Collisionally-activated dissociation spectra of linear peptides in matrix-assisted laser desorption/ionization time-of-flight mass spectrometry. *Rapid Commun. Mass Spectrom.* **9(13),** 1342.

Krutchinsky, A. N., Dolgin, A. I., Utsal, O. G., and Khodorkovski, A. M. (1995) Thin-layer chromatography/laser desorption of peptides followed by multiphoton ionization time-of-flight mass spectrometry. *J. Mass Spectom.* **30(2),** 375.

Kupke, T., Kempter, C., Jung, G., and Gotz, F. (1995) Oxidative decarboxylation of peptides catalyzed by flavoprotein EpiD: determination of substrate specificity using peptide libraries and neutral loss mass spectrometry. *J. Biol. Chem.* **270(19),** 11,282.

Lee, H. G., Tseng, J. L., Becklin, R. R., and Desiderio, D. M. (1995) Preparative and analytical capillary zone electrophoresis analysis of native endorphins and enkephalins extracted from the bovine pituitary: mass spectrometric confirmation of the molecular mass of leucine enkephalin. *Anal. Biochem.* **229(2),** 188.

Liao, P.-C. and Allison, J. (1995) Ionization processes in matrix-assisted laser desorption/ionization mass spectrometry: matrix dependent formation of [M + H]⁺ vs [M + Na]⁺ ions of small peptides and some mechanistic comments. *J. Mass Spectrom.* **30(3),** 408.

Li, K. W., Van Minnen, J., Van der Greef, J., and Geraerts, W. P. M. (1994) Direct mass spectrometry of buccal ganglia and nerves reveals the processing and targeting of peptide messengers involved in feeding behaviour of *Lymnaea. Neth. J. Zool.* **44(3–4),** 432.

Lin, R. C., Smith, J. B., Radtke, D. B., and Lumeng, L. (1995) Structural analysis of peptide-acetaldehyde adducts by mass spectrometry and production of antibodies directed against nonreduced protein-acetaldehyde adducts. *Alcohol Clin. Exp. Res.* **19(2),** 314.

Loutelier, C., Marcual, A., Cassier, P., Cherton, J. C., and Lange, C. (1995) Desorption of ions from locust tissues. III. Study of a metabolite of A-destruxin using fast-atom bombardment linked-scan mass spectrometry. *Rapid Commun. Mass Spectrom.* **9(5),** 408.

Loutelier, C., Marcual, A., Cherton, J. C., and Lange, C. (1994) Characterization of cyclodepsipeptide-glutathionyl conjugates by negative-ion fast-atom bombardment linked-scan mass spectrometry. *Rapid Commun. Mass Spectrom.* **8(12),** 957.

Molle, D. and Leonil, J. (1995) Heterogeneity of the bovine κ-casein caseinomacropeptide, resolved by liquid chromatography on-line with electrospray ionization mass spectrometry. *J. Chromatogr. A.* **708(2),** 223.

Nylander, I., Tan-No, K., Winter, A., and Silberring, J. (1995) Processing of prodynorphin-derived peptides in striatal extracts. Identification for electrospray ionization mass spectrometry linked to size-exclusion chromatography. *Life Sci.* **57(2),** 123.

Pasa-Tolic, L., Huang, Y., Guan, S., Kim, H. S., and Marshall, A. G. (1995) Ultrahigh-resolution matrix-assisted laser desorption/ionization Fourier transform ion cyclotron resonance mass spectra of peptides. *J. Mass Spectrom.* **30(6),** 825.

Peter-Katalinic, J., Ashcroft, A., Green, B., Hanisch, F. G., Nakahara, Y., Ilijma H., and Ogawa T. (1994) Potential of electrospray mass spectrometry for structural studies of *O*-glycoamino acids and -peptides with multiple α2,6-sialosyl-Tn glycosylation. *Org. Mass Spectrom.* **29(12),** 747.

Qian, M. G., Zhang, Y., and Lubman, D. M. (1995) Collision-induced dissociation of multiply charged peptides in an ion-trap storage/reflectron time-ot-flight mass spectrometer. *Rapid Commun Mass Spectrom.* **9(13),** 1275.

Rouse, J. C., Yu, W., and Martin, S. A. (1995) A comparison of the peptide fragmentation obtained from a reflector matrix-assisted laser desorption-ionization time-of-flight and a tandem four sector mass spectrometer. *J. Am. Soc. Mass Spectrom.* **6(9),** 822.

Rozenski, J., Samson, I., Janssen, G., Busson, R., Van Aerschot, A., and Herdewijn, P. (1994) Characterization of modification sites during peptide synthesis using liquid secondary ion collision-induced dissociation mass spectrometry and a computer program. *Org. Mass Spectrom.* **29(11)**, 654.

Squire, N. L., Beranova, S., and Wesdemiotis, C. (1995) Tandem mass spectrometry of peptides III: differentiation between leucine and isoleucine based on neutral losses. *J. Mass Spectrom.* **30(10)**, 1429.

Sullards, M. C. and Adams, J. (1995) On the use of scans at a constant ratio of B/E for studying decomposition of peptide metal(II)-ion complexes formed by electrospray ionization. *J. Am. Soc. Mass Spectrom.* **6(7)**, 608.

Summerfield, S. G., Dale, V. C. M., Despeyroux, D. D., and Jennings, K. R. (1995) Charge remote losses of small neutrals from protonated and group 1 metal-peptide complexes of peptides. *Eur. Mass Spectrom.* **1(2)**, 183.

Tomlinson, A. J. and Naylor, S. (1995) Enhanced performance membrane preconcentration-capillary electrophoresis/mass spectrometry (MPC-CE-MS) in conjunction with transient isotachophoresis for analysis of peptide mixtures. *J. High Res. Chromatogr.* **18(6)**, 384.

Tseng, J. L., Yan, L., Fridland, G. H., and Desiderio, D. M. (1995) Tandem mass spectrometry analysis of synthetic opioid peptide analogs. *Rapid Commun. Mass Spectrom.* **9(4)**, 264.

Tull, D., Miao, S. C., Withers, S. G., and Aebersold, R. (1995) Identification of derivatized peptides without radiolabels: tandem mass spectrometric localization of the tagged active-site nucleophiles of two cellulases and a β-glucosidase. *Anal. Biochem.* **224(2)**, 509.

Van Dongen, W. D., De Koster, C. G., Heema, W., and Haverkamp, J. (1995) The diagnostic value of the *m/z* 102 peak in the positive-ion-fast-atom bombardment mass spectra of peptides. *Rapid Commun. Mass Spectrom.* **9(9)**, 845.

Wang, J. Y., Guevremont, R., and Siu, K. W. M. (1995) Multiple alkali metal ion complexes of tripeptides. An investigation by means of electrospray tandem mass spectrometry. *Eur. Mass Spectrom.* **1(2)**, 171.

Weber, C., Streckert, H. J., and Muller, D. (1995) Antigenicity and mass spectrometric properties of rotaviral polypeptides. *J. Mol. Struct.* **349**, 357.

Yagami, T., Kitagawa, K., and Futaki, S. (1995) Liquid secondary-ion mass spectrometry of peptides containing multiple tyrosine-O-sulfates. *Rapid Commun. Mass Spectrom.* **9(14)**, 1335.

Yan, L., Tseng, J. L., and Desiderio, D. M. (1995) Detection and characterization of native β-endorphin[1–31] in bovine pituitary using electrospray ionization, liquid secondary ion and tandem mass spectrometry. *J. Mass Spectrom.* **30(4)**, 531.

Zaramella, A., Curcuruto, O., Hamdan, M., Cabrele, C., and Peggion, E. (1995) Investigation of crudes of synthesis of [Leu[31], Pro[34]]- neuropeptide Y by capillary zone electrophoresis/mass spectrometry. *Rapid Commun. Mass Spectrom.* **9(13)**, 1386.

Zhang, H., Andren, P., and Caprioli, R. M. (1995) Micro-preparation procedure for high-sensitivity matrix-assisted laser desorption ionization mass spectrometry. *J. Mass Spectrom.* **30(12)**, 1768.

Index

345